Principles of Systems Programming

Principles of Systems Programming

ROBERT M. GRAHAM

The City College of New York

John Wiley & Sons, Inc.   New York   London   Sydney   Toronto

Library of Congress Cataloging in Publication Data:

Graham, Robert M   1929-
    Principles of systems programming.

    Includes bibliographical references and index.
    1. Electronic digital computers—Programming.
I. Title.

QA76.6.G7   001.6'42   74-19390
ISBN 0-471-32100-1

Printed in the United States of America

10 9 8 7 6 5 4 3 2 1

To my VW

Preface

Systems programming has grown rapidly in the past 15 years. At the end of the 1950s the importance of software was generally recognized, and operating systems existed for most large computers, although they were simple systems compared to today's systems. FORTRAN was an unquestioned success, and ALGOL was stimulating intense interest in compilers and programming language theory. In spite of this, software development accounted for only a minor part of a computer manufacturer's budget. The opposite is true today. Over half of IBM's budget for development of the series 360 machines was allocated to software. Projects such as the operating system for the IBM 360 (OS/360) employed as many as a 1000 systems programmers. Larger, faster, and more complex computers require larger and more complex operating systems. Growth in this direction will most probably continue. As computers and operating systems get more complex, the problems they generate get more complex and difficult. Thus, systems programming has become one of the most challenging and important areas of computer science.

Systems programming is a broad field that encompasses many specialties. A number of texts deal with specialized subjects, such as compilers, time-sharing systems, and management information systems. However, there are not enough texts and other material at an introductory level of systems programming. This textbook was written to fill this void.

Systems programming is a specialized branch of programming. Thus, the reader who wishes to study the subject should have some knowledge of, and experience in, programming—particularly the basic ideas and techniques of programming, such as conditionals, looping, subscripted variables, and subroutines. Elementary texts in computer science, such as Forsythe et al,[1] Rice and Rice,[2] or Conway and Gries[3] provide more than the required background

---

[1] A. I. Forsythe, T. A. Keenan, E. I. Organick, and W. Stenberg, *Computer Science: A First Course,* Wiley, New York, 1969.

[2] J. K. Rice and J. R. Rice, *Introduction to Computer Science,* Holt, Rinehart and Winston, New York, 1969.

[3] R. Conway and D. Gries, *An Introduction to Programming: A Structured Approach Using PL/1 and PL/C,* Winthrop, Cambridge, Mass., 1973.

in basic programming. In addition to knowledge, it is necessary for the reader to have the practical experience that is accumulated by coding several programs in a compiler language such as FORTRAN, ALGOL, or PL/1 and successfully debugging them on a computer. It is impossible to appreciate fully all of the problems in systems programming without some first-hand experience in using a computer. I also assume a basic knowledge of assembly language for some computer. Skill in machine language coding is not required, and it is not dealt with in the text. However, basic knowledge of the character of machine language is required as a background for understanding some of the problems that arise in interfacing an operating system to the computer and translating compiler language into machine language.

The importance of actual experience by the student in studying any branch of programming cannot be overemphasized. For this reason, several case studies are examined. Detailed algorithms for them are given in a machine-independent, informal, high-level programming language. The student should do several of the exercises provided at the end of the chapters. Many of these exercises ask him to describe the modifications to a case study that are required to extend an existing feature or add a new feature. Frequently an exercise depends on previous exercises. Therefore, the exercises should be done in sequence. Of course, the best possible experience for the student is for him to actually code and test the relevant algorithms. To achieve this, the instructor may wish to modify some of the programming exercises so that the student's programs can be tested on a locally available computer. If sufficient laboratory time is available, the student, or a small group of students, should develop working versions of one or more of the case studies.

The text is organized into three major sections: assemblers and loading, compilers, and operating systems. However, each of these sections is relatively independent of the others so that some reordering of the material is certainly reasonable. In addition, any of the major sections could be used as introductory material for a course devoted entirely to that subject.

I have tried to be as independent as possible of a particular computer. However, I think that it is better to use the machine language of an existing computer than to invent a new one. Since the IBM 360 and similar byte-oriented computers are the most widely used machines, I have used a simple subset of the IBM 360 machine language in the examples. In addition, some of the interface problems that arise using the more advanced machines are simpler; thus, the student will have less trouble going from a 360-like structure to other machines than going in the opposite direction. The examples in the text do not depend on the particulars of the IBM 360 in any significant way. It is simple to convert all of the machine language examples into the machine language for any other computer, if desired.

Although COBOL and FORTRAN are the most popular and widely used compiler languages, I use a simplified subset of PL/1 in the case study of compilers because it is a very rich language, containing almost all of the important features found in programming languages. It contains most of the best features from both FORTRAN and COBOL.

This text grew out of an introductory course that I developed at MIT between 1965 and 1967. Although the general form is still similar, the material has been completely revised and updated. Many people have had an influence on the text. This large group includes my co-workers on the Multics project at MIT and my colleagues and students at MIT, the University of California at Berkeley, and the City College of New York. However, the text is basically an individual effort that (perhaps too strongly) expresses my personal view of the subject.

I would like to thank Elliott I. Organick, who read two different versions of the manuscript and offered many detailed comments, Muriel Webber, who typed the drafts of the manuscript, and Maryann Archer, who typed the final version of the manuscript.

New York, 1974   Robert M. Graham

Contents

Principles of Systems Programming

1 | Systems Programming

This text is an introduction to the specialized branch of computer science known as *systems programming*. A systems programmer is concerned with some, or perhaps all, aspects of a computer-based information-processing system. All computer-based information-processing systems are a combination of both *hardware* and *software*. Hardware is physical, consisting of the computer and its attached devices. Software is information—programs and their data—and therefore is conceptual instead of physical. Systems programming is principally concerned with the design, implementation, and maintenance of the programs that form the software part of an information-processing system. A systems programmer usually needs to understand how the hardware operates. However, the depth of understanding required is not great, except when involved in one of a few specialized areas such as the operation of input-output devices. In contrast, *applications programming* is principally concerned with the design, implementation, and maintenance of programs that use the system. The reader should be aware that although the distinction between systems programming and applications programming is useful, it is not at all sharp. Indeed, an applications programmer frequently finds himself faced with systems programminglike problems with the difference being more in viewpoint than in fundamental problems and principles.

1.1 What Is a System?

Before proceeding further, it will be helpful to discuss the idea of a *system*. A typical dictionary defines a system as, "an assemblage of objects united by some form of regular interaction or interdependence." This is a very broad definition, and our world abounds with examples: the solar system, electrical networks, the human body, and society. The first thing to notice is that a system is composed of objects, which we will call *components*. Second, these *components* interact with one another according to some set of rules. In the solar system the components are the sun and the planets. Their interaction is described by the law of gravitation. In a simple electrical network, such as those found in our homes, the components are the switches, lights, appliances, the electric company that supplies the power, and the connecting wires. Their interaction is described by laws of physics such as Kirchoff's law.

The human body is much more complex. The components of this system are the various body organs, nerves, blood, skin, and so forth. Their interaction can be partially characterized by laws of chemistry and physics. However, we do not know of any rules that completely describe this interaction. Society is perhaps the most complex system of all. It has components such as people, institutions, and ideas. Very little is understood about how these components interact, and even a partial description of the interaction between them is difficult to formulate.

While the interactions in a household electrical system can be easily, precisely, and completely expressed by mathematical equations relating idealized elements such as resistances (toasters) and inductances (mixer motors), we are unable to describe, with similar precision and completeness, more complex systems such as information-processing systems, the human body, and society. Some scientists hold the view that every system can eventually be described by a set of mathematical equations. Others feel that systems such as complex biological and social systems can never be completely described by mathematical equations. Whatever the outcome of this debate, currently there is no formalism, such as mathematics, for completely and precisely describing the interaction between the components of an information-processing system. Therefore, in this text we are forced to use English, supplemented with diagrams and examples.

One aspect of a system that is not mentioned in the dictionary definition will be quite important in our study: the interaction of a system with its *environment*. A system's environment consists of all objects in the universe that are not part of the system; for example, the user of an information-processing system is an object in its environment. A system built by humans is usually built for a purpose. To achieve this purpose the system must affect its environment in some noticeable way, usually in response to some stimulus from the environment. In an information-processing system the receiving of stimuli from its environment is called *input*, and the affecting of its environment by the system is called *output*.

Actually, information-processing systems are a subset of a more general class of systems called *information systems*. Communication systems are also information systems, since they are principally concerned with information. The major distinction between these two types of information systems is that in a *communication system* the emphasis is on the transmission of information, while in an *information-processing system* the emphasis is on the transformation and storage of information. This distinction is becoming more and more blurred with the development of systems such as airline reservation systems, which have terminals throughout the country, each communicating with the same central computer. There are many types of information-processing systems, and even more names for them. For example, all of the following are names for some kind of information-processing system: operating system, data management system, time-sharing system, airline reservation system, air traffic control system, management information system, Apollo guidance system, and information retrieval system. This list, which is far from complete, should give the reader some idea of the great variety of computer-based information-processing systems.

Not all information-processing systems are computer based. Airline reservation systems were information-processing systems before they were computerized. A library is an example of an information-processing system that is not yet computerized. It is important to realize the generality of information-processing systems, since many such systems not yet computerized will become so in the future, and the reader may become involved in such an effort.

1.2 Operating Systems

In this text we will focus on an important subset of information-processing systems—operating systems—which are also the most common type of information-processing systems. It is unnecessary and beyond the scope of this text to consider all types of information-processing systems. The system programming aspects of all types of information-processing systems are very much the same. Thus, the problems and concepts encountered in operating systems and the applicable design and implementation techniques are easily adaptable to other kinds of information-processing systems.

An operating system is software (programs) that is used to assist in the operation and use of computer hardware. From the user's point of view the function of an operating system is to make it easier to obtain a solution to his problem. Thus, a primary objective of an operating system is to make it easier to code, test, and execute programs. From the administrator's point of view the function of an operating system is to obtain the most efficient use of the computer. Thus, another primary objective of an operating system is to manage the use of the hardware efficiently.

The components of an operating system are programs. Some programs (for example, compilers, assemblers, loaders, report generators, sort-merge programs, file maintenance programs, program testing aids, and input-output programs) are principally for the purpose of making it easier for the users to use the computer. Even though the computer is not part of an operating system as we have defined it, some of the software is so intimately involved with the computer that we extend our definition of an operating system to include the hardware components of the computer. These additional system components include control processors, primary memory, secondary memory devices such as disk units, and input-output devices such as card readers and line printers.

The reader's view of an operating system can probably be represented by the diagram in Figure 1.1. The input is a user's program for the solution of his problem, written in some language such as FORTRAN or PL/1, along with its data. This information is either typed on a typewriter terminal that is connected to the computer or punched on cards and submitted to the computer center for reading by a card reader that is connected to the computer. Some time later the user receives some output that is typed on the typewriter terminal or printed by a line printer. This output consists of the answers to his problem or, if some trouble occurred, comments explaining what trouble occurred.

We assume that the reader has little, if any, information about what happens inside the box. He puts his program and data into the system, and some results come out. One purpose of this text is to help the reader understand what goes on inside. Inside the box is an information-processing system consisting of the computer hardware and the operating system software. Our dis-

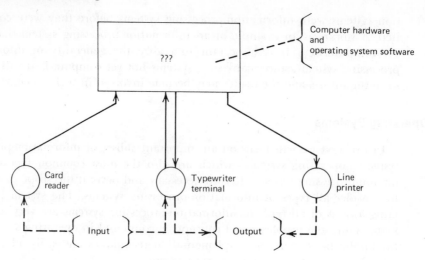

Figure 1.1 Black box view of an operating system.

cussion of the hardware will be limited to the minimum required to understand the software. The text will mostly be concerned with the software,
examining some of the software components in considerable detail.

There are several desirable, if not essential, properties that any useful operating system must have. The system must be composed of a collection of useful
programs such as those previously mentioned. Most contemporary systems
contain a substantial number of such software components, since one measure
of a system's usefulness is the variety of services that it offers. The components
of the system must be organized into some meaningful structure. Conventions
must exist for communication between the components. Finally, the system
should be easily accessible, that is, it should make communication between
the user and the system as easy as possible. These properties are what make an
operating system useful to a user with a problem for which a solution is
desired. A major purpose of an operating system is to aid the user in obtaining
a solution. Assuming the user has in mind an algorithm that will give him the
answers he desires, the system should mechanize as much as possible the
preparation, testing, and execution of a program that carries out his algorithm.
Some very advanced systems even assist the user in discovering an algorithm
to solve the problem.

A telephone system is a good example of a system having properties that
are close analogies of the major properties of a good operating system. The
telephones are analogous to the programs of an operating system. First,
the telephone system is organized. Individual telephones are connected and
grouped together by exchanges. Exchanges are connected together with long-
distance lines. The telephones are organized into a network in such a fashion
that communication between individual telephone subscribers is not only possible, but extremely easy. Second, all telephones follow the same communication conventions. Thus, any two telephones in the system can be connected
together easily to permit communication between them. Last, the telephone
system is easily accessible. The task of making a connection between two telephones has been completely mechanized with the advent of direct dialing.
Dialing a number is such a simple operation that a child can easily do it.

Currently, the telephone company is beginning to computerize the telephone system, which will result in even more simplification in the mechanics of making a connection. In the future a user will be able to store convenient mnemonic codes for frequently used numbers in the telephone system and use these short mnemonics instead of dialing a long number.

1.3 Objectives

This text introduces the reader to systems programming. Systems programming differs from other kinds of programming in that it is concerned with the design and implementation of systems, especially operating systems. Thus, it is essential that a systems programmer have a good undertanding of the problems and concepts of operating systems. In pursuit of this objective we will attempt to answer such questions as the following ones.

What are the purposes and objectives of an operating system? What are its main functions? What are its major components and how are they structured? What are the constraints within which it might have to function? What are the problems of managing the use of the computer's control processor, memory, and attached input-output devices? How does the user communicate with the system? How does the system "understand" a user's program written in a language such as PL/1? How can more than one user use the system at the same time?

Systems programming, being programming, requires the development of standard programming techniques such as the use of independent, separately compiled or assembled procedures and the use of structured data such as arrays and queues. We will find that certain techniques are used very frequently. These techniques will be explored when they occur naturally in the discussion of the operating system components.

Development of the reader's understanding is more difficult than increasing his knowledge. We will approach the problems of each system component in an organized way. We will try to isolate the problem and discuss the inherent requirements that any solution must satisfy. In addition, we explore the implications of the system's objectives. This is illustrated by discussing a specific solution. In all cases we attempt to be as independent of any particular computer as is possible. Whenever it is reasonable, alternate solutions are explored. However, it is impossible, here, to conduct an exhaustive review of all known solutions to any problem.

We would like to help the reader develop to the point where he is able to derive the logical implications of a problem and predict the effect of the choices he makes in the design and implementation of a system component. We use a systematic methodology in the development of our examples and hope the reader will thus develop a feeling for this methodology and then use it when doing the exercises included in this text. Only by doing can the reader develop true understanding.

1.4 A Preview

This text divides into two major parts. The first (Chapters 3 to 12) is concerned with the translation of a user's program from the language in which he wrote it into machine language and with the loading of this translated program into the computer's memory so that it may be executed. The second part (Chapters 13 to 17) is concerned with the management of the system resources such as the memory, the control processors, and the input-output devices.

Chapter 2 discusses the computer hardware and a simple operating system. Our intent here is not to be exhaustive, but to provide an overview of an operating system and a context for the more detailed discussions in the remaining chapters. The operating system described, while simple, is realistic and workable. Examination of this simple system shows where some of the major problems occur in an operating system.

Chapters 3 to 6 cover assemblers and loading. Coding in actual machine language is unreasonable, and all machine language programs are written in assembly language. The functions and problems of an assembler are explored, and a typical assembler for basic assembly language is studied. We then examine the problem of loading an assembled program into the computer's main memory so that it can be executed. Since a user's program may consist of a collection of separately assembled procedures, we must consider the problem of linking these procedures together to form a single program.

Compilers are studied in Chapters 7 to 12. We begin with a discussion of programming languages and some important related concepts. These concepts are illustrated by examples from a subset of PL/1. The functions and problems of a compiler are discussed, and a simple compiler is described. This discussion establishes the context for a more detailed treatment of several of the major problems in compilers in the succeeding chapters.

The remaining chapters consider the resource management aspects of operating systems. We begin with a discussion of the function of an operating system. We explore the properties of an operating system as they relate to the system's objectives. The reader will see that the principal difference between various types of operating systems, such as batch and time sharing, is the degree to which the system has a given property, rather than the presence or absence of the property. The four major topics that conclude the text are process control, memory management, input-output, and the protection of user programs and data.

Process control is concerned with managing the execution of user programs. In particular we consider the major problem of scheduling the use of the control processor when several users are attempting to use the system at the same time. Memory management is concerned with the allocation of the computer's memory for the storage of both user programs and data. We consider both allocation of primary memory to executing programs and allocation of secondary memory for long-term storage of user files. Input-output is concerned with the management of the computer's attached devices that communicate with the outside world. In considering protection of user programs and data we are especially concerned with the problems encountered in this area when the system has many users and permits sharing of data and programs among them.

2 | A Simple Operating System

In this chapter we will explore a simple example of an operating system. This example will be a traditional, *sequential batch system*. This type of system will process a collection of jobs belonging to many different users. A *job* consists of a user's program and its data. The collection of jobs is called a *batch*. The jobs in the batch are processed *sequentially*, that is, each job in the batch is completely processed before any other job is considered. Sequential batch systems have been in common use for more than 10 years. Even though many modern systems are more sophisticated, sequential batch systems are still in wide use, especially with small computers.

This chapter is not principally a discussion of the problems of operating systems. These problems are treated in Chapters 13 to 17. We have two primary goals in exploring an example of an operating system at this point. Even though our example is simple, it contains most of the major internal functions of an operating system. Much of the complexity found in contemporary operating systems is merely elaboration of a few basic functions. By looking at a complete system we hope to give the reader a sense of completeness, a sense of the cohesiveness required to make a collection of components into a system. This will set a context for the more detailed discussions in the following chapters. The reader must have this perspective, or the significance of many of the problems raised in the following chapters will be lost.

Our second goal in this chapter is to introduce the reader to a design methodology. In exploring the example system we will actually design it. Our design process will use a methodology without discussing it in detail. Design should proceed from objectives. Given a set of functional objectives for a system, we explore the implications of these requirements with respect to what basic tasks the system must be able to do internally. In order to perform these basic tasks the system will have to maintain certain information. This process defines the basic building blocks of the system; the tasks are performed by procedure components, and the information is stored in data components. The reader should keep these comments in mind as he reads this chapter.

2.1 A User's View of the System

As the first step in developing the design of the system we must clearly understand what the system is to do. That is, we must understand the *functional objectives* that our design is to satisfy. Since we are building the system to assist the user of the computer, it is important that we look at the objectives from his viewpoint. In the simplest case the user has some program that will cause the computer to perform a desired computation or other processing of his data. He submits the program and his data to the system, which sees to it that the computer performs the computation and then returns the results to the user.

This is too simple a picture. Programs do not grow on trees. At least some of the users will be developing programs. A program is usually a collection of procedures, each procedure being more or less independent of the others. These procedures are always written in some language other than machine language, frequently in languages such as FORTRAN, PL/1, COBOL, or assembly language. It is not necessary that all procedures in a program be written in the same language. For example, a program may consist of three procedures written in PL/1 and one procedure written in assembly language. However, a computer can execute a program only if all of the procedures in the program are expressed in that computer's machine language. Thus, each procedure must be translated from the language in which it was written into machine language. During the development of a program, the job that a user submits to the system may be a collection of procedures some of which have already been translated into machine language, while others still need to be translated.

When a user submits a program and its data to the system, he must also include directions called *commands* (or *control cards* in many systems), which indicate what is to be done with each of the procedures in his program, that is, what translator is required in order to translate each procedure into machine language. The program, data, and commands together form a *job*. Logically the job consists of a sequence of *job steps*, each job step being defined by a single command. "Translate a FORTRAN procedure into machine language" or "execute the program" are typical commands that may define a job step.

We can now formulate the user's view as the following sequence. The user collects the procedures that make up his program and the data he wishes this program to use when it executes. He organizes them into job steps, arranges them in the proper sequence, and adds the required directions. He now has a job. He then submits his job to the system. The system accepts the job, performs the actions specified in each of the job steps, and prints the results of this processing. The results are than returned to the user.

So far we have concentrated on functional objectives that are obvious from the user's point of view. There is also a class of objectives that, even though not obvious to most users, are equally important because they are reflected in lower cost per unit of computing. These objectives are called *performance objectives*. They can be expressed as constraints under which the system must operate. Specifying that the system must perform a particular action in no more than a given time or that the system must be able to work using no more than a specified amount of memory are typical performance objectives. One such objective is important in our simple system. We require that our system is to process one job after another without stopping, with no operator action required either between jobs or during a job except in unusual circumstances.

2.2 Hardware Components of the System

In the preceding section we examined the objectives of our system. Before discussing the implications of these objectives we must look at the computer hardware that the system must use. The structure and characteristics of this computer will also have implications in the design of the system.

The computer that we will use is a simplified version of a typical modern computer. It is composed of three major units: a control (and computational) processor, an I/O (input-output) processor, and a primary memory (see Figure 2.1). In addition, a number of I/O devices (an operator's console, a printer, a card reader, a card punch, a disk storage unit, and several magnetic tape units) are connected to the I/O processor. A computer organized in this fashion is called a *memory-based system*, since both processors have direct access to the primary memory.

The primary memory is used to store programs and data. The I/O processor controls all of the I/O devices and is responsible for all I/O. The disk unit and the magnetic tape units are the computer's secondary storage and are principally used for long-term storage of programs and especially of data. The remaining devices are used for communication with the outside world. A user's program and data are initially entered into the system by reading cards from the card reader. The output from a user job is printed by the printer or punched on cards by the card punch. The operator's console is used to communicate with the operator of the computer. The control processor performs the computation or data processing specified by a user's program.

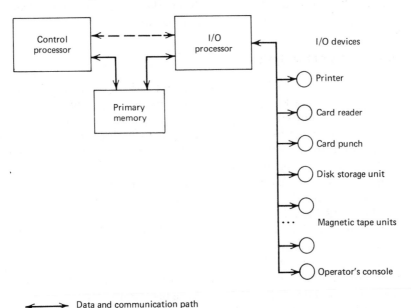

Figure 2.1 Structure of computer hardware.

Detailed knowledge of the operation of the hardware components is not required for the development of our simple system. However, several properties of the hardware do affect the design of our system and must be considered in the following discussion. Both the control and I/O processors execute programs. In order for either processor to execute a program, that program must be stored in primary memory and must be expressed in the computer's machine language.

The control and I/O processors operate *asynchronously* with respect to each other. This means that both processors can be executing programs simultaneously, essentially independent of each other. Direct communication between the two processors (illustrated by the broken line in Figure 2.1) is limited. The control processor can start and stop the operation of the I/O processor. Once the I/O processor starts executing a program, it operates independently of the control processor until it has finished executing its program or encounters some trouble. In order for any I/O to take place the control processor must tell the I/O processor to start. The I/O processor then executes a program that causes the desired I/O to take place. The I/O processor notifies the control processor when the requested I/O has been completed. The I/O processor has very minimal computational and decision-making ability, and all decisions of any complexity must be made by the control processor.

The primary memory of most computers is relatively small compared to the size of secondary memory. In most computers the storage capacity of the disk unit is one or two orders of magnitude larger than that of the primary memory, and the storage capacity of a magnetic tape unit is essentially unlimited. However, secondary storage has an access time that is correspondingly greater. For example, to move a word of data from the disk unit to the primary memory may take 1000 times longer than to move a word of data from primary memory to either of the processors. In addition, the control processor has no direct access to information stored in secondary memory.

2.3 An Insider's View of the System

In this section we will develop the implications of both the objectives discussed in Section 2.1 and the hardware properties described in the preceding section. These objectives and hardware properties are fixed, and we cannot change them. Designing the system consists of finding an algorithm that satisfies the objectives and that will also execute on the given computer. Some implications are immediate consequences of the functional objectives. For example, if the user can include in his job a procedure written in FORTRAN and specify that this procedure be translated, then the system must be able to translate FORTRAN procedures.

Other implications are less immediate. Since a user's job may contain several procedures that are not expressed in machine laguage, all of these procedures must be translated before the user's program can be executed. These procedures are translated one at a time. Thus, the system must save the result of each translation until all procedures have been translated.

In order for any program to execute it must be in the primary memory. The system itself is a program that the computer must execute, so at least the part of the system that is executing must be in the primary memory. Thus, the system must manage the primary memory, that is, it must insure that the currently executing procedure, be it system or user, is in primary memory.

Let us follow the scenario that guides the system in processing a user's job. A job consists of a number of job steps. The system processes each job step in sequence until the final step has been completed. After processing the final step of a job the system will have to clean up any "loose ends" from the now completed job and prepare for processing the next job.

The command in a job step may specify translation of a procedure. We assume the primary memory is not large enough to hold all of the system. This is generally true, even for computers with very large primary memories. Thus, the translator program that is needed to translate the procedure is stored in secondary memory instead of in primary memory. Therefore, the system must find a place in primary memory to store the translator and then move it from secondary memory into primary memory. In doing this the system will have to displace some other program, which is currently in primary memory, in order to find a place for the translator. (If it never had to do this then primary memory would be large enough to hold all of the system, contradicting our assumption.)

Once the translator is in primary memory it can begin to execute. It now translates the user's procedure. In doing so it must be able to find the procedure that is to be translated. The translated procedure, which is expressed in machine language, must be saved somewhere so that it can be found later when the system is ready to let the user's program execute.

After all of the procedures in the job that need to be translated have been translated, the system can allow the user's program to execute. However, one more thing must be done first. The program still consists of a collection of separate procedures. Even though they are all expressed in machine language, they are like pieces of a jigsaw puzzle that must be fitted together to form a whole program before execution is possible. The system must find a place in primary memory for the resultant program, and then execution of the user's program commences.

In order to satisfy the objective that the system be able to process one job after another without operator intervention, the system must be able to recognize the beginning and end of each job and each job step. This it does by reading and examining the commands. Analysis of the commands is also required to find out what action is desired for a job step. Even though the operator is not required to act for normal job processing, he must act in abnormal situations. Consequently, he must be kept informed as to the status of the system. Thus, the system should print a message on the operator's console at the beginning of each job and job step identifying the job or job step. If anything goes wrong during the processing of a job, such as an illegal command, both the operator and the user should be informed. In addition the system should recover from the problem and go on to the next job or job step. Since computer time is not free, the system must keep track of how much time it takes to process each user's job and bill the user for this amount of computer time.

We can group the tasks that the system must perform into three major functions: *job control, memory management,* and *I/O control.* Job control consists of interpretation of the commands, which includes invocation of the appropriate system programs required to perform the action specified by the commands, sequencing from job to job and from job step to job step, time accounting, and some error detection and recovery. Memory management keeps track of which memory areas are in use and which are not. It also puts together the procedures in a job to form a whole program. I/O control does the reading and writing, such as reading cards and printing, as well as generally controlling all I/O devices. In addition, it manages the use of secondary memory. These are the three major internal functions of any operating system.

2.4 Software Components of the System

In this section we begin to give structure to the operating system by determining what software components are implied by the discussion in the preceding section. These components include both procedures and data. The relationship of the procedure components to each other is shown in Figure 2.2 and will be discussed in more detail later. The principle of functional modularity, which is almost universally used in all fields of engineering, dictates that each

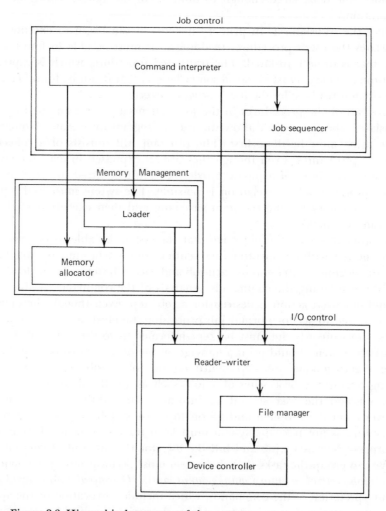

Figure 2.2 Hierarchical structure of the system.

substantial function of the system should be isolated into a separate procedure. The data should be organized into collections of related information. To the extent possible each such collection should be utilized for only a single function. The process of isolating functions is iterative; usually several iterations, each resulting in a new level of increased detail, are required in the course of designing a system. The first level was described in Section 2.2 and consisted of three functional components: job control, memory management, and I/O control. This section is the second iteration, wherein we define the second level by developing each of the first-level components in more detail.

We divide job control into two components: the *command interpreter* and the *job sequencer*. These correspond to the two major functions of job control: interpretation of commands and sequencing of jobs and job steps. The command interpreter analyzes the information in a command to find out what system action is being requested. It then sees to it that this action is performed. This usually consists of initiating execution of a system command program such as a language translator. The principal data required to support the command interpreter is a table of legal commands called the *command table*. This table will have an entry for each command containing the name of the command and information specifying where the corresponding command program is located in primary or secondary memory.

The job sequencer is invoked at the end of each job step to accomplish the transition to the next job step or the next job. The job sequencer also keeps the operator informed as to the progress of the job and "cleans up" after each job step and job. It also performs the accounting function by keeping a record of the charges to be made to the user who submitted the job. The job sequencer also attempts limited error recovery in case an illegal command is encountered or some other difficulty is encountered by a system program. The principal data required to support these functions, in addition to the job itself, is called the *job description*. This data includes the user's identification, the charges he has accumulated so far, and constraints on the processing of the job, such as the maximum time the user wishes his program to execute.

The two major components of memory management are the *memory allocator* and the *loader*. The function of the memory allocator is to keep a list of the portions of primary memory that are currently being used. We call this list the *use list*. Using this data the memory allocator can find space for new programs when requested to do so. The *loader* takes all of the procedures in a job and puts them together so that they form a single, complete program that can be executed by the computer. This activity is called *loading*. Ordinarily the only data required by the loader is the set of procedures that is to be loaded.

I/O control does the reading and writing of data, controls the operation of all I/O devices, and manages the use of secondary storage. The *reader-writer* is responsible for reading data from devices into the primary memory and writing data from primary memory onto devices. The *file manager* manages secondary memory. Information is stored in secondary memory in large units called *files*. The file manager maintains a use list for each secondary memory device and allocates space on these devices. It also keeps a record of where each file is located in secondary memory. This record is called the *file directory*. Both devices and files are addressed in a uniform way using symbolic names. The reader-writer must know if it is being asked to read or write a file or a device. It must also know which device corresponds to each symbolic device name. All of this information is kept in the *attachment table*.

In order to actually carry out I/O the I/O processor must execute an I/O program. The reader-writer and file manager are responsible for constructing the appropriate I/O program, but operation of the I/O processor is initiated by a third component, the *device controller*. It must maintain a record of the status of all current I/O activity so that it can respond with suitable action in case of trouble or when some I/O operation is completed. The information for each I/O operation is maintained in its *I/O status block*.

It should be pointed out that division of the operating system into the three major components that we have described is not the only possible grouping of functions. However, it seems quite natural, since the three major components correspond quite closely to the three major hardware components: memory management to primary memory, I/O control to the I/O processor, and job control to the control processor. Even more choice exists with respect to the second level of detail, which consists of seven procedure components and more than half a dozen data components. This division into components is where art and experience influence the design activity. Principles such as functional modularity help, but they cannot automatically decide how a system should be divided into components.

2.5 The Command Language

In order to proceed further in the design of our system we need a more detailed definition of the system's functions. A complete definition of the commands will satisfy this requirement, since they are the system's interface to the user and thus define what the system does as far as the user is concerned. The complete set of commands constitute a *command language*. In some systems the command language is called the *Job Control Language*, or *JCL*. In many systems, including ours, commands are punched on cards. In other systems they may be typed on a typewriter terminal.

The JOB command defines a job and has the form:

//JOB   *user_name,project_id,time,lines,cards*

It must be the first command in a job. A job consists of a sequence of job steps that follow the JOB command. Each job step specifies a single action, such as translation of a user procedure, loading of his program, or execution of his program. All of the procedures that a user includes in his job are part of a single program. After they all have been translated this program may be loaded and executed as the last two steps of the job. If only translation of procedures is desired these last two steps may be omitted from a job.

In the JOB command "JOB" is a key word. In our example command language and the IBM OS/360 JCL key words are written entirely in uppercase. In Multics they are written entirely in lowercase. The remaining information in the command is a set of arguments that are similar to the arguments in a procedure call. The first two arguments identify the user and the project that is to be charged for processing this job. The remaining arguments specify limits on the execution time, number of lines of printed output, and the number of punched cards. These limits are chosen by the user and apply only to execution of his program, not system programs. This allows the user to limit his liability in case his program does not operate properly, since the system will terminate the job as soon as any one of these limits is exceeded.

Each job step calls for the execution of either a system command program or the user's program. The command specifying this execution has the format:

//EXECUTE   program_name,arg_1,...,arg_n

The number and meaning of the arguments depend on which program is to be executed. A job step consists of an EXECUTE command followed by any data required by that command. The job step is terminated by an END command that has the format:

//END   any_comment

If the EXECUTE command specifies a language translator, the user procedure to be translated is the data that the command requires. If the EXECUTE command specifies execution of the user's program, the data required by the command is whatever data the user's program requires.

The only other kind of EXECUTE command in our system is one that specifies loading the collection of procedures that constitutes a user's program. The data required by this command is the set of procedures that makes up the user's program. In this case all the procedures that were not translated in any of the preceding job steps of this job are included as data following the EXECUTE command. These procedures must be in machine language, that is, they must have been translated by some other job and punched on cards.

Any job step may be preceded by an arbitrary number of comments (comment commands) that have the form:

//*any_comment

All of the commands, including comments, are printed on the operator's console and included in the printed output, which is returned to the user. Notes indicating serious errors and unusual events are also printed in both places. Thus, the operator is aware of the progress and status of the current job. Since all commands are printed for the operator, a user may include comments to communicate to the operator information on any unusual events that the user expects will occur. In addition, when a system command program executes, it may include information concerning its execution in the printed output that is returned to the user. Since all of this output to the user occurs in the sequence in which things happen, he is provided with a log that records all that happened when his job was processed.

Our example system contains two language translators, a PL/1 compiler and an assembler. They translate a procedure written in PL/1 or assembly language, respectively, into machine language. The PL/1 or assembly language version of a procedure is called the source procedure, and the translated machine language version is called the object procedure. A translator normally produces three kinds of output: a detailed listing of the object procedure, the object procedure punched on cards, and the object procedure stored in secondary memory for later use by the loader. Any of these outputs may be suppressed by including the proper argument in the command that specifies the translation. This command may have from zero to three arguments in addition to program_name. The argument "NOLIST" suppresses the detailed listing of the object procedure, the argument "NOPUNCH" suppresses the punched object deck, and the argument "NOOBJECT" suppresses the copy of the object procedure that is saved for the loader.

When a translator is finished, it returns to the system. Prior to returning it sets a status code indicating whether the translation was successful or errors were detected. If any translator reports an error the system will not process a job step calling for either loading or execution of the user's program. The loader also sets this status code before returning to the system and, if it indicates an error in loading, the system will not process the job step calling for execution of the program that was being loaded.

Neither loading nor execution of a user program is required, that is, a job may consist only of job steps calling for translation. However, if loading and execution are desired, they must be the next to last and last job steps, respectively.

Figure 2.3 is an example of a job that our system will process. The job consists of three job steps that compile and execute a simple PL/1 program. The first job step calls for the PL/1 compiler to translate the program that follows the EXECUTE command and is terminated by the first END command. The EXECUTE command includes the argument NOPUNCH, which specifies that no object deck is to be punched. The card containing "END;" is required by PL/1 and will be processed by the PL/1 compiler. The second job step calls for the loader to prepare the translated PL/1 program for execution. Since

```
// JOB   GRAHAM, CLASS, 1, 20, 0
//* DEFINE JOB: CHARGE TO PROJECT CLASS; LIMIT EXECUTION
//*   TO ONE SECOND, LINES PRINTED TO 20, AND NO CARDS
//*   WILL BE PUNCHED
//* FIRST JOB STEP FOLLOWS: IT SPECIFIES TRANSLATION OF
//*   THE PL/1 PROCEDURE WHICH FOLLOWS THE EXECUTE
//*   COMMAND
// EXECUTE   PL/1, NOPUNCH
SPEED: PROCEDURE;
FIRST: GET LIST (DISTANCE, TIME);
       AVERAGE_SPEED = DISTANCE/TIME;
       PUT LIST (DISTANCE, TIME, AVERAGE_SPEED);
       GO TO FIRST;
       END;
// END     THIS COMMAND ENDS THE FIRST JOB STEP
//* SECOND JOB STEP FOLLOWS: IT LOADS THE PROCEDURE
//*   WHICH WAS JUST TRANSLATED
// EXECUTE   LOADER
STR SPEED     LOADER DATA CARD WHICH BEGINS EXECUTION
// END     THIS ENDS THE SECOND JOB STEP
//* THIRD JOB STEP FOLLOWS: IT EXECUTES THE PROGRAM
//*   JUST LOADED; THE DATA REQUIRED FOR EXECUTION FOL-
//*   LOWS THE EXECUTE COMMAND
// EXECUTE   OBJECT
   12.65  8.21;  9  20;  .52  .58
// END     THIS ENDS BOTH THE THIRD JOB STEP AND THE JOB
```

Figure 2.3 Example job consisting of three job steps.

this EXECUTE command is immediately followed by an END command, no other procedures are to be included in the program. The third and final job step calls for execution of the loaded program. The data to be used by the program when it is executing follows the EXECUTE command and is terminated by the last END command.

The printed output, which is returned to the user after the job has been processed, is shown in Figure 2.4. The JOB command that defines this job is printed, followed by the EXECUTE command for the first job step, which is the PL/1 translation. Following this is the detailed listing produced by the PL/1 compiler, since this listing was not suppressed by an argument in the EXECUTE command. Next, the END command that terminated this job step, all of the following comments, and the EXECUTE command that begins the next job step, are all printed. The second job step calls for the loader. Any information that the loader wished to print follows next. For example, a list of all the procedures in the loaded program will be printed. Following the commands for the third job step is any output that was printed by the PL/1 program while it was executing.

Preceding and following the output described above is sign-on and sign-off information supplied by the system. The sign-on information includes the date and time of day when the system began processing this job and a summary of the latest additions to the system or any other important news the user should know. The sign-off information includes the date and time of day when the

```
      {sign-on information}
// JOB   GRAHAM, CLASS, 1, 20, 0
//* DEFINE JOB: CHARGE TO PROJECT CLASS; LIMIT EXECUTION
//*    TO ONE SECOND, LINES PRINTED TO 20, AND NO CARDS
//*    WILL BE PUNCHED
//* FIRST JOB STEP FOLLOWS: IT SPECIFIES TRANSLATION OF
//*    THE PL/1 PROCEDURE WHICH FOLLOWS THE EXECUTE
//*    COMMAND
// EXECUTE   PL/1, NOPUNCH
      {program listing produced by PL/1 compiler}
// END      THIS COMMAND ENDS THE FIRST JOB STEP
//* SECOND JOB STEP FOLLOWS: IT LOADS THE PROCEDURE
//*    WHICH WAS JUST TRANSLATED
// EXECUTE   LOADER
      {loading information such as a list of procedures loaded}
// END      THIS ENDS THE SECOND JOB STEP
//* THIRD JOB STEP FOLLOWS: IT EXECUTES THE PROGRAM
//*    JUST LOADED; THE DATA REQUIRED FOR EXECUTION FOL-
//*    LOWS THE EXECUTE COMMAND
// EXECUTE   OBJECT
   12.65   8.21   1.54;   9.00   20.00   0.45;   0.52   0.58   0.90
// END      THIS ENDS BOTH THE THIRD JOB STEP AND THE JOB
      {Sign-off information}
```

Figure 2.4 Output from the example job shown in Figure 2.3.

system finished processing this job, comments on why the job was terminated, and the charges for processing this job.

2.6 The System Building Blocks

Except for job control, this section contains the final iteration of our design. We will specify in detail the function of each of the procedure components identified in Section 2.4, with particular attention to how each interfaces with the other components. Our basic concern here is with the interaction between the components of the system. Referring to Figure 2.2, observe that the command interpreter occupies a position of special importance in the system. It is the component that directly or indirectly controls the actions of all of the other components in the system. For example, the reader-writer will read or write only when called directly by the command interpreter, the job sequencer, or the loader, each of which is called by the command interpreter. The command interpreter performs its task continuously as long as user jobs (which contain commands to interpret) are supplied to the card reader by the operator.

Job control (the job sequencer and the command interpreter) will be discussed in detail in the next section. Here we will focus on the remainder of the system. One fundamental decision we must make is how primary memory is to be used. In the interest of simplicity, but not necessarily efficiency, we choose a very simple policy for memory use. Job control, memory management, and I/O control will all be permanently stored in the primary memory. We call these components the *resident system*. The remainder of the system, consisting of the command programs, will be stored on the disk unit (secondary memory). Whenever a command program or user program is to be executed, it is loaded into that part of primary memory not used by the resident system (see Figure 2.5). Command and user programs are essentially treated the same.

No attempt is made to save a command program from one use to the next, that is, if the PL/1 translator is requested in two consecutive job steps it is copied from the disk unit into primary memory twice. With this simple policy for the use of primary memory, the memory allocator is so trivial that we can dispense with it by absorbing its function into both the command interpreter and the loader. The use list of the memory allocator is simply the number K from Figure 2.5. In more sophisticated systems memory usage policies cannot be this simple, and the memory allocator can become quite complex. Chapter 15 will consider the problems of memory management in greater depth.

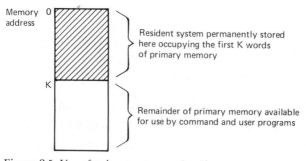

Figure 2.5 Use of primary memory by the system.

I/O control is solely responsible for all of the devices, including the disk unit. Other parts of the system as well as user programs must call the reader-writer to transmit any information to or from any device. Symbolic names are used to reference both the I/O devices, such as the card reader and printer, and the files that are stored in secondary storage. By convention, certain names are reserved for the names of commonly used I/O devices and files. These names and their definitions are listed in Figure 2.6. Any name not listed is assumed to be a user-defined name. The first seven entries in the attachment table are entries for the names in Figure 2.6. These will be permanent entries reflecting the fact that these names are reserved.

It is the duty of I/O control to insure that read or write requests using one of these reserved names are addressed to the indicated device or file. For example, all write requests that reference PRINT will cause the transmitted information to be printed on the printer as part of the job's output, which is given to the user. Thus, if the system components (both the resident components and the command programs) and the user's program always write on PRINT whenever there is information for the user, all such information will always reach the user.

The reader-writer component of I/O control has four entry points that are called by other procedures (both system and user) that wish to use I/O devices or files. Externally there is no distinction between I/O devices and files. Before a file or I/O device can be read or written, it must be *opened* for either reading or writing, but not both. When a file's or I/O device's use is concluded, it is *closed*. Once closed it can be opened again for a different use. For example, a file can be opened for writing, information then written into the file, the file closed, and then opened for reading so that the information that was previously written into the file can now be read.

The call

> **open** (*sname, direction, status*);

will open the file or I/O device *sname* for reading or writing depending on the value of *direction*. Any information already contained in a file being opened for writing is destroyed. Upon return to the calling procedure the value of *status* will indicate success or failure and the reason for the failure. **open** also performs any initialization and repositioning required to ready the file or I/O device for reading or writing as requested. **open** also updates the entry for *sname* in the attachment table.

Name	Device or file and its use
IN	Card reader: jobs to be processed by the system
PRINT	Printer: printed output from the job
PUNCH	Punch: punched output from the job
OPERATOR	Operator's console: comments to the operator
PL/1_CMD	File containing the PL/1 compiler
ASSEMB_CMD	File containing the assembler
OBJECT	File in which object procedures resulting from translation during current job are saved

Figure 2.6 Definition of reserved names for devices and files.

The call

$$\text{close } (sname);$$

will close the file or I/O device *sname*. **open** must be called before *sname* can be read or written again.

The call

$$\text{read } (sname, count, location, status);$$

will read the next record from *sname* into primary memory beginning at *location*. Files and I/O devices contain a sequence of *records*, each record being a convenient unit of information. For example, on the card reader or card punch a record is equal to a card, which is 80 characters, while a record on the printer would be a print line, which is from 1 to 120 characters. After reading the record *count* is set equal to the number of characters in the record. The value of *status* will be set to indicate success or failure. If a record was successfully read *status* will equal zero. If the file or I/O device is empty, that is, there are no more records, *status* will equal one. Other nonzero values of *status* indicate failure to read a record for other reasons.

The call

$$\text{write } (sname, count, location, status);$$

writes *count* characters from primary memory, beginning at *location*, to *sname* as its next complete record. The value of *status* is set to indicate the result.

The system must guarantee that PRINT, PUNCH, and OPERATOR are permanently open for writing and that IN is permanently open for reading. This means that the reader-writer must not honor any calls to **close** or **open** that refer to IN, PRINT, PUNCH, or OPERATOR. In addition, the translators may assume that OBJECT is open for writing; therefore the system must open it when the job begins and close it before calling the loader.

Before an I/O device can be used it must be associated with a symbolic name, *sname*. Once this association has been made, *sname* will be *attached* to the I/O device. The call

$$\text{attach } (sname, device, status);$$

will attach the symbolic name *sname* to the I/O device *device*, unless it is already attached to some other I/O device. Upon return to the calling procedure the value of *status* will indicate success or failure and the reason for failure. The call to **attach** may fail because *sname* is already attached to some other device or because *device* does not exist (is not connected to the computer). Successful execution of a call to **attach** will result in an entry for *sname* being added to the attachment table. The call

$$\text{detach } (sname);$$

will delete the entry for *sname* from the attachment table, if one exists.

There are two entries to the file manager for the creation and destruction of files. The call

$$\text{create } (sname, status);$$

creates an empty file whose name is *sname*. All this really amounts to is making an entry in the file directory. Disk space is not allocated to the file until information is actually written into the file.

The call

$$\text{delete } (sname);$$

destroys the file *sname*. The disk space allocated to it is returned to the file manager for use by other files. The entry for *sname* is also removed from the file directory.

It is the responsibility of the file manager to keep a record of the used and unused space on the disk. Disk space for a file is allocated only as needed, that is, whenever a record is written into the file and thus onto the disk. When a file is deleted or opened for writing, all of the disk space it currently occupies is released for reuse. The reader-writer is called for use of both files and I/O devices. It has the responsibility of determining from the attachment table which one is being referenced. If it is an I/O device the reader-writer takes care of things; however, if it is a file, the reader-writer must call the file manager. The device controller is called by both the reader-writer and the file manager to actually read or write an I/O device or the disk unit.

A detailed description of I/O control, especially the file manager and the device controller, is very complicated. There are many problems that we have not even caught a glimpse of yet. There is more than one view as to how I/O devices and secondary memory should be managed. In Chapters 15 and 16 we consider two different views of secondary memory management. In Chapter 16 we consider the management of I/O devices and the I/O processor in some depth.

The loader is part of the resident system and is used to load a user's program. All of the object procedures that are the result of translations specified in previous job steps and all of the user-supplied object procedures are put together to form a complete user program that is ready to execute. This program will be stored in primary memory beginning at location K in Figure 2.5. The loader calls the reader-writer to read the object procedures that make up the program. The object procedures that are the result of previous translations in this job will be in the file OBJECT. The user-supplied object procedures are included in the job step that specifies loading between the EXECUTE command beginning the job step and the END command ending the job step. The loader reads these object procedures from the device IN. In fact, all command programs read IN to obtain user-supplied data. A typical loader and the problems of loading are discussed in depth in Chapter 6.

2.7 Job Control

In this section we examine in detail the components of job control. As we have previously pointed out, the command interpreter is the component that controls the action of all the other components of the system. Its structure, shown in Figure 2.7, is essentially a single endless loop. We will assume that the operator always has at least one unprocessed job in the card reader. This means that the card reader never becomes empty, and the system never runs out of work to do. Under this assumption, once the system has been started the command interpreter can go into an endless loop, interpreting one command after another.

Each job begins with a JOB command. We will assume that the operator has visually checked each job for the presence of a JOB command before putting the job in the card reader. With this assumption the end of one job and the beginning of the next job are both signaled by a JOB command. The beginning of a job step is signaled by an EXECUTE command. Any comments preceding the EXECUTE command are ignored, except for writing them as part of the user's output and on the operator's console.

The algorithm for job control in Figure 2.7 is also shown in Figure 2.8,

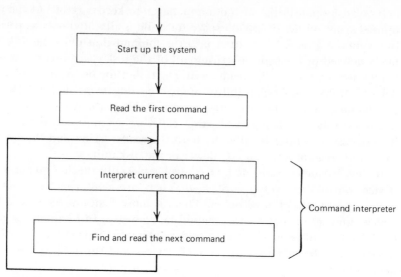

Figure 2.7 Overview of job control.

```
procedure job_control;              {procedure statement}
   Start up the system;             {simple imperative statement}
   Read the first command;
repeat                              {opening bracket of loop statement}
   Interpret current command;       {first statement in loop}
   Find and read next command;      {last statement in loop}
until forever;                      {closing bracket and termination
                                       condition of loop}
end.                                {closing bracket of procedure
                                       definition}
```

Figure 2.8 Algorithm of Figure 2.7 expressed as a procedure definition in the informal programming language.

expressed in an informal programming language (IL). This language has a format similar to a procedure written in PASCAL,[1] PL/1, or ALGOL.[2] However, in order to avoid obscuring the basic structure and significant operations, English-like statements are used frequently instead of writing a procedure that is complete enough to actually be translated and executed. Flowcharts will occasionally be used to show the overall structure of an algorithm, as in Figure 2.7. However, we will use the informal programming language format to present most algorithms, especially detailed ones.

There are two major reasons for this approach. We assume that the reader is interested in improving his skills in programming, particularly in systems programming. In this case continued practice in the reading and writing of

---

[1] See N. Wirth, *Systematic Programming: An Introduction*, Prentice-Hall, Englewood Cliffs, N.J., 1973.

[2] See J. E. Sammet, *Programming Languages: History and Fundamentals*, Prentice-Hall, Englewood Cliffs, N.J., 1969, for descriptions of both PL/1 and ALGOL.

programs and program fragments is essential. Second, we have found that extensive use of flowcharts, especially in designing algorithms, often unwittingly leads the designer into the construction of algorithms that have unnecessarily complex structure when translated into some programming language (which must be done if the algorithm is going to be executed by a computer). This complexity is the cause of errors in the program that must be detected and eliminated during program testing. Both the complexity and the errors can usually be significantly reduced and sometimes even completely avoided by expressing the algorithm in programming language form from the very beginning.

In the language used in Figure 2.8 a procedure is defined as a sequence of statements that are assumed to be executed in the sequence in which they are written. Most simple imperative statements are expressed in informal English and printed in ordinary type. Special statements, such as loop statements and declarative statements, use key words that are printed in boldface. Proper names of procedures and variables are printed in a distinguishable type font. Comments are enclosed in braces and may be inserted between any two statements. Indentation of the additional lines of a statement when it is longer than a single line is an essential feature of the language. This feature is a significant aid in showing clearly the structure of an algorithm.

A procedure definition has the form:

procedure *name*; *body* **end**.

This defines a procedure whose name is *name*. The procedure's action is defined by *body*, which is a sequence of statements that are called the *body of the procedure*. Each statement is terminated by a semicolon. In Figure 2.8 a procedure named job_control is defined. Note that proper names may not contain any blank spaces. Compound names are formed by using the underscore "_" as a connector. In our example the body of the procedure consists of three statements. The first two statements are *simple imperatives*. The third statement is a *loop statement*.

A loop statement has the form:

repeat *body* **until** *condition*;

The statements in *body* are called the *body of the loop* and are repeatedly executed until the value of *condition* is true. This condition is called the *termination condition* and is tested after execution of the body of the loop. Therefore, the statements in *body* get executed at least once. In Figure 2.8 the body of the loop consists of two simple imperatives. The key word "**forever**", which is used as the termination condition, results in a nonterminating loop, since its value is always false.

Figure 2.8 also illustrates the indentation feature. Both the procedure definition and the loop statement are *compound statements*, that is, they consist of one or more statements enclosed by some kind of brackets. The key words "**procedure**" and "**end**" are brackets for a procedure definition and the key words "**repeat**" and "**until**" are brackets for a loop statement. The indentation rule is: all statements in the body of a compound statement are indented two spaces further than the brackets of the compound statement. This applies to a compound statement, which is a statement in the body of another compound statement. Thus, in Figure 2.8.

Start up the system;

is indented two spaces while
Interpret current command;
is indented four spaces.

The design of job_control continues by considering each of the simple imperative statements in Figure 2.8 and expanding them into several statements, thereby introducing more detail. First, we consider the statement
Start up the system;
We assume that the resident system has been loaded into primary memory and that the system command programs are stored on the disk unit. Starting up the system is principally initialization. Figure 2.9 shows more details. The execute, print, and punch limits are set to extremely large (essentially unlimited) values whenever any part of the system is executing. We assume that the system is reliable enough so that no limits are required. In the interest of good accounting practice, whenever the computer is being used its use must be charged to some account. Therefore, the system start-up is executed as part of a special system start-up job, and the charges that that job incurs are billed to system operation overhead.

After starting up the system, the first command, which will be the JOB command for the first job, is read from the card reader. More details of this operation are shown in Figure 2.10. A procedure call in our informal language is written as:
procedure_name (arguments);
where *arguments* is a list of argument values for the procedure, separated by commas. If there are no argument values for the procedure, both *arguments* and the enclosing parentheses are omitted.

The call to read in Figure 2.10 reads the next record from IN, which is the card reader, into an array named cur_cmd, which must be able to hold the 80 characters from a card. The first argument of read is a character string, which is the name of the device or file to be read. When read returns, the value of status indicates success or failure.

In Figure 2.10 the statement used to test the value of status is the if statement, which is another compound statement. Its general form is:
if *condition* then *statement_1* else *statement_2*

Set the execute, print, and punch limits to system values;
Set the job description to indicate system start up job is being processed;
Set permanent entries in the attachment table for the devices IN, PRINT,
   PUNCH, and OPERATOR and the files PL/1_CMD and ASSEMB_CMD;
Initialize other system variables and tables;

Figure 2.9 Expansion of "Start up the System;" from Figure 2.8.

read ('IN', count, cur_cmd, status);
if status $\neq$ 0 then
  begin Write error message on OPERATOR;
    Stop for operator action;
  end;

Figure 2.10 Expansion of "Read the first command;" from Figure 2.8.

Only one of the two statements is executed; *statement_1* is executed if the value of *condition* is true, and *statement_2* is executed if the value of *condition* is false. The variant of the **if** statement used in Figure 2.10 omits the **else** clause. Its form is:

if *condition* **then** *statement*

where *statement* is executed only if the value of *condition* is true. Otherwise this variant of the **if** statement has no effect. The statement following the key word "then" in Figure 2.10 is another kind of compound statement called a *block*. Its form is:

begin *body* **end**;

The sequence of statements in *body* are called the *body of the block*. As with all compound statements it can be used anywhere that a simple statement can be used.

Figure 2.11 is the command interpreter proper. Interpretation of a command depends on its type. The current command is in **cur_cmd**. The type of this command is used to select for execution one of a set of statements that forms a new kind of compound statement, the **case** statement. Its form is:

case *index* **of**
label_1: *statement_1*
label_2: *statement_2*
•••
label_n: *statement_n*
otherwise *statement*
end;

where the **otherwise** clause is optional. At most one of the statements is executed. If the value of *index* matches one of the labels, the first statement with a matching label is executed. If none of the labels match, then the statement preceded by the key word "otherwise" is executed, unless the **otherwise** clause is omitted, in which case no statement is executed. The labels used may be any constant consistant with the data type of *index*.

In Figure 2.11, the type of a command is determined by examining what follows the two characters "//" that begin a command. If it is one of the key words "JOB", "EXECUTE", or "END", the type of the command is 'job', 'execute', or 'end', respectively. If an asterisk follows, the command type is

Write contents of **cur_cmd** on both PRINT and OPERATOR;
type := type of command which is in **cur_cmd**;
case type of
   'job': **begin** Terminate current job;
       Begin next job;
     end;
   'execute': Execute appropriate command program;
   'end': **null**;
   'comment': **null**;
   otherwise message ('Illegal command type.');
end;

Figure 2.11 Expansion of "Interpret current command;" from Figure 2.8.

'comment'. If none of these follow, the type is undefined. Another new statement appears in Figure 2.11, the **null** statement, which does nothing.

The **otherwise** clause is used to catch all commands that are not one of the four legal types. It guarantees that one of the statements in the case statement will always be executed. The proper action for the command interpreter to take is to print an error message for both the user and the operator. The procedure **message** is used to do this. Its definition is in Figure 2.12. This procedure has a single argument, a character string, which is the message to be written on both IN and OPERATOR (the line printer and the operator's console). Notice that in the definition of a procedure, if the procedure has any arguments, dummy variables corresponding to these arguments follow the name of the procedure in the **procedure** statement at the beginning of the procedure definition.

After a command has been interpreted, we must find the next command before we can interpret it. The details of this action are shown in Figure 2.13. Any records in the input that are read before a command is found are simply ignored. Actually, this is part of the job sequencer, which we identified as one of the components of job control. In this iteration of the design we have changed our mind and broken the job sequencer up into three parts. Job-step sequencing is now an integral part of the command interpreter. It occurs as part of finding the next command (Figure 2.13) whenever the next command happens to be an EXECUTE command. The remainder of the job sequencer component is the block:

'job': **begin** Terminate current job;
　　　　　　Begin next job;
　　　end;

from Figure 2.11. The complete definition of job control is shown in Figure 2.14.

The first part of job sequencing, termination of the current job, is shown in Figure 2.15. The action taken here is relatively straightforward. The procedure **delete** is called to delete OBJECT, just in case it still exists, as it might if there were any problems in processing the job. The accounting card that is punched is saved for later use as data for the computation center's accounting programs. These programs will be executed periodically to compute and print bills for the users of the system.

The second part of job sequencing begins the next job; it is shown in

```
procedure message (text);
    count := number of characters in text;
    write ('PRINT', count, text, status);
    write ('OPERATOR', count, text, status);
end.
```

Figure 2.12 Definition of procedure to write error messages.

```
repeat Read next record from IN into cur_cmd;
until first two characters in cur_cmd = '//';
```

Figure 2.13 Expansion of "Find and read next command;" from Figure 2.8.

```
procedure job_control;
    Set the execute, print, and punch limits to system values;
    Set the job description to indicate system start up job is being processed;
    Set permanent entries in the attachment table for the devices IN, PRINT,
        PUNCH, and OPERATOR and the files PL/1_CMD and ASSEMB_CMD;
    Initialize other system variables and tables;
    read ('IN', count, cur_cmd, status);
    if status ≠ 0 then
        begin Write error message on OPERATOR;
            Stop for operator action;
        end;
    repeat
        Write contents of cur_cmd on both PRINT and OPERATOR;
        type := type of command that is in cur_cmd;
        case type of
            'job': begin Terminate current job;
                    Begin next job;
                end;
            'execute': Execute appropriate command program;
            'end': null;
            'comment': null;
            otherwise message ('Illegal command type.');
        end;
        repeat Read next record from IN into cur_cmd;
        until first two characters in cur_cmd = '//';
    until forever;
end.
```

Figure 2.14 Second level of detailed specification of job control obtained by substituting the expansions in Figures 2.9, 2.10, 2.11, and 2.13 for the appropriate statements in Figure 2.8.

```
delete ('OBJECT');
Compute charges for this job;
Punch accounting card containing user's name, project identification, and
    charges for this job;
Write sign-off information on both PRINT and OPERATOR;
```

Figure 2.15 Expansion of "Terminate current job;" from Figure 2.11.

```
Copy the user name, project identification, and user limit values from the
    JOB command in cur_cmd into the job description;
Write sign-on information on both PRINT and OPERATOR;
create ('OBJECT');
open ('OBJECT', 'write', status);
execute_flag := true;
```

Figure 2.16 Expansion of "Begin next job;" from Figure 2.11.

Figure 2.16. The information in the JOB command, which defines the job, must be saved in the job description for later use. A file named OBJECT is created and opened for writing. This is the file in which each translator will store a copy of the object procedure that it produces. By convention translators expect to find this file open for writing. The value of execute_flag is initially set equal to **true**. If any translator or the loader encounters an error, it will set the value of execute_flag equal to **false**. Loading and execution of the user's program will be suppressed unless the value of execute_flag equals **true**.

The details of executing the command or user program specified by an EXECUTE command are shown in Figure 2.17. If the appropriate program is not the loader, which is part of the resident system, or the user's program, which will already have been loaded into primary memory by the loader, it will have to be copied from the disk unit into primary memory before it can be executed. The procedure load_cmd does this. It is shown in Figure 2.18. Each command program is stored in a separate file in a form that is ready for execution. By convention a command program's entry point is always the first word of the command program. This will be memory address K after the command program has been loaded into primary memory. The first three arguments of the call to a command program indicate if an object listing is to be made, a copy of the object procedure saved for loading, or the object program to be punched. The fourth argument, execute_flag, will be set to the value **false** if any errors occur when translating the source procedure.

Returning to Figure 2.17, the loader is part of the resident system and may

```
begin
    name := command name from execute command in cur_cmd;
    case name of
        'pl/1': load_cmd ('pl/1');
        'assembler': load_cmd ('assemb');
        'loader': begin close ('object');
                        if execute_flag
                            then loader (K, execute_flag, start);
                            else message ('Program loading suppressed.');
                    end;
        'object': if execute_flag then
                        begin Set limits to user-limit values that are stored in the
                                job description;
                            Execute a procedure call to the user's program at the
                                entry address that is stored in start;
                            {User program returns here when finished.}
                            Reset limits to system values;
                        end;
                        else message ('Program execution suppressed.');
        otherwise message ('Unknown command name.');
    end;
end;
```

Figure 2.17 Expansion of "Execute appropriate command program;" from Figure 2.11.

procedure load_cmd (cmd_name);
    Find entry for cmd_name in the command table;
    From this entry get name of file in which corresponding command program
       is stored;
    Read contents of this file into primary memory beginning at location K;
    Set values of list, object, and punch to **false** if 'NOLIST', 'NOOBJECT', or
       'NOPUNCH', respectively, are arguments of the command in cur_cmd,
       otherwise their value is **true**;
    Execute a call to location K with arguments (list, object, punch,
       execute_flag)   {By convention the entry point of a command program
       is always at location K.}
end.

Figure 2.18 Definition of procedure to load and execute a command program.

be called directly when it is the program specified by an EXECUTE command. Before calling the loader OBJECT is closed. The loader will open it for reading in order to get the object procedures to be loaded. The loader is called only if the value of execute_flag is true, otherwise loading is suppressed. The first argument of the call to the loader indicates the primary memory location at which to begin loading. The loader returns the address of the entry point to the loaded user program as the value of the third argument.

When the command name in an EXECUTE command is "OBJECT", the user's program is to be executed, but only if the value of execute_flag is **true**. Before actually calling the user's program, the execute, print, and punch limits are set to the values that the user specified in his JOB command. After the user's program has finished and returned to the system, the limits are reset to the special system values. A user program always returns to this part of the command interpreter, even if the program gets into trouble or exceeds any of the limits that the user specified. The system programs that enforce the limits know how to achieve this.

This concludes the detailed description of our simple example of an operating system. Our example includes only a very limited number of commands for the user and does very little error checking. For example, after most of the calls to reader-writer, the value of the status argument is not even examined and no check is made to verify that loading and user program execution are the last two job steps in a job. The reader can easily find additional examples. In practice even a minimal system will include much more complete error checking. Also, many more commands will be included, especially commands dealing with I/O devices and files. In the exercises at the end of this chapter the reader will be asked to correct some of these deficiencies.

The omissions that we have just pointed out are intentional. Our purpose in this chapter was not to study a typical operating system in detail, but to describe the general structure and functioning of a system so that the reader is aware of how the various components interface with each other (fit together) and the role they play in the total system. Much of what we omitted is irrelevant conceptually, but not irrelevant practically. We have also not discussed many significant problems. These will be considered in later chapters.

2.8 Resource Management

Each of the basic functions described in the preceding sections is actually a particular example of a very general function—*resource management*. Our understanding of a complex system is greatly aided if a single idea can be used to explain all of the functions of the system. Resource management is such an idea. In order to apply this idea to our example system we must view the elements of the system as resources. There are four major groups of resources: the control processor, memory (both primary and secondary), I/O devices (and the I/O processor), and the operating system itself (in particular the command programs such as the PL/1 translator).

The heart of the resource allocation view is that each job requires the use of certain resources in order to perform the requested actions and computations. In order to process a job the operating system must make the required resources available to the job. The function of making required resources available is *resource management*. There are two different and independent aspects of resource management that we should be careful to keep separate: *policy* and *mechanics*. For example, the decision whether a job may use a certain device is policy, or rather, policy determines the outcome of this decision. The mechanics of actually making the device available to the job once it has been decided that the job may use the device is independent of the allocation policy.

Let us apply the resource management view to our example system. The first resource is the control processor. The jobs in a batch are processed in sequence, one after the other. Each job has complete use of the control processor from the initiation of the job until the termination of the job. The only constraint will be a maximum execution time, which is specified by the user. This simply limits his liability in case his program does not execute correctly. This particular policy is a basic factor in the system design and, in fact, greatly simplifies the operating system. Policies that share the use of the control processor during the processing of a job and the mechanics of processor allocation will be explored in Chapter 14.

All of the primary memory, except the area occupied by the resident system, is allocated to each job step. Nothing is retained in primary memory from job step to job step except the resident system. All of secondary memory except the portion used to store the system command programs and permanent user files is allocated to the job. Users may permanently store information in the system, but no details of this were discussed. Policies that permit permanent storage of user information and sharing of primary memory are explored in Chapter 15.

The disk unit is permanently reserved for file storage. The card reader, card punch, and printer are permanently reserved for job input and job output. The user's job may use these devices for these purposes and no other. Thus, even though they are allocated to the job, their use is controlled. The remaining devices are all allocated to the job for the duration of the job. Chapter 16 explores more restrictive policies.

Finally, all of the system command programs and other system services are available to the user's job. However, the job's use of the system is restricted. System procedures and commands may be called by a user program only at designated entry points. System procedures and data bases may not be modified by user programs. No such restriction is placed on the use or modification of the user's procedures and data by his own procedure. Alternate policies are explored in Chapter 17.

EXERCISES

2.1 Describe all of the modifications and additions that are required to add a new command to the system described in this chapter:

a. If the command is another language translator, such as FORTRAN.

b. If the command has a different number of arguments than PL/1 and the assembler, or the arguments, have a different interpretation.

2.2 Write IL programs for the following commands:

a.   // EXECUTE   FILE, *fname*
                                   {data cards}
      // END

This command creates the file *fname* and copies the data cards that follow the command into the file. If the file named *fname* already exists, it is used, and the previous contents are lost.

b.   // EXECUTE   DELETE, *fname*

This command destroys the file named *fname* and removes all record of it from the system, if it exists, otherwise the command does nothing.

c.   // EXECUTE   PRINT, *fname*

This command prints the contents of the file named *fname* on the printer. If no such file exists, the system prints an appropriate comment.

d.   // EXECUTE   COPY, *fname1, fname2*

This command creates a new file named *fname2* and writes into it a copy of the contents of the file named *fname1*. If no file named *fname1* exists, an appropriate comment is printed. If a file named *fname2* already exists, it is used and its previous contents are lost.

e.   // EXECUTE   ATTACH, *sname, device_id*

This command attaches the symbolic name *sname* to the I/O device *device_id*. If the attachment cannot be made, an appropriate comment is printed.

2.3 Assume there is a separate, special directory that contains entries for all of the command programs. In the initialization section of the system eliminate placing entries for the command program files into the attachment table. Modify the remainder of the system to find and open the appropriate file before loading a command program and close it after loading each time a command is executed.

2.4 Assume there is a system directory for command programs as in Exercise 2.3 and, in addition, there is a private directory for each user that is stored in secondary memory along with the files. Describe a mechanism that will find and use the proper directory whenever a file is referenced.

2.5 Modify the system so that it can input a single job step or an entire job from some device other than the card reader or from a file. This should be done in such a way that none of the command programs or user programs have to be modified. New control cards will be needed to specify which jobs and job steps are to be read from what devices and files.

3 | Machine and Assembly Languages

The first major software component that we will study is an assembler. An assembler is a translator. It translates from assembly language to machine language. The language translated from is called the *source language,* and the language translated to is called the *object language.* One problem that arises in the design of all translators is that of understanding the two languages involved. In order to build a translator that will correctly translate from source language to object language, we must understand both of these languages. Thus, our first task is to acquire an understandinig of the character of both assembly and machine languages. Only then can we understand the problems an assembler has to solve.

Assembly language is basically symbolic machine language. In fact, most assembly language statements translate into a single machine language instruction. The structure and other characteristics of assembly language are very similar to the characteristics of machine language. For this reason assembly language is different for different computers, while languages such as FORTRAN (which are sometimes called *compiler languages*) are essentially the same for different computers. The best approach to understanding assembly language is first to look at machine language, exposing the different kinds of instructions and the types of operands that can occur as a part of them. After this, understanding the corresponding assembly language forms will be straightforward.

3.1 Machine Language

One property of machine language makes programming in it very difficult. Machine language instructions are binary numbers in most computers. For example, the binary number

$$01000011100010111010000000000000$$

is a machine language instruction for the IBM 360. A program would be a long sequence of numbers like this. This is so difficult to work with that almost no one would be able to write large programs in machine language. Assembly language was invented to overcome this disadvantage.

There is another property of machine language that makes it difficult to use:

the amount of detail required to express a computation. To specify the computation A = B + C in machine language requires three to six instructions. Each instruction requires specification of four or five components. This means selecting and writing a total of 12 to 30 separate components. This is almost an order of magnitude more than are required in a language like FORTRAN where one simply writes A = B + C, which is a single statement (instruction) with five components. Most of this detail required in machine language is routine and of secondary relevance to the computation. In effect this makes machine language programming an order of magnitude more complex than compiler language programming. Assembly language does little to overcome this disadvantage. For this reason most programming today is done using compiler languages such as FORTRAN, COBOL, and PL/1.

There is one advantage in using machine language (actually assembly language). When programming in machine language the programmer may have access to every instruction of the computer and may use them in virtually every possible combination. This gives him complete, detailed control over the way his computation is carried out. It is not possible to exercise the same degree of control using most compiler languages. This complete control is occasionally necessary, and a machine language program is then required. However, in general, such extreme flexibility is not necessary, nor even desirable, since it makes programming more difficult and the resulting programs are harder to understand.

It is not our intent to study computer hardware structure in any detail. However, there are a few facets of the hardware structure and its operation that are important in developing an understanding of machine language. Figure 3.1 is a slightly expanded version of Figure 2.1. It is still a substantial simplification of the functional diagram for any real computer. Our diagram shows only the major functional units and the data transmission paths between them. The control processor and primary memory have both been subdivided, illustrating the major internal functions. Examination of the operation of the I/O processor is postponed until Chapter 16.

The instructions that the control processor executes and the data that these instructions reference must be stored in primary memory. There are four major steps in execution of an instruction by the control processor.

1. Fetch the (next) instruction from primary memory.
2. See what action is specified by this instruction (*instruction decoding*).
3. Fetch the operands from primary memory, if required.
4. Perform the action specified by the instruction.

The control unit of the processor is in charge of instruction execution. The computation unit actually performs the action specified by the instruction. The registers are used for storage of the instruction currently being executed, its operands, the address of the next instruction, and other information needed by the control processor. There is an exception to the requirement that all data that instructions reference must be in primary memory. Some instructions reference data stored in the registers. The registers actually function as a very small, private memory for the control processor.

The primary memory consists of a number of equal-sized units called *memory cells*. An item of information (instruction or data) is stored in one or more of

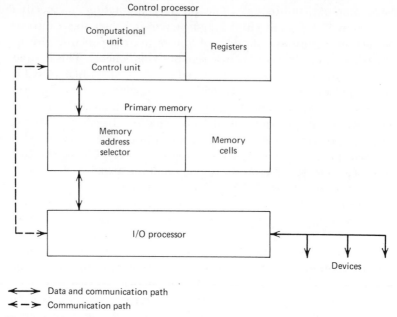

Control processor

Computational unit

Registers

Control unit

Primary memory

Memory address selector

Memory cells

I/O processor

Devices

⟷ Data and communication path
◄ ─ ➤ Communication path

Figure 3.1 Functional diagram of a computer.

these memory cells. Each memory cell has a unique address. To fetch the contents of a memory cell its address must be sent to the memory address selector. To fetch the contents of a contiguous set of memory cells, only the address of the first cell in the set must be sent. In response, the memory address selector sends back the contents of the addressed memory cell or cells. To store information into memory the control unit sends an address and the data to the memory. Clearly, in both cases, the control unit must also send a signal indicating to the memory whether information is to be fetched or stored. The I/O processor references memory in exactly the same way, and the memory address selector distinguishes the source of a request and keeps from getting confused.

A major part of operand fetching is computing the *effective address*. This is the address actually sent to the memory address selector. In most computers the effective address is an integer that is the sum of the contents of an "address" field in the instruction and the contents of one or more special registers that are specified by other fields in the instruction. Effective address computation is exactly the same, whether the instruction specifies a data fetch, a data store, or a control transfer. A control transfer is accomplished by putting the effective address into the register that contains the address of the next instruction. If there is no transfer of control, this register is incremented so that it contains the address of the next instruction in sequence.

An instruction consists of an operation code specifying the action to be performed and other codes specifying zero or more operands. The number of operands depends on the type of the instruction. Instruction decoding consists of determining the number of operands specified in the instruction, the format for the specification of each operand, and the action to be performed. This information is either explicitly or implicitly contained in the operation code. The information on the number of operands and format of their specification is used to guide the effective address computation and subsequent fetching of the operands. Some of the operands may be in the registers. When such an operand is specified, it does not have to be fetched. As soon as any required operand fetching has been accomplished, the control unit instructs the computational unit to carry out the action specified by the operation code in the instruction. In addition to sending the computational unit a code designating the action to be performed, the control unit also sends it information indicating which registers contain the operands involved in the instruction. Any operands fetched from primary memory will have been put into registers. The registers may only conceptually be a part of the control processor. In many computers they are physically a part of the control processor. However, in some computers they are physically part of primary memory.

3.2 Computer Description Checklist

In this section we present a set of questions that form a checklist for the description of a computer. It is based on the conceptual description of a computer discussed in the preceding section. Answering all of the applicable questions on this list for some computer should give one a good understanding of the machine language for that particular computer. In fact, a generalization of these questions forms a good basis for understanding any programming language, including compiler languages. A few of these questions, such as those pertaining to interrupts and the I/O processor, will not make sense at this time and should be ignored until reaching Chapter 16. The questions are grouped into five sections, corresponding roughly to various functional units of a computer.

1. Registers
    —How many are there and for what purposes may each be used?
    —How are they referenced (named)?
    —What is the size of each register?
    —How are they interconnected, that is, from what registers to what registers may data be moved?
2. Memory structure and addressing
    —How is the memory organized?
    —What is the size of a memory cell?
    —How are cells organized into larger, more complex units?
    —How large is the memory, or what is its maximum size?
    —How are the complex units of memory addressed?
    —What are the different ways the address of a memory cell can be specified as an operand of an instruction (effective address computation)?
3. Computational unit
    —What are the types of data on which the computer can operate?
    —What are the formats of these different types of data?
    —What operations can the computer perform on each of the different types of data?
    —What are the formats of instructions that operate on data (e.g., number and location of operands)?
4. Control unit
    —What is the normal sequencing of instructions?

—What different kinds of transfer of control are possible (e.g., absolute, conditional, subroutine call)?

—What conditions can be tested in a conditional transfer?

—How is the return location obtained in a subroutine call?

—What is the format of control instructions?

—What conditions can cause an interrupt?

—Can interrupts be inhibited and if so how?

—What happens to control when an interrupt occurs?

5. I/O processor

—How is input and output in general achieved?

—How is the I/O processor started?

—How is completion of an input or output operation detected?

—How are abnormal or error conditions detected?

—How is the memory location, amount of information, and device specified for an input or output transmission?

—What devices can be attached to the computer?

—How is each device addressed and controlled?

3.3 A Subset of the IBM 360

Machine language examples in this text will be given in terms of a small, simple subset of the machine language for the IBM 360. A complete description of this subset is given in Appendix A. We will refer to this subset as Our 360. In this section we will discuss some of the more significant characteristics of this machine language.

There are 18 registers of interest. Sixteen of these are identical in size and use. They are called *general purpose registers* and are used for a number of different purposes, including effective address computation and arithmetic computation. These general registers are referenced by the integers 0-15 (written G_0-G_{15} in the text). Each general register is 32 bits in size. The remaining two registers are the instruction counter and the condition code. The instruction counter is 24 bits in size and contains the address of the next instruction. The contents of the condition code, which is 2 bits in size, is set by many of the instructions to indicate the outcome of some operation.

The basic memory unit (cell) is called a *byte*, which is 8 bits in size. The maximum memory size is 16,777,216 (2^{24}) bytes, which are addressed by the integers 0, 1, ..., 16,777,215. Each consecutive group of 4 bytes (32 bits), beginning at an address that is evenly divisible by 4, is called a *full word*. A full word is referenced by using the address of the first byte of the full word (the smallest of the four addresses). It should be pointed out that the IBM 360 term *full word* does not have the same meaning as the term *word* used in connection with some other computers. In these computers the basic addressable unit (cell) is called a word. Also, in speaking of computers in general, the term word is commonly used to designate the basic addressable unit. When we use the term full word we will be speaking specifically of the IBM 360 or Our 360. When we use the term word we will be speaking more generally, usually meaning the basic addressable unit.

The effective address is computed as the sum of the contents of a field in the instruction called the *displacement* field and the contents of one or two of the general registers. The displacement field is only 12 bits in size, hence its contents can never be greater than 4096, which is very much smaller than the memory size. It should be clear from this observation that in order to reference most of primary memory, the contents of at least one of the general registers used in effective address computation will have to be nonzero.

Our 360 operates on three different kinds of data: integer, logical, and character. An integer or logical quantity is stored in a full word, while a character is stored in a byte. An integer i is stored in the format

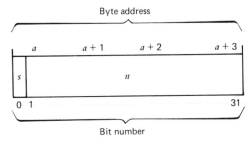

where the address a is evenly divisible by 4. The 32 bits in a full word are numbered as indicated. In this representation n is a 31-bit, unsigned integer in binary representation, while s is a single bit called the *sign bit*. The value of i is determined as follows:

$$\textbf{if } s = 0 \textbf{ then } i := n;$$
$$\textbf{else } i := -(2^{31} - n);$$

This form of representation for negative values of i is called the *two's complement form*. This is because the value of n is equal to the highest possible power of 2 (i.e., 2^{31}) minus the absolute value of i (this is the complement of i).

There is no distinction between any of the bits in logical quantities or characters. A logical quantity is stored in the format

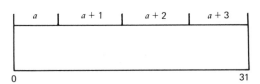

where the address a must be evenly divisible by 4. None of these bits have any meaning other than that given to them by the programmer. A character is stored in the format

where the address a may be any address. The value of n, which is an 8-bit integer, specifies the particular character. The codes used in this text are given in Appendix A. However, only minimal knowledge of them is required.

Operations on integers include addition, subtraction, multiplication, and division. Operations on logical quantities include logical **and**, logical **or**, and shifting. All types of data can be explicitly loaded from primary memory into any of the general registers or stored into primary memory from any of the general registers. Any two items of the same data type may be compared with each other for equality or ordering of their values.

Our 360 normally executes instructions in the sequence in which they are stored in primary memory. The transfer instructions break this normal sequence. The normal transfer instruction can be unconditional or can depend on the contents of the condition code register. There is also an instruction that is used to transfer control to a subroutine entry point. In addition to effecting a transfer of control, this instruction also puts the address of the return point in the calling procedure into one of the general registers.

3.4 A Sample Machine Language Program

It will be easier to explore machine language in more detail if we do it in the context of a specific example. Assume that we wish to write a procedure in machine language that will search a string of characters from left to right for the first occurrence of a space, starting at a specified position in the string. The procedure is to be coded so that it can be called from other procedures. We may assume that the strings used as arguments to our procedure are from cards. Therefore, they are 72 characters in length, and we should not search any further than the 72nd character.

If we let the name of our procedure be find, then it can be called from another procedure by

find (s, k, n)

where s is the string to be searched, k is the index of the character at which to begin searching, and n is the number of characters preceding the first space. The characters in the string s are numbered from left to right, beginning with zero. For example, suppose the first 10 characters of the string s are

"ABC XWYZ K"

After executing

find (s, 4, n)

the value of n will be 4.

Figure 3.2 shows the desired searching algorithm written in our informal

```
1   procedure find (s, k, n);
2       i := k;
3       while s [i] ≠ ' ' do
4           i := i+1;
5           if i > 72 then unloop;
6       end;
7       n := i−k;
8       return;
9   end.
```

Figure 3.2 Algorithm to search for a space character.

programming language. When called, as in the preceding example, i will initially be set to the value 4 and will be the index of "X":

$$\text{"ABC \quad XWZY \quad K"}$$
$$\text{i} = \ 0\ 1\ 2\ 3\ \ 4\ 5\ \ 6\ 7\ 8\ \ 9$$
$$\qquad\qquad\uparrow\qquad\uparrow$$
$$\qquad\qquad\text{k}\qquad\text{final value of i}$$

After finding the first space following "X", i will be equal to 8 and n will be set equal to 4. The **while** statement is another form of loop statement, which has the form:

$$\text{while } \textit{condition } \textbf{do } \textit{body } \textbf{end;}$$

The principal difference between the **while** statement and the **repeat** statement is that *condition* is tested before executing the body of a **while** statement and tested after executing the body of a **repeat** statement. Execution of the **unloop** statement terminates the loop in which it appears, regardless of the value of *condition*. The square brackets are used to enclose a subscript.

Our algorithm calls for testing one character of the string at a time, in sequence. This sequencing is exactly like sequencing through the elements of a one-dimensional array in FORTRAN or PL/1. Subscript notation was used in Figure 3.2 to suggest this. Therefore, the most reasonable way to store a string of characters in memory is in a contiguous block of bytes, with the first character stored in the byte with the smallest address. The address of the first character in the string s, called the *base address* of s, will be supplied to **find** by the calling program. A reference to some character s_i in the string s requires both the address of s and the subscript, or *index,* i. In fact, the address of s_i is equal to the sum of the base address of s and the value of the index i.

Instructions in Our 360 with an X-type format allow the use of two different general registers in computing the effective address. These two registers are called the *base register* and the *index register*. This suggests use of one register for the base address of s and the other register for the index i. The X-type instruction format is

It occupies 4 bytes, and the address of the first byte must be evenly divisible by 2. It is subdivided into five fields, each of which will contain a binary integer that has the following meaning.

OP = the operation code; specifies the action of the instruction and implies the type of the instruction, thus indicating the number of operands and the format of their specification.

R = the number of the general register containing the first operand.

X = the number of the general register containing the index.

B = the number of the general register containing the base address.

D = the displacement; a fixed offset from the base address.

The location of the second operand is specified by the effective address, which is computed as a function of the values contained in the D field and the general registers specified by the X and B fields. If we let G_i designate general register number i, then the effective address ea is defined by

$$ea := (\text{contents of D field}) + base + index$$

where

> if B = 0 then $base := 0$;
>
>   else $base :=$ contents of G_B;
>
> if X = 0 then $index := 0$;
>
>   else $index :=$ contents of G_X;

Therefore, we can easily address a character in the character string **s** by putting the index i into G_X and the base address of **s** into G_B.

For example, the *insert character* (IC) instruction replaces byte 3 (bits 24-31) of the first operand (the contents of G_R) by the contents of the single byte in memory specified by the effective address. As an example, consider the instruction (a binary number)

$$0100001110001011101000000000000000$$

which is clearly difficult to read. Let us rewrite this number in hexadecimal representation:

$$438BA000$$

Hexadecimal is a base 16 representation, so each group of four binary digits becomes a single hexadecimal digit. The 16 digits used in hexadecimal representation are 0, 1, ... , 9, A, B, ... , F. Notice that in hexadecimal the OP field is 2 digits, the R, X, and B fields are each 1 digit, and the D field is 3 digits. In the above instruction the value contained in the X field is B_{hex}, which equals 11 (decimal). Whenever numbers are written in hexadecimal representation in a context where they may be misinterpreted, they will be subscripted with *hex*. Numbers written in decimal representation will not be subscripted. Instructions will normally be written in hexadecimal. The value of the B field in the above instruction is A_{hex}, which equals 10. The effective address is then

$$ea := 0 + G_{10} + G_{11}$$

If G_{10} contains 562 and G_{11} contains 753, the effective address is 1315, and the instruction will insert the contents of byte 1315 into byte 3 of G_8.

An R-type format instruction is used when both operands are in general registers. The R-type format is

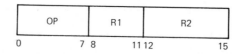

It is stored in 2 bytes, and the address of the first byte must be evenly divisible by 2. It is subdivided into three fields, each containing a binary integer with the following meaning.

OP = the operation code.

R1 = the number of the general register containing the first operand.

R2 = the number of the general register containing the second operand.

For example, the *load register* instruction

$$18A2$$

transfers the contents of G_2 to G_{10}.

Since the procedure we are coding is to be callable by other procedures, it must observe the *system standard calling sequence*. This standard requires that four of the general registers be used as follows.

G_{15} = the address of the *entry point* of the called procedure.

G_{14} = the address of the *return point* in the calling procedure.

G_{13} = the address of the first full word of the *save area*.

G_{12} = the address of the first full word of the *argument list*.

The *argument list* is a contiguous block of full words, one for each argument of the called procedure. Each of these full words contains the address of the corresponding argument. Addresses are stored in the form of logical data, even though the maximum size address requires only 24 bits. The address is stored in bits 8-31, with bits 0-7 equal to zero. The standard calling sequence assumes that the contents of any registers that are used by the called procedure have been saved in the save area and restored before returning to the calling program. The save area is a contiguous block of 17 full words.

The machine language program for the procedure find is shown in Figure 3.3. The function of all registers that are used is shown in Figure 3.4. In Figure 3.3 the machine language instructions (in hexadecimal) appear between the pair of double vertical lines. For easier reading the fields of the instructions are separated into columns. The symbolic operation code for each instruction is written to the left of the instruction. Comments on the action of the instruc-

Address	Symbolic operation code	Instruction OP R X B D					Program line number and comment	
00	STM	90	E	C	D	008	1	save all registers
04	L	58	2	0	C	000	1	get base address of s
08	L	58	3	0	C	004	1	get address of k
0C	L	58	4	0	C	008	1	get address of n
10	L	58	5	0	3	000	2	i := k
14	SR	1B	6	6			3	clear G_6
16	SR	1B	7	7			3	clear G_7
18	IC	43	7	0	F	03E	3	get space character
1C	IC	43	6	5	2	000	3	get s_i
20	CLR	15	6	7			3	$s_i = ''$?
22	BC	47	8	0	F	032	3	branch if =
26	A	5A	5	0	F	040	4	i := i + 1
2A	C	59	5	0	F	044	5	i > 72 ?
2E	BC	47	C	0	F	01C	5,6	branch if ≤
32	SR	1B	5	3			7	i − k
34	ST	50	5	0	4	000	7	n := i − k
38	LM	98	E	C	D	008	8	restore all registers
3C	BCR	07	F	E			8	return to caller
3E		40						the constant ''
3F		XX						unused
40		00	0	0	0	001		the constant 1
44		00	0	0	0	048		the constant 72

Figure 3.3 Machine language program for procedure in Figure 3.2.

Register	Use
2	address of s
3	address of k
4	address of n
5	i
6	s_i
7	byte 3 = space character
12	base address of argument list
13	base address of save area
14	return address
15	base address of find

Figure 3.4 Assignment of general registers for procedure find.

tions and the line number of the corresponding line in Figure 3.2 are written on the right.

We have coded the program as if it were going to be stored in primary memory beginning at address 0. Recalling Chapter 2, we know that no user's program will be loaded beginning at address 0, since the resident system is stored there. As we will see in Chapter 6, it is the responsibility of the loader to resolve this conflict. Thus, we do not need to know where our program will be loaded; we simply assume it will be at address 0. The address of each instruction and constant is written to the left of the instruction or constant. We will not explain every instruction in detail, but will comment on some of the more interesting and potentially confusing instructions in the following paragraphs. The reader should follow the operation of the program in detail to be sure that he understands how the machine language instructions function.

The first four instructions save the registers as required by the standard calling sequence and obtain the addresses of all of the arguments. This is sometimes called the *prologue* of the procedure. The first instruction is a *store multiple* (STM) instruction that stores the contents of the set of registers G_{14}, G_{15}, G_0, G_1, ..., G_{12} into the save area, as shown in Figure 3.5. The R field specifies the first register in the set, and the X field specifies the last. The B field equals 13 (D_{hex}), since G_{13} contains the base address of the save area. The D field is 12, since that is the displacement relative to the base of the save area of the full word in which the contents of G_{14} are to be saved. The contents of G_{12} is the base address of the argument list, which is shown in Figure 3.6. The

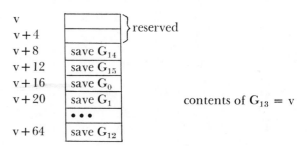

Figure 3.5 Format of save area.

a	base address of s	
a+4	address of k	contents of G_{12} = a
a+8	address of n	

Figure 3.6 Argument list for find.

three *load* (L) instructions put the addresses from the argument list into registers where they can be used to access the arguments.

The three instructions at 14_{hex}-18_{hex} initialize for the loop. The registers used to hold the space character and the character s_i must be cleared to zero since the *insert character* (IC) instruction inserts a character into byte 3 of a register without clearing the remainder of the register's contents, yet the *compare logical register* (CLR) instruction compares the entire contents of the registers. The compare logical register instruction is an R-type instruction and compares the logical quantities in the two registers specified by the R1 field and R2 field of the instruction. As a result of the comparison, the condition code is set as follows:

$$\text{if } G_{R1} = G_{R2} \text{ then CC:} = 0;$$
$$\text{if } G_{R1} < G_{R2} \text{ then CC:} = 1;$$
$$\text{if } G_{R1} > G_{R2} \text{ then CC:} = 2;$$

Normally a compare instruction is immediately followed by a *branch on condition* (BC) instruction that conditionally branches, depending on the value of the condition code.

The insert character instruction at $1C_{hex}$ begins the loop. The termination condition is tested first and is done by the compare logical register instruction and the following branch on condition instruction (at 22_{hex}). The R field of this instruction, instead of indicating a register, contains a mask that identifies the conditions for a branch. The condition code register of the computer is 2 bits in size and may have values 0 to 3. The R field in an instruction is a 4-bit field. Thus the mask is a 4-bit quantity. When a branch on condition instruction is executed, the current value of the condition code is used to select one of the bits in the mask. The bits in the mask are numbered from 0 to 3 in left to right sequence. If the bit selected by the value of the condition code is one, then program control branches to the effective address. For example, the instruction at 22_{hex} will branch to 032_{hex} if the condition code is equal to 0, since the mask is 8.

mask = 8 =

1	0	0	0

condition code =   0   1   2   3

A condition code of 0 corresponds to s_i = ' ' (space character). The instruction at $2E_{hex}$ will branch to $01C_{hex}$ if the condition code is equal to either 0 or 1, since the mask is 12 (C_{hex}).

mask = 12 =

1	1	0	0

condition code =   0   1   2   3

A condition code of either 0 or 1 corresponds to $i < 72$.

The loop ends with the instruction at $2E_{hex}$. After the loop terminates, the value of n is computed and stored at the address supplied in the argument list. This address was put into G_4 by the prologue. Before returning to the calling procedure, the contents of the registers must be restored to what they were when this procedure was called. The *load multiple* (LM) instruction, which is the inverse of the store multiple instruction, reloads the registers. The actual return is accomplished by a *branch on condition register* (BCR) instruction. This instruction functions exactly like the branch on condition instruction, except that it has an R-type format, and the branch address is found in the register specified by the R2 field. When the mask is 15 ($= F_{hex} = 1111_{bin}$), whatever the value of the condition code, the corresponding mask bit is 1. Thus, the instruction at $3C_{hex}$ is an unconditional branch to the address that is the contents of G_{14}.

The constants needed by the program are stored beginning at address $3E_{hex}$. The first is a single byte containing the USASCII code for the space character. The second and third constants are full word integers. They must be stored at addresses evenly divisible by four. In order to do this the single byte at $3F_{hex}$ must be left unused.

3.5 Assembly Language

In studying the machine language procedure in the last section the reader no doubt had some difficulty reading the instructions in Figure 3.3. The difficulty encountered in such a small procedure would be magnified many times if we were dealing with a large machine language procedure. Reading and working with binary, or even hexadecimal, numbers is difficult. Remembering the meaning of numeric operation codes is difficult. Another difficulty in the use of machine language, which was not illustrated in the example, is that the use of actual addresses in the instructions makes it inconvenient to modify a procedure. The instructions in a procedure must be stored in primary memory in sequence without any unused bytes between instructions. Inserting a new instruction will cause the location of all the succeeding instructions and constants to change. This will require a change in each instruction that refers to any of these succeeding instructions or constants.

The major objective of assembly language is to substantially or completely eliminate these difficulties. Assembly language achieves this by allowing the programmer to use symbolic names, called *symbols*, for numbers. Since an instruction is composed of several fields, each of which contains a number, it may be written as a combination of symbols. The number that a symbol names may be an operation code, an address, a displacement, a register number, a mask, or any other number used in the composition of an instruction. The assembly language used for examples in this text is described in Appendix B. This is assembly language for Our 360. A symbol is defined as any combination of from one to eight letters and digits, the first of which must be a letter.

Since symbols name numbers, we define the *value* of a symbol to be the actual number that the symbol names. Every symbol used in an assembly language procedure must have a value. The act of associating a symbol with a value is known as *symbol definition*. The values of symbols that name operation codes are predefined. For example, the symbol "SR" is the name of the operation code for the *subtract register* instruction. All other symbols must be defined within the assembly language procedure written by the programmer.

There are two ways a programmer may define a symbol. If the value of a

symbol is the address of an instruction or datum, the symbol is defined by writing it in front of the corresponding instruction or datum. For example
                    FOUND   SR   5,3
defines the value of the symbol "FOUND" to be the address of the first byte of the subtract register instruction, wherever it may be. In writing assembly language procedures the programmer does not choose the addresses of any of the instructions or data. Instead, he uses symbols for all addresses, defining them in the manner just described. One of the major tasks of the assembler (the program that translates assembly language to machine language) is to choose addresses for all of the instructions and data in the program. The other method of defining a symbol is by using the *equate* (EQU) pseudo instruction. For example, the symbolic instruction
                    K   EQU   6
defines the value of the symbol "K" to be 6. This method of symbol definition is usually used for defining symbols whose values are register numbers and masks.

An assembly language statement has the form
            label   operation   operands   comment
where *label* is optional. If it is present it must be a single symbol that begins in the first column of the card. (We are assuming the operating system described in Chapter 2 where jobs are punched on cards.) The *comment* is also optional. If present it may be any sequence of characters. The components of an assembly language statement are separated by one or more spaces. The *operation* is a single symbol that is predefined as the name of some operation code. The format of the *operands* depends on the type of assembly language statement. There are two types of statements: *machine instructions* and *pseudo instructions*. *Machine instructions* correspond to machine language instructions. *Pseudo instructions* are used to give information to the assembler, generate constants, and reserve space in the procedure. Further details on the format and meaning of assembly language statements will be exposed in the next section.

3.6 An Assembly Language Program

Figure 3.7 shows the machine language procedure from Figure 3.3 coded in assembly language. Both the assembly language source procedure and the (assembled) machine language object procedure are shown. The assembler always assembles a procedure as if it were going to be loaded as a contiguous block in memory beginning at address 0. This block is called a *procedure segment*. The addresses in Figure 3.7 are actually displacements from the base of the procedure segment. The value of a symbol that is defined by writing the symbol as a label of an instruction will actually be a displacement from the procedure segment's base. The START pseudo instruction in line 1 defines the name of the entry point of this procedure to be FIND. This entry point will be at the base of the procedure segment.

Whenever an instruction references other instructions or data that are part of the same procedure, the number of the base register need not be specified, provided the programmer has included a USING pseudo instruction as in line 2. This pseudo instruction tells the assembler that G_{15} will contain the base

Line	Assembled instruction	Label	Operation	Operands	Comment
1		FIND	START		define FIND as name of procedure
2			USING	*,15	G_{15} = procedure base address
3	00 90 E C D 008		STM	14,12,8(13)	save all registers
4	02	S	EQU	2	define G_2 as s
5	04 58 2 0 C 000		L	S,(12)	get base address of s
6	03	K	EQU	3	define G_3 as k
7	08 58 3 0 C 004		L	K,4(,12)	get address of k
8	04	N	EQU	4	define G_4 as n
9	0C 58 4 0 C 008		L	N,8(12)	get address of n
10	05	I	EQU	5	define G_5 as i
11	10 58 5 0 3 000		L	I,(,K)	i := k
12	14 1B 6 6		SR	6,6	clear G_6
13	16 1B 7 7		SR	7,7	clear G_7
14	18 43 7 0 F 03E		IC	7,=C' '	get space character
15	1C 43 6 5 2 000	NEXT	IC	6,(I,S)	get s_i
16	20 15 6 7		CLR	6,7	s_i = ' ' ?
17	22 47 8 0 F 032		BE	FOUND	branch if =
18	26 5A 5 0 F 040		A	I,=F'1'	i := i+1
19	2A 59 5 0 F 044		C	I,=F'72'	i > 72 ?
20	2E 47 C 0 F 01C		BNH	NEXT	branch if $\leq$
21	32 1B 5 3	FOUND	SR	I,K	i − k
22	34 50 5 0 4 000		ST	I,(,N)	n := i − k
23	38 98 E C D 008		LM	14,12,8(13)	restore all registers
24	3C 07 F E		BR	14	return to caller
25	3E 40 X X				
26	40 00 0 0 0 001				
27	44 00 0 0 0 048				
28			END		

Figure 3.7 Assembly language version of machine language program in Figure 3.3.

address of the procedure segment (indicated by the asterisk). (In assembly language numbers are normally written in decimal representation.) The assembler will then use 15 as the base register number in all instructions that are written without one.

In assembly language the store multiple (line 3) and the load multiple (line 23) instructions are written

$$\text{OP} \quad \text{R,X,D(B)}$$

R-type instructions are written

$$\text{OP} \quad \text{R1,R2}$$

and X-type instructions are written

$$\text{OP} \quad \text{R,D(X,B)}$$

If the value of the D field or the X field is 0, then a 0 need not be written, as in line 5. However, all of the punctuation (commas and parentheses) must be written. In addition, there are special forms of the branch on condition and branch on condition register instructions that use symbolic operation codes called *extended operation codes*. Use of these codes allows the programmer to omit the mask. The extended operation codes used in Figure 3.7, their names, branch condition, and mask supplied by the assembler are shown in Figure 3.8.

In writing the assembly language procedure we have used symbolic names for register numbers 2, 3, 4, and 5. This is simply to make the program more readable. It is easier to remember that register S contains the base address of **s** than to remember that register 2 contains it. The same holds for the addresses of **k** and **n** and the index **i**. This use of symbols makes it essentially obvious that the instruction

$$\text{IC} \quad 6,(I,S)$$

inserts character s_i into register 6, whereas the equivalent instruction

$$\text{IC} \quad 6,(5,2)$$

is much less obvious.

Another feature that our assembly language provides is the use of *literals* for referencing constants. In line 18 the notation $=\text{F'1'}$ is called a literal and is a reference to the constant 1, that is, a reference to the constant that literally appears between the pair of quote marks. The "F" means that the constant is to be a full word integer. The "C" in the literal $=\text{C' '}$ in line 14 means that the constant is to be the code for the space character stored in a single byte. Instead of actually writing out a constant in the form of an instruction, as was done in the machine language program in Figure 3.3, the programmer may use

Extended operation code	Used in line number	Mask supplied	Name	Branch condition
BE	17	8	branch on equal	A = B
BNH	20	12	branch on not high	A ≤ B
BR	24	15	branch register	always

A is first operand of preceding compare instruction
B is second operand of preceding compare instruction

Figure 3.8 Definition of extended operation codes.

the literal notation. The assembler will do the rest of the work assigning the constant to an appropriate location at the end of the program.

This concludes our discussion of the assembly language program in Figure 3.7 and assembly language in general. If the reader has referred to Appendix B he will be aware of the fact that even the simple assembly language described there is much richer than the example in this section suggests. We have deliberately used only the most basic assembly language features in the example so as not to obscure the basic idea of symbolic assembly language. We will encounter some of the additional features in the next chapter, where we will study the problems and inner workings of an assembler.

EXERCISES

In the following exercises use only the machine language and assembly language described in Appendixes A and B.

3.1 Write assembly language instructions for the statement

type := type of command that is in cur_cmd;

which appears in Figure 2.11.

3.2 Assume cur_cmd contains a JOB card. Write an assembly language program to locate the user limit values and convert each to a binary integer (32-bit, two's complement form). Each user limit value appears in cur_cmd as a string of characters that are decimal digits. Assume the USASCII codes in Appendix A are used.

3.3 Write an assembly language procedure to convert a binary integer (32-bit, two's complement form) to a string of decimal digits using the USASCII codes that are suitable for printing on the line printer.

4 | Assemblers

In this chapter we will study the problems that must be faced when designing an assembler for basic assembly language. To illustrate these problems we will discuss a simple assembler for the assembly language that was discussed in Chapter 3. (A complete definition of this language appears in Appendix B.) We will assume that the assembler is to be used as the assembler command program in the operating system that was discussed in Chapter 2. This means that the assembler must observe the system standards, particularly the format of object procedures and the use of I/O.

4.1 Design Methodology

In Chapter 2 we used a design method without describing it. Since we intended to use the approach throughout this text, a brief discussion of the method is approprite. The probability of successfully solving a complex problem is much higher if the problem is attacked methodically. Often the particular method used is not as important as the fact that some method is used. The design method used in this text has worked well for designers in many other fields of engineering and in computer system software design.

Our design method is iterative and is essentially repeated application of the following steps.

1. Define and understand the problem. The designer needs a clear understanding and definition of the functions that must be performed by the program being designed. The designer must also have a clear and precise definition of the environment in which the program will operate and any constraints that it must satisfy.
2. Define functional modules. The major functional modules are identified and described. In isolating these functional modules care should be taken to minimize the interconnections between them. These are the major building blocks of the program, and functional modularity demands that the interfaces between them be explicit and described in a standard terminology.
3. Define required data. The data required by each functional module is defined and organized into tables and other data structures.

4. Define the algorithm. The building blocks and data defined in steps 2 and 3 are used to define an algorithm for performing the functions of the program which is being designed. In defining this algorithm watch for redefinitions of the building blocks that will reduce the number of modules, simplify the algorithm, minimize the number of interconnections, or simplify the interfaces. Some of the functional modules may be simple and small enough so that they are combined with other functions into a single procedure segment. However, most of the functional modules should be implemented as separate procedures, physically as well as functionally.

After step 4 has been completed, the design method should be applied to each separate module surviving step 4. Each major functional module becomes a new, relatively isolated subproblem. The design method can be applied to each of these new problems. No knowledge of the other modules is required except a definition of the interfaces. This is the iterative aspect of the method. Each iteration produces a new level of increased detail. Iteration is continued until the structure of each of the procedures in the program is specified in enough detail that coding of the procedure is straightforward and simple. This method was applied to the design of the operating system in Chapter 2, as a review of Figures 2.7 to 2.16 will show.

4.2 Functions of an Assembler

What does an assembler have to do? It must translate an assembly language procedure into a machine language procedure. An example of an assembly language procedure and its translation is shown in Figure 3.7. Another example appears in Figure 4.2.

The procedure in Figure 4.2 is the assembly language equivalent of the procedure defined in Figure 4.1. This procedure takes 1 byte as an input argument, reverses the order of its 8 bits, and returns the reversed bits as output. For example, if $b = 10111010_{bin}$, then the call

reverse (b, rb)

```
procedure reverse (b, rb);
    y := b;
    i := 7;
    repeat
        X [i] := low-order bit of y;
        y := y with low-order bit shifted off;
        i := i−1;
    until i < 0;
    i := 0;
    repeat
        y := y with X [i] appended as its low-order bit;
        i := i+1;
    until i > 7;
    rb := y;
end.
```

Figure 4.1 Definition of procedure to reverse the order of the bits in a byte.

Line	Loc	Assembled instruction	Label	Operation	Operands	Comment
1			REVERSE	START	*,15	name of procedure is REVERSE
2				USING	*,15	G_{15} = procedure base address
3	04		Y	EQU	4	define G_4 as y
4	03		I	EQU	3	define G_3 as i
5	00	90 E C D 008		STM	14,12,8(13)	save all registers
6	04	58 2 0 C 000		L	2,(12)	get address of b
7	08	43 4 0 2 000		IC	Y,(2)	y := b
8	0C	58 3 0 F 068		L	I,SEVEN	i := 7
9	10	8C 4 0 0 001	OFF	SRDL	Y,1	shift off low-order bit
10	14	50 5 3 F 044		ST	Y+1,X(I)	X_i := low-order bit of y
11	18	5B 3 0 F 064		S	I,ONE	i := i-1
12	1C	47 A 0 F 010		BNL	OFF	branch if i $\geq$ 0
13	20	1B 3 3		SR	I,I	i := 0
14	22	58 5 3 F 044	ON	L	Y+1,X(I)	get X_i
15	26	8D 4 0 0 001		SLDL	Y,1	shift into y as low-order bit
16	2A	5A 3 0 F 064		A	I,ONE	i := i+1
17	2E	59 3 0 F 068		C	I,SEVEN	i > 7?
18	32	47 C 0 F 022		BNH	ON	branch if i $\leq$ 7
19	36	58 2 0 C 004		L	2,4(12)	get address of rb
20	3A	42 4 0 2 000		STC	Y,(2)	rb := y
21	3E	98 E C D 008		LM	14,12,8(13)	restore all registers
22	42	07 F E		BR	14	return to caller
23	44		X	DS	8F	reserve 8 full words for X
24	64	00 0 0 0 004	ONE	DC	F'4'	define constants to index
25	68	00 0 0 0 01C	SEVEN	DC	F'28'	... through full word array
26				END		

Figure 4.2 The procedure from Figure 4.1 coded in assembly language.

will set $rb = 01011101_{bin}$. The assembly language version, uses some instructions which were not discussed in Chapter 3. The *shift right double logical* (SRDL) instruction in line 9 shifts right the contents of a pair of registers (in this case G_4 and G_5, since the value of the symbol Y is 4) the number of bits specified by the effective address. What we are interested in here is the fact that the low-order bit (bit 31) of the number in G_4 is shifted into bit 0 of G_5. After shifting, the contents of G_5 are stored into X_i. When the loop terminates, each of the 8 full words in the array X will contain one bit of b in bit 0. The array X is defined by a *define storage* (DS) pseudo instruction in line 23. The operand field, "8F", specifies 8 full words. After the bits of b have been spread out, they are recombined in reverse order by a loop using a *shift left double logical* (SLDL) instruction in line 15.

The *define constant* (DC) pseudo instructions in lines 24 and 25 are used to define two full word constants with values of 4 and 28. These constants are used for loop control. Since the index i in the loop is used as a subscript to reference elements of an array of full words, it should be incremented by 4, since there are 4 bytes in a full word, not 1, as shown in Figure 4.1 (a compiler for the language used in Figure 4.1 would take care of this for us). Likewise, the constant 7 used in the loop termination condition in Figure 4.1 must be equal to 28 (4 times 7). If we define these constants and give them symbolic locations of ONE and SEVEN, the resulting procedure is more readable.

The major task of the assembler in translating an assembly language procedure is the translation of each symbolic machine instruction into its corresponding machine language instruction. The value of each field of a machine language instruction is specified by a symbolic expression in the symbolic machine instruction. Each of these symbolic expressions must be evaluated before the machine language instruction can be assembled. The value of a symbolic expression depends on the value of any symbols in the expression. Thus, all of the symbols in the assembly language procedure must be associated with a value by the assembler. The appropriate value to associate with a symbol is discovered when the definition of that symbol is processed by the assembler. The definition of symbols that are labels of instructions or data require the assignment of memory locations to the labeled instructions or data. Since a literal may also appear as an expression in a symbolic machine instruction, all literals appearing in the assembly language procedure must be assigned a memory location. In addition, the constant in a literal must be converted to binary representation.

The assembler must also process the pseudo instructions appearing in the assembly language procedure. The processing required is different for each of the pseudo instructions. The EQU pseudo instruction defines a symbol. The START pseudo instruction defines the entry point name of the procedure being assembled, and the END pseudo instruction defines the end of the assembly language procedure. The DS pseudo instruction calls for the assignment of a block of memory locations in which the procedure can store data. The DC pseudo instruction also requires the assignment of a block of memory locations, and the conversion of constants to binary representation.

So far, we see that the functions of the assembler are symbol definition, evaluation of symbolic expressions, assignment of memory locations, conversion of constants, and assembly of machine language instructions. In addition, the assembler must output the machine language object procedure and a listing of it. We may select the format of the listing. But the format for the object procedure is specified as a system standard, since the *loader* command program must be able to load the object procedure. This format will be defined in Chapter 6. The standard format for object procedures is often called an *object deck*, even though it may never be punched on cards. The form we choose for the listing will be described later in this chapter.

4.3 Major Modules and Interfaces

The assembler cannot assemble most machine language instructions without evaluating symbolic expressions. The assembler cannot evaluate a symbolic expression until it knows a value for each symbol appearing in the expression. The rules for writing an assembly language procedure do not require that the definition of a symbol precedes its use. These two facts imply that there may be expressions that the assembler cannot evaluate until it has processed all of the assembly language statements in the procedure. For example, the instruction in line 8 of Figure 4.2 contains a symbol whose definition does not occur until line 25. The practical implication of this situation is that the assembler will have to process the assembly language procedure twice before it can accomplish a complete translation. Thus, the assembler must be organized into two *passes*. Each pass reads and processes the entire assembly language source procedure. *Pass one* is principally for defining the values of all symbols that appear in the program. *Pass two* can then assemble the machine instructions, which requires evaluating the expressions designating the values of the fields in the instructions.

Definition of symbols involves associating a symbol with the number that is its value. This association must be remembered so that it can be used later, especially during pass two. A table called the *symbol table* will be used to record symbols and their associated values. In order to obtain the addresses needed for symbol definition pass one will have to assign memory locations to all instructions generated from symbolic machine instructions, constants generated from DC pseudo instructions, and storage reserved by DS pseudo instructions. Since a literal is a special kind of symbol referring to a constant whose value is equal to the literal, pass one will also have to assign memory locations to all of the literals in the program. These addresses will be recorded in the *literal table* along with the binary representation of the constant corresponding to the literal. When a symbol is defined by an EQU pseudo instruction, its value is the value of the expression in the operand field of the pseudo instruction. Thus, pass one must also be able to evaluate an expression.

Pass one processes each of the pseudo instructions according to the pseudo operation code appearing in it. Pass one must be able to distinguish pseudo instructions from symbolic machine instructions. It can do this most easily by reference to the *operation code table*, which has an entry for each symbolic operation code. This entry distinguishes between machine operations and pseudo operations.

Pass two finishes up the translation. The major task is that of assembling the machine language instruction corresponding to a symbolic machine instruction. Each expression in the operand field defines the value of one of the fields in the machine language instruction. These expressions are evaluated separately. The symbolic machine operation code must be translated into its corresponding numeric operation code. This is easily accomplished with the operation code table. Each entry in the table for a symbolic machine operation contains its corresponding numeric operation code. The numeric operation code and the values for all of the fields are then assembled into the proper format for the type of instruction being assembled.

An expression is composed of symbols, numbers, and operators, such as "+" for addition and "*" for multiplication. The value of an expression is computed by replacing all of the symbols by their values and performing the operations indicated by the operators.

Pass two is also responsible for producing the output. Unless suppressed by the command that invoked the assembler, there are three major outputs. Two of the outputs are virtually the same, a punched object deck containing the object procedure and a copy of the object procedure for loading and execution later in the current job. The third output is the listing. This is a printed copy of the source procedure and the assembled object procedure. The listing will also include comments on any errors that were detected during translation, such as multiply defined symbols (a multiply defined symbol is one for which more than one definition occurs in the program).

The preceding discussion suggests that the assembler be composed of four major modules.

1. Pass one.
2. Pass two.
3. Evaluate expression.
4. Convert constant.

The last two modules are subroutines that are needed in both pass one and pass two. Evaluate expression is used by pass one in processing EQU pseudo instructions and by pass two to obtain the value for the fields of machine instructions. Convert constant is used by pass one in processing literals, by pass two in processing constants that appear in a DC pseudo instruction, and by evaluate expression in processing literals. Hence, it makes sense to make them separate modules. Each of these modules is a separate procedure segment and would be compiled or assembled separately (depending on the language in which each was written). The relationship of these modules to each other is shown in Figure 4.3.

The interfaces between these modules are relatively simple. The only input to pass one is the source procedure which is to be assembled. The output from pass one and the input to pass two is:

1. The symbol table.
2. The literal table.
3. A copy of the original source procedure.
4. The length of the object procedure, error flags, and other information needed for preparation of the object deck and listing.

The output from pass two is:

1. A listing of the object procedure.
2. The object deck.
3. A copy of the object procedure for loading and execution.

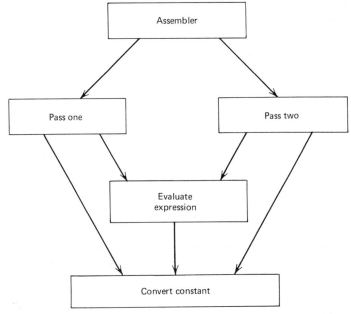

Figure 4.3 Major modules of the assembler and their relationship.

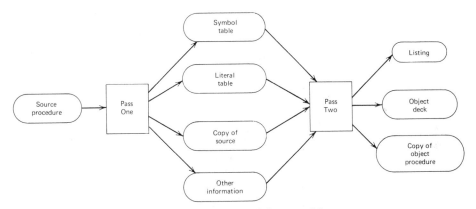

Figure 4.4 Data flow for pass one and pass two of the assembler.

These interfaces are shown in Figure 4.4. The input to evaluate expression is the location, in an assembly language statement, of the first character of an expression. Its output is the value of that expression. The input to convert constant is the location, in an assembly language statement, of the first character of a constant. Its output is the value of that constant in binary form.

In this section we carried out the first iteration in the design of our assembler. We identified four major procedure modules. We also identified several tables and other items of data. In the following sections we will examine each of the four procedure modules in more detail, carrying out the second iteration in the design of our assembler.

4.4 Pass One: Symbol Definition

The major task of pass one is to define all of the symbols used in the program. In order to do this, memory locations must be assigned to:

1. Each instruction generated by a symbolic machine instruction.
2. Each constant generated by a DC pseudo instruction.
3. Each block of storage reserved by a DS pseudo instruction.
4. Each constant generated by a literal.

Except for literals, the language requires that all items be assigned to memory in the sequence in which they occur in the procedure. This assignment is to begin at memory address zero and proceed consecutively without any unassigned memory except the minimum amount required to satisfy *alignment* requirements. For example, an integer must have full word alignment, that is, it is stored in a full word, and its location must be an address that is evenly divisible by four. To accomplish this sequential assignment, the assembler maintains a *location counter* whose value is always the address of the first unassigned byte in memory.

As the assembler processes the assembly language statements in the procedure, the location counter is incremented by an appropriate amount, which depends on the particular statement being processed. A symbolic machine instruction that corresponds to an R-type machine instruction causes the location counter to be incremented by two, since this type of machine instruction is only 2 bytes in length. All other symbolic machine instructions cause it to be incremented by four. A DC pseudo instruction causes the location counter to be incremented by an amount equal to the number of bytes required to store the constants being defined and any *fill* (unused bytes) required to satisfy alignment requirements. A DS pseudo instruction causes the location counter to be incremented by an amount equal to the number of bytes of storage being reserved. None of the other pseudo instructions cause any assignment of memory. All of the literals are assigned to memory in a block immediately following the rest of the procedure, the only unused space being the minimum fill required for alignment.

Assuming that the assembler keeps the location counter up to date, definition of symbols is quite straightforward. A symbol definition is called for whenever a symbol appears in the label field of a symbolic machine instruction, DC pseudo instruction, DS pseudo instruction, or EQU pseudo instruction. In the first three cases the value of the symbol is defined to be the location of the first byte of the generated instruction, the generated constant, or the block of storage reserved. This value is exactly the value of the location counter before it is updated by processing the assembly language instruction, except for adjustment for alignment when the instruction is DC or DS. Therefore, symbol definition in these three cases means entering the symbol from the label field of the assembly language instruction and the value of the location counter into the symbol table as a single entry.

If the symbol appears in the label field of an EQU pseudo instruction, the value that should be entered into the symbol table along with the symbol is the value of the expression in the operand field. In either case, if any entry for the symbol already exists in the symbol table, a multiple definition error is flagged. Any symbols occurring in the operand field expression of an EQU pseudo instruction must have been previously defined. If any undefined symbols are encountered when evaluating this expression, an error is flagged.

A literal is a special symbol referring to a constant. The notation for writing literals allows several different literals to refer to the same constant. For example, the literals $=$C′A′ and $=$X′C1′ both refer to the same constant, namely the binary number 01000001 stored in a single byte. For this reason only one copy of each distinct constant is generated, no matter how many different literals refer to it. In order to detect this duplication each literal must be converted to its binary form. The binary form of each literal is saved in the literal table. The length, in bytes, of the constant and the alignment constraint must also be saved in the literal table so that memory addresses may be assigned correctly at the end of pass one.

Since the processing is different for symbolic machine instructions and the various pseudo instructions, the assembler must be able to distinguish between them. In addition, when processing a symbolic machine instruction, the assembler must know the instruction type in order to know how much to increment the location counter. This information is stored in the operation code table. Each entry in this table corresponds to one operation, machine or pseudo. Any entry contains the symbolic operation code and identifies it as either a machine or pseudo operation. If it is a machine operation the entry also contains the operand type, which implies the length of the instruction, and the numeric operation code (for use in pass two).

The source program is read one record at a time by pass one, using the **read** entry of the system's I/O control. Each record contains one card and thus one assembly language instruction. Since a card contains 80 characters, the record is read into a block of 80 contiguous bytes, each byte containing a single character of the assembly language instruction. This contiguous block of bytes is treated as a character string, which is scanned by the assembler to locate the label, operation, and operand fields of an assembly language statement.

The definition of pass one is shown in Figure 4.5. The first few statements are initialization for the loop statement that follows. The loop statement is repeated until a END pseudo instruction is encountered. The file s_COPY is used to save a copy of the source procedure and error flags identifying any errors detected by pass one. The contents of this file will be used by pass two. The body of the loop processes a single assembly language instruction. This instruction is read from the card reader by calling the **read** entry of I/O control with the device name "IN". After locating the three fields of the symbolic instruction, the type of the operation code in the instruction is determined. The processing appropriate to that type of operation code is selected by a **case** statement.

The fields in the symbolic instruction are located by scanning the character string in **S**. A label, if present, must begin with the first character of S (column 1 on the card). However, the operation and operand fields do not have to begin at any fixed position in the instruction. Furthermore, the length of each field is variable, the first blank terminating the field. The result of locating the label, operation code, and operand field will be a pair of integers for each field. The first integer is the index, in S, of the first character of the item. The second integer is the number of characters in the item. The first integer for the label field will always equal zero, and the second integer will equal zero if there is no label. These pairs of integers are obtained by three successive calls to a procedure similar to the procedure **find** defined in Figure 3.2. Since the fields may be separated by one or more spaces, all of these spaces must be skipped

```
procedure pass_one;
   loc_ctr := 0;
   Create file s_COPY and open it for writing;
   repeat
      Clear error flags;
      Read next card of source procedure from IN into S₀, . . . ,S₇₉;
      Locate label, operation, and operand fields;
      type := type of operation code;
      case type of
         'machine': Process symbolic machine instruction;
         'EQU': Process EQU pseudo instruction;
         'DC/DS': Process DC or DS pseudo instruction;
         'END': Process END pseudo instruction;
         'START': null;
         'USING': null;
         otherwise Set error flag;
      end;
      Write contents of S and error flags into file s_COPY;
   until type = 'END';
   close ('s_COPY');
   return;
end.
```

Figure 4.5 Definition of pass one.

over in order to determine the first integer of the next pair. Once this integer is determined, the second integer of the pair could be obtained by calling find as it is presently defined.

Figure 4.6 describes the processing of a symbolic machine instruction in detail. Any symbol appearing in the location field is defined by calling the procedure define_symbol with the value that is to be associated with the symbol as the argument. In this case the value of the symbol is the current value of the location counter. This procedure is defined in Figure 4.7. If the length of the label (the second integer in the pair of integers that locates the label) is zero, there is no symbol to be defined. Otherwise, an entry that defines the symbol is made in the symbol table, unless there is already an entry for the symbol. If there is, a multiple definition error is flagged.

```
define_symbol (loc_ctr);
if operand field contains a literal then
   begin Convert constant following '=' by calling convert_constant, which
            returns triple = (converted constant, length, alignment);
      Search literal table for entry that matches triple;
      if no match then Enter triple into literal table;
   end;
k := length of machine instruction from operation code table;
loc_ctr := loc_ctr + k;
```

Figure 4.6 Expansion of "Process symbolic machine instruction;" from Figure 4.5.

```
procedure define_symbol (value);
    if second integer in locator pair for label field = 0
        then return;
    Search symbol table for an entry that matches the symbol in the label field;
    if no entry found
        then Enter symbol and value into symbol table;
        else Set multiple definition flag;
    return;
end.
```

Figure 4.7 Definition of procedure to define value of a symbol.

If the operand field of the symbolic machine instruction contains a literal, the literal must be processed. Our assembly language allows a literal only as the second subfield of the operand field. Thus, the presence of a literal can be verified easily by looking to see if an equals sign immediately follows the first comma in the operand field. The first step in processing a literal is to convert the constant to its corresponding binary representation. This is done by calling convert constant, which is one of the four major modules in the assembler. The input argument of this procedure is the index, in S, of the first character following the equals sign. Convert constant returns a triple of values consisting of the converted constant, its length in bytes, and its alignment constraint. The literal table is searched to see if it already contains an entry for this triple. If it does not, the constant, its size, and its alignment constraint are entered in the literal table as a single entry. Unless all three values in the triple match, an additional entry must be made. This is because two constants may have equal values, but different lengths or alignment constraints.

The final step in processing a symbolic machine instruction is to increment the location counter by an amount equal to the length of the machine language instruction that will be generated from the symbolic machine instruction. This length is deduced from the operand type, which is found as part of the entry for the symbolic operation code in the operation table. If it is an R-type instruction, its length is two. All other instructions have a length of four.

The processing of the EQU pseudo instruction is described in Figure 4.8. This pseudo instruction defines the symbol in the label field of the instruction to have a value equal to the value of the expression in the operand field. This expression is evaluated by calling evaluate expression. This procedure is another of the four major modules in the assembler. The input argument to the procedure is the index, in S, of the first character of the expression, that is, the first integer of the pair of integers that locates the operand field. Evaluate expression returns the value of the expression. Once the expression has been evaluated, the symbol in the label field of the EQU instruction can be defined by calling the symbol definition procedure with the value of the expression as an argument.

```
Evaluate expression in operand field by calling evaluate_expression;
define_symbol (value returned by evaluate_expression);
```

Figure 4.8 Expansion of "Process EQU pseudo instruction;" from Figure 4.5.

Determine alignment requirement for type of first item in operand field;
k := number of bytes of fill needed for required alignment;
loc_ctr := loc_ctr + k;
define_symbol (loc_ctr);
i := index in S of first character in operand field;
repeat
    Determine alignment requirement for type of item beginning at S [i];
    k := amount of fill for required alignment + length of item;
    loc_ctr := loc_ctr + k;
    i := index of first comma or space after S [i] that is not enclosed by single
        quote characters;
    i := i+1;
until S [i−1] = ' ';

Figure 4.9 Expansion of "Process DC or DS pseudo instruction;" from Figure 4.5.

Processing is the same for both the DC and DS pseudo instructions, and is defined in Figure 4.9. Both of these pseudo instructions have the same format, except for the presence of constants in DC pseudo instructions. These constants do not need to be converted until pass two, since only the amount of space required to hold them need be known in order to assign locations. The operand field of a DC pseudo instruction consists of a sequence of *items* of the form

$$t'constant' \text{ or } nt'constant'$$

where t is a letter that identifies the *type* of the constant and n is a nonnegative integer that is the *repetition count*, that is, the number of repetitions of the constant. The operand field of a DS pseudo instruction has the same form, except that *'constant'* is omitted. The operand field of a DS pseudo instruction specifies the amount of memory to be reserved by describing the types of the data that will be stored there. Since a DS pseudo instruction only reserves space, no actual constants need be written.

The type t of an item implies both the item's alignment and its length.

t	Alignment	Length in bytes	Type of *constant*
F	Full word	$n*4$	Decimal integer
C	Byte	$n*length(constant)$	Characters
X	Byte	$ceiling((n*length(constant))/2)$	Hexadecimal integer
A	Full word	$n*4$	Symbolic Address

Full word alignment means that the address of the first byte in the item must be evenly divisible by four. Byte alignment means the first byte of the item may have any address. The basic size of both F- and A-type constants is a full word, therefore, the length of such an item is four times the repetition count. In an A-type item, *constant* must be an expression whose value is an address.

The value of the function length(*constant*) is the number of characters in the *constant*. The value of the function ceiling(*i*) is the smallest integer greater than or equal to *i*. For example, if length(*constant*) = 4, ceiling(length(*constant*)/2) = 2 and if length(*constant*) = 9, ceiling(length(*constant*)/2) = 5. When *'constant'* is omitted, we define length(*constant*) = 1. The number of bytes of memory required by an item in the operand field of a DC or DS pseudo instruction is equal to the sum of its length and the amount of fill required to satisfy its alignment. If $n = 0$ the number of bytes of memory required exactly equals the required fill. A repetition count of zero would be used to insure alignment of a C- or X-type constant as a full word.

In Figure 4.9 the location counter must be updated by the amount of fill required by the first item in the operand field before the label symbol is defined. The value associated with this symbol must be the address of the first byte of the constant in the first item. The number of bytes of fill needed is zero if the alignment is byte, otherwise it is equal to the remainder resulting from dividing the current value of the location counter by four. The loop processes a single item. After using the *nt* part of the item to determine the required fill and length of the item, the following *'constant'*, if any, is skipped. An item is terminated by either a comma, which indicates at least one more item follows, or a space, which indicates the end of the operand field. Any commas or spaces that are part of a C-type constant must be ignored. Figure 4.10 shows some examples of DC and DS instructions.

Figure 4.11 shows the processing of the END pseudo instruction. This pseudo instruction signals the end of the assembly language procedure being translated. At this point all constants corresponding to literals that appeared in the procedure will have been entered into the literal table. Memory addresses can now be assigned to these constants. The literal table entry for a constant includes its alignment constraint and its size. When assigning an address to a constant, the location counter first has to be incremented by an amount equal to the number of bytes of fill needed for alignment of the constant. The address assigned to the constant is equal to the value of the location counter after it has been updated for the constant's alignment constraint. This value is entered into the entry for the constant in the literal table. The location counter is then incremented by an amount equal to the size of the constant to account for the memory where the constant will be stored. After all of the constants have been assigned addresses, the value of the location counter is equal to the length of the program. This is true since address assignment started with address zero and the address counter contains the address of the first unassigned memory location.

Before finishing our discussion of pass one, a few comments about functional modularity and procedures are appropriate. We began by trying to define all of the major functions of pass one (the functional modules). These were then organized into an algorithm for pass one. These functions were simple enough so that we did not consider making them separate procedures as we did expression evaluation. We then proceeded to a detailed description of several of the functional modules. In doing this we discovered a function that was required in several different places, symbol definition. For this reason we defined this as a new functional module and made it into a separate procedure.

Source instruction	loc_ctr value	Item	Bytes of fill	Length of item	Locations assigned to item	Assembler generated contents
DC C'AB'	4	C'AB'	0	2	4–5	AB
DC 0F,C'AB'	4	0F	0	0	4–5	
	4	C'AB'	0	2	4–5	AB
DC C'AB'	2	C'AB'	0	2	2–3	AB
DC 0F,C'AB'	2	0F	2	0	2–3	unused
	4	C'AB'	0	2	4–5	AB
DS 3C,F,3X	0	3C	0	3	0–2	nothing
	3	F	1	4	3–7	nothing
	8	3X	0	2	8–9	nothing
DC 3C'A',F'13',3X'F'	0	3C'A'	0	3	0–2	AAA
	3	F'13'	1	4	3	unused
					4–7	$0000000C_{hex}$
	8	3X'F'	0	2	8–9	$FFF0_{hex}$

Figure 4.10 Examples of DC and DS pseudo instructions.

```
i := 0;
repeat
    k := number of bytes of fill required for alignment of constant
        in literal_table [i];
    loc_ctr := loc_ctr + k;
    Enter value of loc_ctr into literal_table [i] as location of constant;
    k := size of constant from literal_table [i];
    loc_ctr := loc_ctr + k;
    i := i+1;
until i > index of last entry in literal_table;
procedure_length := loc_ctr;
```

Figure 4.11 Expansion of "Process END pseudo instruction;" from Figure 4.5.

4.5 Pass Two: Assembly of Instructions

The principal task of pass two is to assemble and output numeric machine instructions from symbolic machine instructions. In order to assemble an instruction, a value has to be determined for each of the fields in the instruction. The value for the OP field is found in the operation code table entry corresponding to the symbol that appears in the operation field of the symbolic instruction. This entry in the operation code table also indicates the operand type. From the operand type pass two can tell how many and which fields require values. Using this information as a guide, pass two can evaluate the expressions in the operand field of the symbolic instruction and identify the field in the machine instruction into which the resultant value is to be placed. If there is no expression in the symbolic instruction corresponding to a field, the value of that field will be set to zero. If the symbolic instruction contains an extended operation code, the value of the mask (R field) to be used in the numeric instruction is found in the operation code table entry for the extended operation code. After obtaining values for all of the fields in the instruction, these values are packed into 2 or 4 bytes, depending on the instruction type, in the proper fields and dispatched for inclusion in the object deck, object file, and program listing.

The constants in a DC pseudo instruction must be converted and assembled. After a constant has been converted to its binary equivalent, using convert constant, it is packed into the number of bytes specified by its size. The assembled constant is then dispatched for output. Any fill required in order to satisfy the alignment constraint for the constant must be accounted for by the output routines, that is, *unused space* must also be dispatched for output. The dispatching of unused space is also required for a DS pseudo instruction, since the purpose of that instruction is to reserve space in the object program segment. In order to get the correct fill for the constants in a DC pseudo instruction and the total space reserved by a DS pseudo instruction, the operand fields of these instructions need to be processed in virtually the same way that they were in pass one.

Since there are as many as three different places output may go, depending on the options specified in the command that invoked the assembler, all output is sent to the *dispatcher*, which routes it to the appropriate places. In addition, the dispatcher keeps track of the location address associated with each instruction and constant that it receives. It does this by keeping a location counter. Accompanying a constant or instruction sent to the dispatcher is its length in bytes. After dispatching this output, the dispatcher increments its location counter by an amount equal to the number of bytes just dispatched. When unused space is dispatched for output, the dispatcher simply increments its location counter.

Output for the punched object deck and the object file (a copy of the object deck is saved in OBJECT for later use by the loader) has the same format. This format is specified as part of the system standards to which all translators must adhere. The unit of output is a record (which corresponds to a card) containing from 1 to 56 bytes. Each record also contains a count of the number of bytes in the record and the location address of the first byte in the record. This address is called the *loading address*. The location addresses of all remaining bytes in the record follow consecutively. The same functional module can process output for both the object file and the object deck. Its function is to collect bytes until it has a full record of 56 bytes or until a gap occurs in the output due to fill or a DS pseudo instruction. In either case the current record is written into the object file and punched, depending on the options specified in the invoking command, by calling the **write** entry of the system I/O control. A new record is then started using the address of the next byte as its loading address.

Output for the listing must be converted to character form and merged with a copy of the source procedure before it can be written on the printer. Any error flags that were included with the copy of the source program by pass one are printed beside the symbolic instruction in which the error was found. Unfortunately, there is not a one-to-one relationship between the source instructions and the contents of the object procedure. A DC pseudo instruction may output many bytes of information in the object program corresponding to the constants in the operand field, the DS pseudo instruction outputs only unused space, and the EQU pseudo instruction outputs nothing. The format of the listing is:

Error flags	Location address	Object instruction or 1-4 bytes of constant	Copy of source instruction

Figure 4.2, without the first column containing line numbers, is exactly the listing that the assembler would print for the procedure **reverse**. If there are no errors in the corresponding source instruction, the first column of the listing is blank. The second column contains the location address of the first byte in the third column. The third column contains the instruction generated by a symbolic machine instruction or the data generated by a DC pseudo instruction. If this data is longer than 4 bytes, it is continued on succeeding lines, 4 bytes per line. The information in the second and third columns is printed in hexadecimal representation. The fourth column contains a copy of the entire source language instruction that produced the output printed in the third column. If the source instruction was a DS or EQU pseudo instruction, the third column will be blank, but the second column will contain the address of the block of storage reserved or the value of the symbol, respectively.

```
procedure pass_two;
    loc_ctr : = 0;
    Open file s_COPY for reading;
    default_base : = 0;
    while forever do
        Read next assembly language instruction and its associated error flags
            from s_COPY into S and flags;
        Set all bytes in print_line to blanks;
        Put copy of S and flags into print_line;
        p_out : = true;
        p_cnt : = 0;
        Locate label, operation, and operand fields;
        type : = type of operation code;
        case type of
            'machine': Assemble machine instruction;
            'EQU': Print EQU pseudo instruction;
            'DC': Assemble DC constants;
            'DS': Reserve DS storage;
            'END': End assembly;
            'START': Define object procedure;
            'USING': Set default base;
            otherwise begin write ('PRINT', 132, print_line, status);
                        p_out : = false;
                    end;
        end;
        if p_out then write ('PRINT', 132, print_line, status);
    end;
end.
```

Figure 4.12 Definition of pass two.

Pass two is defined in Figure 4.12. The copy of the source procedure saved
in s_COPY is used in pass two, so this file must be opened for reading. The value
of default_base is the number of the register to use as the base register for all
references to locations within the procedure being assembled, for example, in
lines 8, 11, and 18 of Figure 4.2. This is initially set to zero and will be reset
by the USING pseudo instruction. The loop in Figure 4.12 is similar to the
loop in Figure 4.5. After the next assembly language instruction and its flags
have been read from s_COPY, they are copied into print_line, which is an array
of 132 characters. Each line of the listing is built up in print_line before it is
printed. The flag p_out is true if print_line is not empty. Object output that
is generated when an assembly language instruction is processed is also put into
print_line. As soon as a line of the listing is complete, it is printed. A listing
line is complete as soon as there are 4 bytes of information for printing in the
third column of the listing. The counter p_cnt indicates the number of these
bytes that have been accumulated. Some assembly language instructions generate more than one listing line, and the last line may not be a complete one,
in which case the write entry of I/O control is called to print it before reading
the next source instruction.

Assembly of a machine language instruction is defined in Figure 4.13. The

j := index in operation_code_table of entry for operation in current
  assembly language instruction;
OP := numeric operation code from operation_code_table [j];
type := operand type from operation_code_table [j];
i := index in S of first character in operand field;
case type of
  'R': Process R-type operand field;
  'X': Process X-type operand field;
  'M': Process M-type operand field;
  'S': Process S-type operand field;
end;
if type = 'R'
  then begin
        assembled_instruction := OP, R, and X packed into 2 bytes;
        length := 2;
      end;
  else begin
        assembled_instruction := OP, R, X, B, and D packed into
          4 bytes;
        length := 4;
      end;
dispatcher (length, 'info', assembled_instruction);

Figure 4.13 Expansion of "Assemble machine language instruction;" from Figure 4.12.

major task in assembling a machine instruction is evaluation of the expressions
defining the values of the various fields of the instruction. Unfortunately, these
expressions are arranged differently in the operand field for each of the four
operand types. The operand type from the operation code table entry is used
to determine how the operand field will be processed. After processing the
operand field, the values obtained are combined with the numeric operation
code and packed into 2 or 4 bytes, depending on the operand type. The assembled instruction can then be dispatched for output. The dispatcher has three
arguments. The first is the length in bytes of the item being dispatched. The
second argument indicates whether information ('info') or unused space
('space') is being dispatched. If information is being dispatched, the third
argument is the location of this information.

Figures 4.14 to 4.17 define the processing of the different operand field types.
Actual computation of a value for an expression is done by evaluate expression.
This procedure is defined and discussed in Section 4.6. It is called with two
arguments:

evaluate_expression (i,value)

The value of the first argument, i, is the index in S of the first character of a

evaluate_expression (i, R);
i := i+1;
evaluate_expession (i, X);

Figure 4.14 Expansion of "Process R-type operand field;" from Figure 4.13.

```
if symbolic operation code is an extended operation code
    then R := mask from operation_code_table [j];
    else begin evaluate_expression (i, R);
             i := i+1;
         end;
evaluate_expression (i, D);
if S [i] ≠ '('
    then begin B := default_base;
             X := 0;
         end;
    else begin i := i+1;
             evaluate_expression (i, X);
             if S [i] ≠ ')' then
                 begin i := i+1;
                     evaluate_expression (i, B);
                 end;
                 else B := default_base;
         end;
```

Figure 4.15 Expansion of "Process X-type operand field;" from Figure 4.13.

```
evaluate_expression (i, R);
i := i+1;
evaluate_expression (i, X) ;
i := i+1;
evaluate_expression (i, D);
if S [i] ≠ '('
    then B := default_base;
    else begin i := i+1;
             evaluate_expression (i, B);
         end;
```

Figure 4.16 Expansion of "Process M-type operand field;" from Figure 4.13.

```
evaluate_expression (i, R);
i := i+1;
X := 0;
evaluate_expression (i, D);
if S [i] ≠ '('
    then B := default_base;
    else begin i := i+1;
             evaluate_expression (i, B);
         end;
```

Figure 4.17 Expansion of "Process S-type operand field;" from Figure 4.13.

67

symbolic expression. When the procedure returns, value has been set equal to the value of that expression. In addition, the value of i has been updated so that it is equal to the index in S of the delimiter character that immediately follows the expression. The characters ",", "(", ")", and " " are expression delimiters.

The form of an R-type operand field is:

R,X

There are six possible forms for an X type operand field:

$$R,D(X,B) \qquad D(X,B)$$
$$R,D(X) \qquad D(X)$$
$$R,D \qquad D$$

The second three forms are used only with extended operation codes, which imply a value for the R field. When evaluate expression is called, if S_i is already a delimiter, the procedure immediately returns with value set equal to zero. This will be the case when an expression is omitted. Thus, none of the variations

$$R,(X,B) \quad R,(,B) \qquad R,(X,) \quad R,(,)$$
$$R,D(X,) \quad R,D(,B) \quad R,D(,)$$
$$R,$$

needs to be processed as a special case, since a field value is interpreted as being equal to zero when its corresponding expression is omitted. The second and third of each set of forms for an X-type operand field are used to reference a location in the procedure being assembled. Therefore, the value of the B field is equal to the value of default_base which contains the number of the register that contains the base address of the procedure segment as specified in a USING pseudo instruction. When the symbolic operation code is an extended operation code, the value for the R field is found as part of the operation's entry in the operation code table.

There are two forms for an M-type operand field,

R,X,D(B)
R,X,D

and two forms for an S-type operand field,

R,D(B)
R,D

The processing for M-type and S-type operand fields is straightforward.

The EQU pseudo instruction does not generate any output for the object procedure. However, as Figure 4.18 shows, the value of the symbol that is defined by the EQU pseudo instruction is printed along with the source language instruction by the procedure lister.

DC and DS pseudo instructions are processed almost exactly as they were in pass one, except that the constants in the operand field of a DC pseudo instruction must be converted to their binary form and sent to the dispatcher for output. Figures 4.19 and 4.20 define this processing. Conversion of the constants

Find entry in symbol table for symbol in label field;
Extract the value of the symbol from this entry;
Call lister to print this value along with the source instruction;

Figure 4.18 Expansion of "Print EQU pseudo instruction;" from Figure 4.12.

i : = index in S of first character in operand field;
while S [i] ≠ ′ ′ do
   k : = amount of fill for item beginning with S [i];
   dispatcher (k, ′info′, zero);
   n : = repetition count from current item;
   i : = index of type character following repetition count;
   convert_constant (i, converted_constant);
   Fill block with n repetitions of converted_constant;
   k : = number of bytes in block;
   dispatcher (k, ′info′, block);
   i : = i+1;
end;

Figure 4.19 Expansion of ″Assemble DC constants;″ from Figure 4.12.

Call lister to print the DS pseudo instruction and address of first
   byte reserved by it;
k : = 0;
i : = index in S of first character in operand field;
while S [i] = ′ ′ do
   k : = k + (amount of fill for item beginning with S [i]);
   k : = k + (length of item beginning with S [i]);
   i : = index of next ′,′ or ′ ′ in S;
   i : = i+1;
end;
dispatcher (k, ′space′);

Figure 4.20 Expansion of ″Reserve DS storage;″ from Figure 4.12.

is done by convert constant. This procedure is called with the index in S of
the type character of a constant as an input argument. It returns the constant
converted to binary representation. The value of the index is also updated so
that it is the index of the terminator immediately following the constant.

If n is the repetition count, n copies of the converted constant must be output as part of the object procedure. These repetitions must not have any
unused space between them. For example, the pseudo instruction

$$DC \quad 3X′7F6′$$

should output the 5 bytes

$$7F\ 67\ F6\ 7F\ 60_{hex}$$

However, three separate calls to dispatcher of the form

$$dispatcher\ (2,\ ′info′,\ converted\_constant)$$

will result in output of the following 6 bytes

$$7F\ 60\ 7F\ 60\ 7F\ 60_{hex}$$

since the dispatcher does not accept units of information smaller than a byte.
Therefore, to achieve the desired output, the converted constant is copied n
times into a block of memory with no unused space between copies. Then one
call to the dispatcher outputs the entire block of n repetitions.

Notice also that the fill, which is required for proper alignment, is set to zero
by the call:

$$dispatcher\ (k,′info′,zero)$$

The third argument **zero** is a 3-byte constant whose value is zero. The amount of fill needed is always 1, 2, or 3 bytes, that is, $k \leq 3$. Since the fill is unused space, it is permissible to put anything we wish in these bytes. If we put nothing there, that is, dispatch unused space, the contiguity of the output information is interrupted. This results in a new output record being started. If this happens often, which is not uncommon, then an excessive number of short records will be written. This inefficiency can be avoided by actually writing out something for the fill bytes. We choose to write zeros.

The END pseudo instruction signals the end of the assembly language procedure. As Figure 4.21 shows, each of the constants in the literal table is sent to the dispatcher for output. Preceding each constant, from 1 to 3 bytes equal to zero are dispatched to account for the fill required for alignment of the constant. After all of the constants in the literal table have been dispatched, pass two must force termination of the object file, punched object deck, and listing.

The START pseudo instruction begins an assembly language program. In pass one no processing was required. The pass two processing is defined in Figure 4.22. The procedure's length, determined in pass one, and the name of the procedure, found in the label field of the START pseudo instruction, are used to make up a standard procedure definition card as required for the object deck. Processing of the USING pseudo instruction is straightforward, as shown in Figure 4.23. The procedure **convert_integer** is used to convert the decimal integer in the second subfield (following the comma) of the operand field to its binary representation. This is the number of the register to use as a base register when none is specified in a symbolic machine instruction.

The dispatcher is defined in Figure 4.24. If unused space is being dispatched,

```
i := 0;
repeat
    constant := constant from literal_table [i];
    k := amount of fill required for alignment of constant;
    dispatcher (k, 'info', zero);
    k := size of constant;
    dispatcher (k, 'info', constant);
    i := i+1;
until i > index of last entry in literal_table;
Call object_out to write standard end of object deck card into
    file OBJECT and onto PUNCH;
Call lister to print end of assembly information;
close ('OBJECT');
delete ('s_COPY');
```

Figure 4.21 Expansion of "End assembly"; from Figure 4.12.

```
name := symbol from label field of START pseudo instruction;
Call object_out to write standard procedure definition card containing name
    and procedure_length into OBJECT and onto PUNCH;
Call lister to print title and other heading information;
```

Figure 4.22 Expansion of "Define object procedure;" from Figure 4.12.

i := index in **S** of first character following first ',' in operand field;
convert_integer (i, default_base);

Figure 4.23 Expansion of "Set default base;" from Figure 4.12.

```
procedure dispatcher (n, type, X);
    if type = 'info' then
        begin card_out (loc_ctr, n, X);
            lister (loc_ctr, n, X);
        end;
    loc_ctr := loc_ctr + n;
    return;
end.
```

Figure 4.24 Definition of procedure that dispatches object output.

```
procedure card_out (add, n, X);
    if both NOPUNCH and NOOBJECT options in assembler command
        then return;
    if add > loading_address + buff_cnt then
        begin card := loading_address and first buff_cnt bytes of c_buff
                in object deck card format;
            if ¬ NOPUNCH option then Write card into OBJECT;
            if ¬ NOOBJECT option then Write card onto PUNCH;
            buff_cnt := 0;
            loading_address := add;
        end;
    i := 0;
    repeat
        c_buff [buff_cnt] := X [i];
        i := i+1;
        buff_cnt := buff_cnt+1;
        if buff_cnt ⩾ 64 then
            begin card := loading_address and all 64 bytes of c_buff
                    in object deck card format;
                if ¬ NOPUNCH option then Write card into OBJECT;
                if ¬ NOOBJECT option then Write card onto PUNCH;
                buff_cnt := 0;
            end;
    until i ⩾ n;
    return;
end.
```

Figure 4.25 Definition of procedure to output object deck.

only the location counter needs to be updated. If information is being dispatched, then this information must be sent to **card_out**, which formats it into object deck form and writes it into OBJECT, and onto PUNCH, and sends it to **lister**, which prints it as part of the output listing.

The fundamental idea used for output of the object program, shown in Figure 4.25, is to accumulate the output in a buffer until it is full (64 bytes)

```
procedure lister (add, n, X);
    if NOLIST option in assembler command then return;
    Convert add to hexadecimal and insert into print_line;
    if n = 0 then
       begin Write print_line onto PRINT;
          p_out := false;
          return;
       end;
    i := 0;
    repeat
       Convert X [i] to hexadecimal and put into print_line;
       i := i+1;
       p_cnt := p_cnt+1;
       if p_cnt ⩾ 4 then
          begin Write print_line onto PRINT;
             print_line := blanks;
             p_cnt := 0;
             p_out := false;
          end;
    until i ⩾ n;
    return;
end.
```

Figure 4.26 Definition of procedure to print listing.

or until a gap appears in the output. A gap in the object procedure output is detected by comparing the address of the first byte of the new output (add) with the sum of the loading address for the current object deck card and the number of bytes (buff_cnt) currently in the card buffer (c_buff). The loading address, which is required by the loader as part of each card, is the location address of the first byte currently in c_buff. If a gap has occurred, the present contents of the buffer is written out to PUNCH and OBJECT. Each record written must be in the standard object deck card form, which contains a loading address and a count of the number of bytes of the object procedure in the record as well as the actual object procedure bytes. After writing out the buffer the new output can be treated as if no gap occurred when the loading address and buffer count are set properly. In either case the new output is now added to the buffer, beginning immediately after the last byte currently in the buffer. As each new byte is added to the buffer, a test is made to see if the buffer is full. Whenever it is, its contents are written out to PUNCH and OBJECT.

Before they can be written out, the output records must be formatted as required by the loader. As we saw in Chapter 2, the contents of OBJECT will be processed by the loader later in the job when the user's program is loaded for execution. The records written onto PUNCH will be punched on cards to make the object deck. The object deck may be included as input to the loader in some other job. The format of these records and the additional information required in each record will be discussed in Chapter 6, which is a discussion of loaders.

The procedure for printing the listing is defined in Figure 4.26. It must convert the output that it receives to hexadecimal form before this output can be printed. If the number of bytes n is equal to zero, only the address add will be printed. This feature is used in processing the EQU and DS pseudo instructions. The converted output is then merged with the copy of the source line that produced the output and any associated error flags. Both of these are already in the print buffer, print_line. At most, 4 bytes in converted form can be printed in each line. The first record contains the error flags, the converted address of the first byte, the converted object instruction or data, and the source instruction that generated the output. In all remaining lines, the error flag and source line portion of the record is blank.

4.6 Expression Evaluation

Expression evaluation is done by evaluate_expression, which is one of the major modules of the assembler. Its input argument is the index in the source instruction of the first character of an expression. The expression is terminated by the first comma, right parenthesis, left parenthesis, or blank. If the value of the input argument is initially the index of one of these characters, zero is immediately returned as the value of the expression. Otherwise, the expression is evaluated and the value of the input argument is updated so that it points to the character that terminated the expression.

An *expression* is either a single literal or consists of one or more *terms* separated by plus or minus signs, for example,

$$t + t - t + t$$

A *term* is one or more *primaries* separated by asterisks (meaning multiplication) or slashes (meaning division), for example,

$$p * p / p$$

A *primary* is either a symbol or an integer (in decimal form). Thus, the expression evaluater might encounter an expression such as:

$$COUNT * 3 - TOP * 4 + BOTTOM$$

By definition the multiplications and divisions are to be done before additions and subtractions. Otherwise, all operations are to be done in left to right sequence.

The action of the expression evaluator is shown in Figure 4.27. If the expression is a literal its value is the location address assigned to the literal. In order to find this value the literal must be converted as in pass one, and the resulting triple used to find the entry for the literal in the literal table. The expression value, e_value, is initialized to zero and the value of each term, t_value, is added to it or subtracted from it after the term has been evaluated. The operator preceding the term must be remembered while the term is being evaluated. The first term is treated as if it were prefixed with a plus operator. If the character that terminated the evaluation of a term is not a plus or minus, e_value is the value of the expression and is returned to the caller.

The same character index is used throughout evaluate_expression, including evaluation of primaries, thus it will point to the proper character on return to the caller. This index must be adjusted before entering the loop in order that the loop work properly. Evaluation of a term is identical to evaluation of

```
procedure evaluate_expression (i, value);
    if S [i] ε (',', '(', ')', ' ') then
        begin value := 0;
            return;
        end;
    if S [i] = '=' then
        begin Convert constant following '=' by calling convert_constant,
                which returns triple = (converted constant, length, alignment);
            Search literal table for entry that matches triple;
            value := address field of matched entry;
            return;
        end;
    i := i−1;
    e_value := 0;
    e_op := '+';
    repeat
        t_value := 1;
        t_op := '*';
        repeat
            i := i+1;
            evaluate_primary (i, p_value);
            if t_op = '*' then t_value := t_value * p_value;
                else t_value / p_value;
            t_op := S [i];
        until t_op ≠ '*' and t_op ≠ '/';
        if e_op = '+' then e_value := e_value + t_value;
            else e_value := e_value − t_value;
        e_op := S [i];
    until e_op ≠ '+' and e_op ≠ '−';
    value := e_value;
    return;
end.
```

Figure 4.27 Definition of procedure to evaluate an expression.

an expression, except that t_value is initialized to one instead of zero, and the initial primary in the term is treated as if it were preceded by a multiply operator.

Evaluation of a primary is relatively straightforward, as shown in Figure 4.28. A primary may be either a symbol or an integer in decimal form. A symbol begins with a letter, while an integer begins with a digit. A symbol is a string of letters and digits beginning with a letter. All of the consecutive letters and digits up to the first character that is something else are collected to form the symbol. The value of the symbol is obtained from the location address field of the symbol table entry for the symbol. This value is the value of the primary.

An integer is converted by calling the procedure convert_integer defined in Figure 4.29. The integer is in decimal form, each digit being represented as the 8-bit character code for one of the digits 0-9. These character codes are 00110000_{bin} through 00111001_{bin}, respectively (see Appendix A). The binary

```
procedure evaluate_primary (i, value);
    type : = type of S [i];
    case type of
        'letter': begin symbol : = S [i] . . . S [k] where S [k + 1] is first character
                    past S [i], which is neither a letter or a digit;
                i : = k + 1;
                value : = value item from symbol table entry for symbol;
            end;
        'digit': convert_integer (i, value);
    end;
    return;
end.
```

Figure 4.28 Definition of procedure to evaluate a primary.

```
procedure convert_integer (i, value);
    value : = 0;
    while S [i] is a digit do
        value : = 10 * value + S [i] − 30_{hex};
        i : = i + 1;
    end;
    return;
end.
```

Figure 4.29 Definition of procedure to convert an integer from decimal form to binary form.

form for the value of a digit can be obtained by subtracting 00110000_{bin} ($= 30_{hex}$) from the character code. For example, the code for 6 is 00110110_{bin}. Thus, $00110110_{bin} − 00110000_{bin} = 00000110_{bin} = 6$. A decimal integer, abc, is shorthand for $a*10^2 + b*10^1 + c*10^0$, which can also be written as $(a*10 + b)*10 + c$. Thus, an N-digit integer $d_1 d_2 \ldots d_N$ can be written as $((d_1 * 10 + d_2)*10 \ldots + d_{N-1})*10 + d_N$. We can convert an integer from its decimal form to the corresponding binary integer by actually evaluating this expression, which is what the loop in Figure 4.29 does. For an excellent discussion of the general problem of conversion of numbers from one base to another, see Knuth.[1]

4.7 Conversion of Constants

The fourth major module of the assembler is convert_constant. It is called by pass one for converting literals and by pass two for converting constants in a DC pseudo instruction. When the procedure is called, its argument is the index in S of the first character in the constant. This character will be the constant's type code. There are four types: F, which is an integer written in decimal form; X, which is a sequence of bytes written in hexadecimal form; C, which is also a sequence of bytes, but written in character form; and A,

[1] D. E. Knuth, *The Art of Computer Programming, Volume 2: Seminumerical Algorithms*, Addison-Wesley, Reading, Mass., 1969.

which is an address written as a symbol. After the constant is converted, its binary equivalent, its length in bytes, and its alignment are returned to the caller. Type F and A constants are always 4 bytes in length. Type X and C constants have a length equal to the number of bytes in the constant.

Type C constants require no conversion, since the binary form of the characters is exactly what is wanted. Type F constants are converted using convert_integer, described in the preceding section. The binary equivalent of an A-type constant is just the value of the symbol in the constant. This value is obtained from the symbol table. An X-type constant specifies bytes in hexadecimal form. Two hexadecimal digits are required to specify the full range of values for a byte. The hexadecimal digits are 0 to 9 and A to F. The binary character representation of these digits is 00110000_{bin} to 00111001_{bin} and 01000001_{bin} to 01000110_{bin}, respectively. Each hexadecimal digit specifies 4 bits of a byte. To obtain these 4 bits in binary subtract 00110000_{bin} from the hexadecimal digit if it is 0 to 9 or subtract 00110111_{bin} if the digit is A to F. The low-order 4 bits of the resulting values are then packed two per byte to get the binary form of the constant. If there are an uneven number of hexadecimal digits, the last 4 bits of the last byte are set equal to zero.

4.8 Table Maintenance

Our design of the assembler requires three different tables: the symbol table, the literal table, and the operation code table. Conceptually these three tables are exactly the same; their similarities are significant, while their differences are insignificant. A table is a special class of *information structure*. An information structure, often called a *data structure*, is a collection of information (data) whose individual items are related to each other in some fashion. We will meet a wide variety of information structures in this text, and we will discuss general information structures in a later chapter. In this section we will explore the structure and management of tables.

A *table* is composed of a number of *entries*. Each entry is composed of a number of *items* of information (frequently called *fields*) that are related to each other. Usually one of the items is distinguished as a *key*, since one of the most common operations on a table is to find the entry containing a given key. For example, an entry in the symbol table consists of two items, a symbol and its value. The symbol is the key, since the table is always used by looking for the entry containing a given symbol. An entry in the operation code table contains four items: a symbolic operation code, its type, its numeric operation code, and its operand type. An operation code table entry has the symbolic operation code as its key.

All of the entries in a table should have unique key values, otherwise a reference to the table can be ambiguous. This is why we considered multiple definitions for a symbol to be an error. Conceptually there is no reason why the entries in a table cannot have more than one key. This is the case with the literal table. There are four items in a literal table entry: a constant, its address, its length, and its alignment constraint. The constant, its length, and its alignment are all keys. As we saw in earlier sections, all three of these keys must match when we are searching the literal table. Searching on multiple keys is only slightly more complicated than searching on a single key. For example, to find an entry in the literal table, we search for a given value of (*constant, length, alignment*). Search the table for a match on *constant*. Check that entry to see if *length* and *alignment* also match. If they do, we have found the desired entry. If not, continue searching for the next match on *constant*.

Several basic operations were needed by the assembler in its use of the tables. We call the building and use of tables *table maintenance*. The table maintenance operations needed by the assembler are:

1. *Find* the entry corresponding to a given key.
2. *Add* a new entry to the table.
3. *Modify* an item or items in an existing entry.
4. *Copy* an item or items from an existing entry.

There is a fifth operation that completes the set of basic table maintenance operations; *delete* an existing entry. This operation was not required by the assembler.

So far we have not said anything about how the tables are stored in memory. It is extremely important to distinguish between two separate concepts: the conceptual relationship between the entries and items in a table and the way these items are stored in actual memory. The way the items of a table are stored in memory is called the *representation* of the table. The representation of a table is what a procedure actually manipulates when it maintains the table. Conceptually we can visualize tables as just that. Figure 4.30 shows examples of parts of the tables that the assembler uses. Each entry is drawn as a single line in the table. This is the form of table with which we are all familiar.

Any valid representation must in some way capture the essence of the conceptual relationships between the items of the table. In picturing the tables in Figure 4.30, we implied that the entries in the table had a definite order, that is, there was a first entry, a second entry, and so forth. Actually, this is an unnecessary constraint, since the assembler never makes use of this relationship between entries of the table. All references to the symbol table and operation table were made using a value for the key. Even when assigning addresses to the constants in the literal table or sending them to the dispatcher for output, the particular order was not important. All we cared about was getting them all; the order in which they were obtained was not significant. If we do not insist that unused structural relationships be preserved in the representation, we will have more choices for a valid representation. As we will see in the next chapter, ordered tables are needed in practice. Thus, tables really fall into two subclasses, *ordered* and *unordered*.

Once a representation for a table has been chosen, we can write the *find, add, modify, copy*, and *delete* functions for that table. The reader should realize that if all references to a table are made using these functions, his entire program is independent of the table representation, except for the implementation of these five functions. The only requirements for the representation of a simple, unordered table are that the above five functions work as follows. Given a value for the key, *find* must be able to either find the proper entry for the key value or be certain that no such entry is in the table. That is, if *add* is used to make an entry, *find* must be able to find it later and, if an entry was never made or has been removed by *delete*, *find* must not be able to

Symbol Table

Symbol	Value
N	4
I	5
NEXT	1C

Literal Table

Constant	Address	Length	Alignment
40	3E	1	byte
00000001	40	4	full word
00000047	44	4	full word

Operation Code Table

Symbol	Operation type	Numeric code	Operand type
L	machine	580	X
LR	machine	180	R
LM	machine	980	M
B	extended	47F	X
EQU	pseudo		

Figure 4.30 Example fragments of the assembler's three tables for assembly of procedure in Figure 3.7 (all numbers in hexadecimal).

find it. In addition, for an entry know to exist in the table, *modify* and *copy* must be able to modify or copy all of the items in that entry.

EXERCISES

4.1 Modify the assembler described in this chapter so that assembly language programs may include two new data types in both literals and DC and DS pseudo instructions: binary $(t = ''B'')$ and half-word $(t = ''H'')$. In

$$B'constant'$$

constant is a string of binary digits, 0 or 1. The corresponding constant is aligned on a byte. In

$$H'constant'$$

constant is a decimal integer, n, such that $-2^{15} \leq n < 2^{15}$. The corresponding constant is aligned on a half word, that is, on an address that is evenly divisible by two. (The IBM 360 has a set of instructions for arithmetic with half-word quantities.)

4.2 Recall that a literal is a special symbol whose value is the relative address of the corresponding constant, for example, the value of $=$X'F6' is the relative address of a byte in the procedure segment that contains the constant $F6_{hex}$. We define a new special symbol written:

$$t'constant'$$

where t is a type letter and *constant* is a constant consistent with t, that is, the new symbol is like a literal without the equals sign. We call this a *self-defining symbol*. Its value will be the value of *constant*. For example, the value of the symbol X'F6' is 246 ($F6_{hex}$), the value of B'110' is 6, and the value of C'A' is 65 (41_{hex}), which is the USASCII code for the letter A. Describe the modifications to evaluate_expression that are required in order that it accept self-defining symbols as a primary in an expression.

4.3 The IBM 360 has many additional instructions, some with formats that are different from R-, X-, M-, and S-types. For example, the compare logical immediate (CLI) instruction has the format

95_{hex}	I	B	D

0   7 8   15 16   19 20   31

The assembly language form for this instruction is

            CLI   D(B),I

The move characters (MVC) instruction has the format

$D2_{hex}$	L	B1	D1	B2	D2

0   7 8   15 16   19 20   31 32   35 36   47

with an assembly language form of

            MVC   D1(L,B1),D2(B2)

Describe modifications to the assembler so that it can assemble these two new types of instructions.

4.4 Write the procedure convert_constant in IL.

4.5 In addition to the default base register, an assembly language procedure often uses additional base registers for referencing various data areas. The assembler can manage all of these base registers as it does the default base register by redefining the USING pseudo instruction as follows:

            USING   e,k

where e is any expression whose value is a relative address, A, and k is an integer, $0 \leqslant k \leqslant 15$. The new USING instruction indicates to the assembler that G_k is to be used as a base register and that it will contain the absolute address corresponding to the relative address A. The assembler may then use G_k as a base register until a corresponding DROP pseudo instruction of the form

            DROP   k

is encountered. No such DROP pseudo instruction need appear, in which case G_k is used as a base register for the remainder of the assembly language procedure.

   More than one register at a time may be specified as a base register by including several USING pseudo instructions before any DROP pseudo instructions occur. In pass two, whenever a symbolic machine language instruction that does not explicitly specify a base register is assembled, the assembler chooses one register from those that are currently specified for use as base registers, according to the following rule. The expression that defines the value of the D field is evaluated. Call this value B. The base register selected is that one of the current base registers whose corresponding value A (specified in the USING pseudo instruction that defined it as a base register) is closest to, but not greater than, B. The actual value used in the D field of the final instruction will then be equal to B − A.

Describe the modifications to the assembler that are required to implement multiple default base registers as defined above.

4.6 Write an assembly language procedure that will search the symbol table for a given symbol. The procedure should work in the context of **evaluate_ primary**, defined in Figure 4.28. The procedure should have three arguments. The first is the symbol whose entry is to be searched for. The second is a variable whose value will be set equal to the value of the symbol, if an entry for it can be found. The third is a variable that will be set equal to zero if any entry for the symbol was found; otherwise its value will be set equal to one. Choose a representation for the symbol table that is appropriate for the search algorithm that you select.

5 | Macros and Macro Processing

Virtually every modern assembler has "macros". A *macro instruction* is a user-defined assembly language instruction that generates one or more other assembly language instructions. Macro instructions allow the programmer to extend the assembly language. In a sense he may define the machine he wishes he really had by defining a set of macro instructions, which he then uses as the symbolic machine instructions of his new machine. While this is perhaps the most common use of macros, the macro capability of many assemblers is powerful enough to be able to use it for such diverse applications as text and character string manipulation.

Macro capabilities are not restricted to conventional assemblers. PL/1 has a macro capability called the *compile time facility*. In fact, the range of languages that have some kind of macro capability is quite wide, as is the range of application of these languages. In addition, macro-processing functions often show up as a subtask of some larger task. In a later chapter we will see how macro processing can be used inside of a compiler to accomplish one of its major tasks.

5.1 Definition and Use of Macros

Our discussion of macros will center on a macro capability for the assembly language and the assembler, which were discussed in Chapter 4. The simplest form of macro instruction is just an abbreviation for a fixed sequence of symbolic assembly language instructions. For example, if comparison of a blank space with another character, as in Figure 3.7, is required in several places in the program, the programmer can define a macro instruction TESTCHR, which is an abbreviation for the three instructions:

```
        IC      7,=C' '      get first character (space)
        IC      6,(I,S)      get second character (Sᵢ)
        CLR     6,7          compare the two characters (Sᵢ=' '?)
```

He can then write the instructions:

```
        TESTCHR
        BE              FOUND        branch if =
```

and the assembler will output exactly the same machine language instructions as if the programmer had written:

```
IC    7,=C'  '    get first character (space)
IC    6,(I,S)      get second character (Sᵢ)
CLR   6,7          compare the two characters (Sᵢ='  '?)
BE    FOUND        branch if =
```

The assembler must equate the macro-operation code, TESTCHR, with the three instructions for which it is an abbreviation. This is called *macro definition*. Then each occurrence of the TESTCHR macro instruction must be replaced by the corresponding set of three instructions. This is called *macro expansion*. Both of these actions together are called *macro processing*.

The ability to make the simple kind of abbreviation just illustrated is much too weak. Therefore, almost every macro capability includes additional features that add a significantly greater power to the macro capability. Unquestionably one of the most important features is the ability to allow parameters in the macro definition for which substitutions will be made during macro expansion. This feature makes it possible to have a different set of instructions generated in each different place the macro instruction occurs.

Using this feature, we can redefine the TESTCHR macro instruction to include the branch instruction, which normally follows the comparison instruction. We make the branch address and the base and index registers that define the location of the second character all parameters. This gives us a very general macro instruction, since the character that is compared with a space and the location to which the procedure branches if the two characters are equal can both be different each time the macro instruction is used. Our new definition is:

```
IC    7,=C'  '    get first character (space)
IC    6,(A,B)      get second character
CLR   6,7          compare two characters
BE    C            branch if =
```

When the macro instruction is used it will have to specify values for the parameters. Assuming that values for the parameters A, B, and C are specified in that order, writing the single macro instruction

TESTCHR   I,S,FOUND

is equivalent to writing the sequence of instructions:

```
IC    7,=C'  '    get first character (space)
IC    6,(I,S)      get second character
CLR   6,7          compare two characters
BE    FOUND        branch if =
```

If this macro instruction is written with different values for the parameters then a different set of instructions is generated. For example, the macro instruction

TESTCHR   ,2,REPEAT

expands to:

```
IC    7,=C'  '    get first character (space)
IC    6,(,2)       get second character
CLR   6,7          compare two characters
BE    REPEAT       branch if =
```

The parameter values in a macro instruction are separated by commas. Parameter values may be omitted, as in the preceding example, but the commas must remain so that the parameter values can be matched to the proper parameters. As is true of all assembly language instructions, the first blank terminates the operand field. A parameter value may be any character string not containing either a blank, except if enclosed by a pair of quote marks, or a comma, except if enclosed by either a pair of quote marks or a pair of parentheses. This generality of parameter value permits us to simplify and increase the scope of application of the TESTCHR macro instruction. We can define a more general version that compares any two characters. We need only three parameters in the new definition:

```
IC    7,A        get first character
IC    6,B        get second character
CLR   6,7        compare two characters
BE    C          branch if =
```

Now when we use the new version of the TESTCHR macro instruction, we can specify in a single parameter value the complete specification of the D, B, and X fields of each of the IC instructions. For example, the macro instruction

```
             TESTCHR   = C' ',(I,S),FOUND
```

expands to:

```
IC    7,= C' '   get first character
IC    6,(I,S)    get second character
CLR   6,7        compare two characters
BE    FOUND      branch if =
```

and the macro instruction

```
             TESTCHR   PAT(3,),25(,5),FAIL
```

expands to:

```
IC    7,PAT(3,)  get first character
IC    6,25(,5)   get second character
CLR   6,7        compare two characters
BE    FAIL       branch if =
```

A macro definition must specify the name of a macro, that is, the macro-operation code. It must also identify the parameters, the sequence in which their values are to appear in the macro instruction, and the symbolic instructions that are to be generated when the macro instruction is expanded. A macro definition has the following format:

$$m\_op\_code \quad \text{MACRO} \quad p_1, \ldots, p_n$$

$$\left. \begin{array}{c} \cdots \\ \cdots \end{array} \right\} \quad body$$

$$\text{MEND}$$

The symbol m_op_code is the name of the macro being defined and will be used as the symbolic operation code in a macro instruction. The symbols p_i are the n parameters of the macro. Between the MACRO pseudo instruction and the MEND pseudo instruction is the body of the macro definition. This is the set of assembly language instructions to be generated when the macro instruction is expanded.

Any of the parameters may appear in any of the symbolic instructions in the body where it is legal for a symbol to appear, including the label field and the operation field. In addition, a parameter may appear as an item in a DC or DS pseudo instruction. When the macro instruction is expanded, the value specified for each parameter in the macro instruction is substituted for all occurrences of the parameter in the body of the macro definition. Parameter values in the macro instruction must be specified in the same sequence as the parameters are listed in the MACRO pseudo instruction.

5.2 Overview of Macro Processing

Two major extensions to the assembler are required in order to implement the macro capability described in the preceding section. The assembler must be able to recognize and expand a macro instruction. In order to do this it must have first recognized and saved the macro definition corresponding to the macro instruction. A definition is recognized by encountering the MACRO pseudo instruction. The definition is terminated by the first MEND pseudo instruction that follows the MACRO pseudo instruction. The simplest way to recognize a macro instruction is to add an entry to the operation table for each macro instruction that is defined. This entry will contain the name of the macro as a symbolic operation code with its type identified as macro instead of pseudo or one of the machine instruction types. The numeric operation code item will contain a pointer to the corresponding macro definition. The assembler can then easily identify a macro instruction and locate its definition so that the macro instruction can be expanded.

The body of a macro definition may contain parameters in the label field of some of the symbolic instructions. After any expansion of the corresponding macro instruction, these label fields will contain symbols that must be defined by pass one of the assembler. Thus, the expansion of macro instructions must be done before or during pass one. By requiring that a macro definition precede any use of its corresponding macro instruction, we will be able to expand macro instructions during pass one. A macro instruction is expanded by making a copy of the body of the corresponding macro definition, substituting parameter values for all occurrences of parameters in the copy, and inserting the modified copy into the source program immediately following the macro instruction. The inserted symbolic instructions are then processed by pass one of the assembler just as if the programmer had actually included them in his assembly language procedure. The macro instruction is left in the source program. Pass two will treat it as if it were a comment that is printed as part of the listing to help the programmer identify which instructions he wrote and which instructions resulted from expansion of macro instructions.

5.3 Implementation of Macro Definitions

Two tasks must be performed to process a macro definition. A new entry must be added to the operation table, and the macro definition must be saved. Macro definitions will be saved in the *definition table*. The entire macro definition will be stored in the definition table, including both the MACRO and MEND pseudo instructions. Figure 5.1 is a new definition of pass one that shows how processing of a macro definition fits into its overall structure. The new definition is identical to the definition in Figure 4.5, except for the addition of two new cases in the **case** statement and initialization of the definitition table index **def_idx** and the macro instruction expansion flag **m_expand**.

Figure 5.2 shows the details of how a macro definition is processed. Figure 5.3

```
procedure pass_one;
   loc_ctr := 0;
   def_idx := 0;
   m_expand := false;
   Create file s_copy and open it for writing;
   repeat
      Clear error flags;
      Read next card of source procedure from IN into S₀, . . . ,S₇₉;
      Locate label, operation, and operand fields;
      type := type of operation code;
      case type of
         'machine': Process symbolic machine instruction;
         'EQU': Process EQU pseudo instruction;
         'DC/DS': Process DC or DS pseudo instruction;
         'END': Process END pseudo instruction;
         'START': null;
         'USING': null;
         'MACRO': Define a macro;
         'macro': Expand macro instruction;
         otherwise Set error flag;
      end;
      Write contents of S and error flags into s_copy;
   until type = 'END';
   close (s_copy);
   return;
end.
```

Figure 5.1 New definition of pass one (from Figure 4.5) with macro processing.

```
Add new entry to operation code table consisting of (macro name from label
   field of instruction in S, 'macro', def_idx, ' ');
definition_table [def_idx] := S;
def_idx := def_idx + 1;
repeat
   Write contents of S into file s_copy;
   Read next assembly language instruction from device IN into S;
   definition_table [def_idx] := S;
   def_idx := def_idx + 1;
until operation field of S = 'MEND';
```

Figure 5.2 Expansion of "Define a macro;" from Figure 5.1.

shows the contents of the operation code table and the definition table after processing the macro definition:

```
TESTCHR   MACRO   A,B,C
          IC      7,A      get first character
          IC      6,B      get second character
          CLR     6,7      compare two characters
          BE      C        branch if =
          MEND
```

Operation Code Table

Symbol	Operation type	Numeric code	Operand type
TESTCHR	macro	k	

Definition Table

k	TESTCHR	MACRO	A,B,C	
$k+1$		IC	7,A	get first character
$k+2$		IC	6,B	get second character
$k+3$		CLR	6,7	compare two characters
$k+4$		BE	C	branch if =
$k+5$		MEND		
$k+6$				

Figure 5.3 Operation code table and definition table entries for TESTCHR macro definition.

Entries in the operation code table consist of four items: a symbolic operation code, its type, its corresponding numeric code, and its operand type. If an operation's type is 'pseudo', the last two items are not relevant. Since a macro instruction has no corresponding numeric operation code, we can use that item to store a pointer to the corresponding macro definition in the definition table. This pointer is the index, in the definition table, of the entry for the corresponding macro definition. The last item (operand type) is not relevant for a macro instruction. As noted in Chapter 4, the operation code table is not an ordered table, so the place where the new entry goes is determined by the particular representation for the operation code table that we choose in our implementation.

Once the macro definition has been added to the definition table, the only function ever performed on that entry is to copy its contents. Even the find function is never used. Since the index of each entry is stored in the corresponding entry of the operation code table, the definition table never needs to be searched. Only the operation code table is searched. Therefore, the simplest representation for the definition table is to store the macro definitions, and the symbolic instructions within them, in the sequence in which they occur in the source procedure. Thus, after the entry in the operation code table has been made for a new macro being defined, the definition is completed by copying all of the symbolic instructions of the definition into the definition table. This table is an array of character strings, each string being 80 characters in length. In the example of Figure 5.3, the first unused element in the definition table has an index of k, so this index value is put into the numeric code item of the entry for TESTCHR, which is added to the operation code table. The symbolic instructions of the definition for TESTCHR are then copied into elements k through $k+5$ of the definition table.

5.4 Expansion of Macro Instructions

To expand a macro instruction a copy must be made of the body of the corresponding macro definition. All occurrences of parameters in the body must be replaced by their corresponding values, which appear in the operand field of the macro instruction being expanded. This modified copy must be processed by pass one of the assembler immediately following the macro instruction being expanded. The method used in our assembler is to modify the input procedure so that symbolic instructions can come from either the system's input device or from a macro definition in the definition table. In a sense, during expansion of a macro instruction, the assembler is tricked into thinking that the modified copy of the macro-definition body is actually being read in from the system input device, just as if the programmer has included it in the source program immediately following the macro instruction.

The body of the macro definition is not copied all at once, but one line at a time. Each line is processed through pass one before the next one is copied. Figure 5.4 shows the expansion of the input statement of pass one that is needed to accomplish this. When the expansion flag m_expand is true, a macro instruction is being expanded. In this case, the next symbolic instruction of the definition for the macro instruction being expanded is copied from the definition table into the source input area S. If the expansion flag is false, the next symbolic instruction of the source procedure is read from IN into the source input area by calling the **read** entry of I/O control.

From that point on, pass one does not know where the source statement actually came from. The expansion flag is set to true when a macro instruction is recognized and processed, as shown in Figure 5.5. The expanded input statement recognizes the end of expanding a macro instruction by testing each symbolic instruction in the macro definition before it is copied into the source input area. A MEND pseudo instruction signals the end of the definition and the expansion flag is set to false.

As each symbolic instruction is copied from the definition table into the source input area, every occurrence of a parameter in the instruction is replaced by its corresponding value from the macro instruction being expanded. The correspondence between parameters and their values is recorded in the *parameter table*. Each entry in this table contains two items, the parameter and

```
if m_expand then
    begin i := i+1;
        if operation code of instruction in definition_table [i] = 'MEND'
            then begin m_expand := false;
                    read ('IN', count, S, status);
                end;
            else Copy definition_table [i] into S replacing all occurrences
                    of parameters by their corresponding values from
                    the parameter table;
    end;
    else read ('IN', count, S, status);
```

Figure 5.4 Expansion of input statement "Read next card of source procedure from IN into $S_0, \ldots, S_{79}$;" from Figure 5.1 to handle macro expansion.

m_expand : = **true**;
i : = value of numeric operation code item from operation code table entry
   for the macro instruction in **S**;
   {this is the index in definition_table of its definition}
Copy parameters from MACRO pseudo instruction in definition_table [i]
   into parameter_table;
Copy parameter values from operand field of macro instruction in **S**
   into parameter_table;

Figure 5.5 Expansion of "Expand macro instruction;" from Figure 5.1.

its corresponding value. This table is built during the processing of the macro instruction. The parameter table for our example is shown in Figure 5.6. The parameters are obtained from the MACRO pseudo instruction of the macro definition in the definition table, and their values are obtained from the operand field of the macro instruction. The correspondence is positional, that is, the first value in the operand field of the macro instruction corresponds to the first parameter in the MACRO pseudo instruction, the second value to the second parameter, and so forth.

When a symbolic instruction is copied, only symbols need to be tested for parameter value substitution. Literals, integers, and separators such as "+", "(", or a string of spaces cannot be parameters of a macro definition. The symbolic instruction is copied one primary or separator at a time. For example, the instruction

IC   7,A   get first character

is copied in seven steps: a string of spaces, the symbol "IC", another string of spaces, the integer "7", the separator ",", the symbol "A", and finally the comment with the spaces preceding it. As each symbol in the instruction is copied, it is tested against the parameters in the parameter table. If the symbol matches one of the parameters, it is not copied; instead, the corresponding value of the parameter is copied out of the parameter table into the source input area. Copying is resumed in the symbolic instruction immediately following the symbol that was a parameter. All symbols in the symbolic instruction must be tested, since there is no restriction on the number of parameters that may occur in the instruction. In fact, it is possible for every symbol in the instruction to be a parameter.

5.5 Additional Macro Features

The macro capabilities of most assemblers include additional features.[1] There is a great variety of additional features, and it is not practical even to attempt to mention them all or to explain any of them in great detail. The macro facilities in the assemblers of most major operating systems are considerably more elaborate than what we discuss here. Our purpose in this section is to describe briefly a few additional features that we will make use of in later chapters.

[1] P. J. Brown, "A Survey of Macro Processors," *Annual Review in Automatic Programming*, *6*, Part 2, Pergamon Press, Oxford, 1969.

Parameter	Parameter value
A	=C' '
B	(I,S)
C	FOUND

Figure 5.6 Parameter table for expansion of the macro instruction
TESTCHR =C' ',(I,S),FOUND.

All of the features we are interested in here contribute to one important functional extension of the macro capability: the ability of a single definition to generate a varying number of instructions when it is expanded. The basic feature needed for this ability is some form of conditional. Our conditional has the form

$$\text{IF} \quad relation, k$$

where k is a decimal integer and $relation$ has the form

$$opnd_1 \; rel \; opnd_2$$

The operands $opnd_1$ and $opnd_2$ are either null or primaries, as defined in Chapter 4, that is, a symbol, literal, or integer, and rel is one of the six relation characters

$$= \; \neq \; < \; \leq \; > \; \geq$$

The source lines in a macro definition are numbered from 0 to N, the MACRO pseudo instruction being number 0 and the END pseudo instruction being number N. If the value of the relation is true, the expansion of the macro instruction will continue with source line number k in the macro definition. The integer k must lie within the range 1 to N. If it is N, expansion of the macro instruction is terminated. For example, we redefine the TESTCHR macro instruction as:

0	TESTCHR	MACRO	A,B,C
1		IC	7,A	get first character
2		IC	6,B	get second character
3		CLR	6,7	compare two characters
4		IF	C=,6	omit BE if 3d parameter value omitted
5		BE	C	branch if =
6		MEND		

(Note that the numbers written in the left column are not part of the macro definition. They are written to help the reader follow the action of the IF instruction.) The BE instruction will not be generated when expanding any TESTCHR macro instruction in which the third parameter value is omitted. If the third parameter value is omitted the null character string will be substituted for the third parameter, C, when the macro instruction is expanded. Thus, the IF pseudo instruction in the expansion will be

$$\text{IF} \quad =,6$$

and the two expressions being compared for equality have the same value.

A potential problem arises with respect to the value of the third parameter when we use the TESTCHR macro instruction. If the third parameter value is not omitted it is to be a symbol that will be a label to be used in a branch instruction. Thus, it may not be defined at the time the macro instruction is being expanded. Our intended interpretation of the value of a relation is that its value is to be determined using the values of the operands as defined for the evaluation of expressions in Chapter 4. The value of an integer is simply its value. The value of a symbol that is a label or a literal is its location address. What is this value if a symbol has not been defined yet? For the purpose of evaluating relations, we define the value of an undefined symbol to be a special value that is not equal to any value that a defined symbol can possibly have.

One exception to this interpretation is the value of a null operand. In Chapter 4 this value was equal to zero. In the context of evaluation of relations, the value of a null operand is defined to be another special value that is different from the value of any symbol, defined or undefined. With these conventions, there will be no problem in determining the desired value for the relation in

$$\text{IF} \quad C =, 6$$

If C is any symbol, its value will not equal the value of the null operand, and the value of $C=$ will be false. Its value will only be true if no parameter value for C appeared in the macro instruction.

The next feature we want is one that allows looping within the macro definition during expansion of a macro instruction. What we need for this is the SET pseudo instruction, which allows the value of a symbol to be redefined an arbitrary number of times. The form of the SET pseudo instruction is

$$\text{symbol} \quad \text{SET} \quad \text{expression}$$

The value of the symbol in the label field is set equal to the value of the expression in the operand field.

For example, we can define a macro that raises a number X to an arbitrary positive integer power N and stores the result in Y.

0	POWER	MACRO	X,N,Y	
1		IF	N=0,8	exit if N=0
2	I	SET	1	I:=1
3		L	3,X	get X
4		IF	I⩾N,9	exit if I⩾N
5		M	2,X	multiply by X
6	I	SET	I+1	I:=I+1
7		IF	=,4	repeat loop
8		L	3,=F'1'	$X^N:=1$ if N=0
9		ST	3,Y	store X^N
10		MEND		

When this macro is expanded it will generate instructions to set the value of Y equal to X^N, even if N=0. It does this by generating $N-1$ multiply instructions. If N=0 the constant 1 is stored in Y and if N=1, the value of X is stored in Y. Values for the parameters X and Y in a use of the POWER macro instruction may be any valid address, however, N must either be an integer or a symbol whose value has already been defined as an integer. Note that

$$\text{IF} \quad =, 4$$

is equivalent to an unconditional branch, since the relation is always true.

As examples of the expansion of the POWER macro consider

$$\text{POWER} \qquad 8,0,Z$$

which expands to

L	$3, = F'1'$	$X^N := 1$ if $N = 0$
ST	$3,Z$	store X^N

and

POWER	A,3,B

which expands to

L	$3,A$	get X
M	$2,A$	multiply by X
M	$2,A$	multiply by X
ST	$3,B$	store X^N

The final feature we need is the ability to have a variable number of parameter values in a macro instruction. We do this by allowing a list of parameter values to correspond to a single parameter and allowing the parameter to be subscripted in the macro definition. A list of parameter values is a sequence of parameter values separated by commas and enclosed in parentheses. For example, $(3,XVALUE,Y+8,K(I,4))$ is a list of four parameter values. If A is the parameter that corresponds to this list of values, then:

A	refers to	$(3,XVALUE,Y+8,K(I,4))$
A.(1)	refers to	3
A.(2)	refers to	XVALUE
A.(3)	refers to	$Y+8$
A.(4)	refers to	$K(I,4)$
A.(n)	for $n > 4$ refers to the null string	

The period separating the parameter from its subscript is needed so that a parameter appearing in the D field of an instruction is not confused with a base or index reference. The subscript inside the parentheses may be any expression that results in a positive integer. Any subscript value larger than the number of parameter values in the list results in a reference to the null string.

For example, the following macro definition defines a macro instruction that adds together an arbitrary set of numbers.

1	ADD	MACRO	R,X,SUM	
2		IF	$X.(1) = ,10$	exit if empty parameter value list
3		L	$R,X.(1)$	get first number
4	I	SET	2	$I := 2$
5		IF	$X.(I) = ,9$	exit if parameter value list exhausted
6		A	$R,X.(I)$	add next number
7	I	SET	$I+1$	$I := I+1$
8		IF	$=,5$	repeat loop
9		ST	R,SUM	store sum
10		MEND		

Expansion of the ADD macro instruction will generate a sequence of add instructions until the list of parameter values corresponding to the parameter X is exhausted, at which point X.(I) will refer to the null string. For example, the macro instruction

ADD $3,(TOP,Z(5,),(K,8)),(9)$

expands to

L	3,TOP	get first number
A	3,Z(5)	add next number
A	3,(K, 8)	add next number
ST	3,(9)	store sum

Macro features such as those described in this paragraph allow an assembly language procedure to be written that can easily be adapted to different situations by changing a few EQU pseudo instructions and assembling the procedure. The EQU instructions are used to define symbols that are effectively parameters in macro definitions. Using conditional macros, the set of instructions that actually gets assembled can be made to depend on the values of some symbols that are defined by EQU instructions. This makes it easy to "customize" a procedure. One common application in which this technique is used is in system generation. The procedures in an operating system that depend on the computer hardware configuration are parameterized using EQU instructions and conditional macros. Versions of the operating system for different hardware configurations are then easily generated by simply changing EQU instructions and assembling.

5.6 Implementation of the Additional Features

We will not describe in detail the implementation of the additional macro facilities described in the preceding section. We will only mention briefly the main points that have to be considered when implementing these features. We leave to the student the task of working out the details as an exercise. The context of this discussion will be the basic macro-processing capability whose implementation we discussed earlier in this chapter.

Since the IF pseudo instruction may affect the sequencing through the macro definition when it is being expanded, it must be interpreted by the macro-processing portion of the input section (Figure 5.4). The relation in the IF pseudo instruction must be evaluated in order to determine which line of the macro definition to expand next. This requires little more than evaluation of the primaries that are the operands of the relation, taking into account the special values for null expressions and undefined symbols. If the value of the logical expression is true, the next line of the macro definition to be copied into the source input area is specified by the integer that is the second subfield of the operand field of the IF pseudo instruction. This integer is converted to binary form, added to the base index of the macro definition, and stored as the value of the index i, which is used to obtain the next line of the macro definition (see Figure 5.4). If the value of the logical expression is false, the value of the index i is sequenced normally.

Processing of the SET pseudo instruction is basically the same as the processing of the EQU pseudo instruction. If an entry for the symbol already exists in the symbol table, no multiple definition error is flagged, and the previous value is simply replaced by the new one. As an additional error check, a third item could be added to the symbol table entries indicating whether the symbol was defined by a SET pseudo instruction or by some other means. Then a mixing of the two types of definition for the same symbol could be flagged as an error.

Handling subscripted parameters and lists of parameter values is the most complicated of the three new features. It is not conceptually difficult, just messy. When copying a source line from the macro-definition table to the source input area, the input routine has to check for a subscript whenever it finds a parameter in the source line. If the parameter has a subscript, which is indicated by the characters ".(" immediately following the parameter, then the expression inside the parentheses must be evaluated. The expression evaluation procedure can be used for this evaluation. After obtaining a value for the subscript, say i, the ith parameter value in the parameter list must be extracted from the list. The list is scanned from left to right until the ith value is found, or the list is exhausted. The values in the list are separated by commas that do not occur inside any paired quote marks or parentheses except the outermost parentheses, which enclose the entire list of values. Once the ith value in the list is located, it is copied into the source input area in place of the parameter and its subscript. If there are fewer than i values in the list nothing is copied, neither a value nor the parameter and its subscript.

EXERCISES

5.1 Let the macro MAX be defined by:

```
MAX    MACRO   A,B,C
       L        2,A     get A
       C        2,B     compare with B
       BNL      SKP     skip if A ⩾ B
       L        2,B     get B
SKP    ST       2,C     store largest in C
       MEND
```

If MAX is called more than once in the same program, the symbol SKP will be multiply defined. Choose a notation for symbols of this type. Modify the macro processor so that whenever a special symbol of this type is encountered during macro expansion, a unique symbol is generated by the macro processor and used in its place. For example, define MAX as:

```
MAX    MACRO   A,B,C
       L        2,A
       C        2,B
       BNL      #SKP
       L        2,B
#SKP   ST       2,C
       MEND
```

Suppose that

```
                MAX   X,Y,Z
```

is the first macro call, then its expansion might be:

```
       L        2,X
       C        2,Y
       BNL      #SKP1
       L        2,Y
#SKP1  ST       2,Z
```

Then, if

$$\text{MAX   P,Q,R}$$

is the second macro call, its expansion might be:

```
                      L      2,P
                      C      2,Q
                      BNL    #SKP2
                      L      2,Q
           #SKP2      ST     2,R
```

Is there any way to prevent programmers from using any of these special symbols (e.g., #SKP2) in ordinary instructions, that is, in instructions that are neither in a macro definition nor in a macro expansion?

5.2 Modify the macro processor to implement the IF instruction (conditional) in a macro definition.

5.3 Modify the results of Exercise 5.2 to allow arbitrary Boolean expressions in place of the relation in an IF instruction. The new form is:

$$\text{IF   }B\_expression,k$$

where B_expression consists of one or more B_terms separated by "||" (logical or). A B_expression:

$$t_1 \,||\, t_2 \,||\, \ldots \,||\, t_n$$

is true if any B_term, t_i, is true. A B_term consists of one or more B_primaries separated by "&" characters (logical and). A B_term

$$p_1 \,\&\, p_2 \,\&\, \ldots \,\&\, p_m$$

is true only if all of the B_primaries $p_1, p_2, \ldots, p_m$ are true. A B_primary has one of the forms

$$relation$$
$$\neg\, relation$$
$$(\,B\_expression\,)$$
$$\neg\,(\,B\_expression\,)$$

where relation is as defined in Section 5.5.

5.4 Modify the macro processor so that a list of parameter values may correspond to a single parameter in a macro definition. Individual values in this list may be referenced by subscripting the parameter as described in Section 5.5.

5.5 Modify the macro processor so that a macro call may occur in the body of a macro definition. The definition of the called macro may also contain a macro call. How can an indefinitely deep nesting of such calls be handled?

6 | Loaders

In this chapter we will study the problems that arise in trying to get a program into the primary memory of the computer and ready for execution. We mentioned in Chapter 2 that the computer can only execute instructions that are in its primary memory. We also pointed out that programs are usually composed of more than one procedure. These procedures may be written in different languages, and each one is translated separately, at different times. Thus, if a program is to be executed, its constituent procedures must be loaded into the primary memory and connected together to make a program. This is the function of a *loader*.

It should be mentioned that there is a simple but impractical solution that avoids the need for a loader. The program could be written as a single procedure, say in assembly language. If the desired memory address for the first instruction in the program were included in the START pseudo instruction of the source program, the assembler could then assemble the object program directly into the desired memory locations. Directly after assembling the program, it could be executed. This is a simple solution, and it eliminates the need for a loader.

However, its disadvantages are so severe that it is not a practical solution. The entire program would have to be written in a single language and translated every time it was to be executed, even after the program was completely checked out and working perfectly. This would be an excessive, unnecessary cost. Another disadvantage lies in the requirement that the entire program is a single procedure translated all at once. A large, complex program would be difficult to work with, eventually becoming so large that it could not be translated by the translator. However, there is one situation where this solution is worth considering: the translation and execution of small programs that are exercises in elementary programming classes. A few of the compilers for student use actually use a solution similar to this. The reason this is practical is that such student programs are always compiled when they are submitted to the system. Once the program works, the student is finished, and the program is not executed again.

The practical solution to this problem is to use some form of loader. The loader accepts as input data one or more procedures, which it loads into primary memory, and it then connects them together to make a program that is ready to execute. This solution requires that each translator in the system package the object procedure in a standard form, which is known to the loader. An object procedure in this form will be called an *object deck* because it is often punched as a deck of cards. The object deck also contains more information than just the instructions and constants in the object procedure. For example, it also contains the length of the object procedure. The relationship between a source procedure, its object deck, the assembler, the loader, and the final executable program can be diagrammed as shown in Figure 6.1. As we saw in Chapter 2, the object deck need not be punched out, but may be passed to the loader via the system object file for execution as part of the current job. If the object deck is punched it may be used as part of any future job. The information included in the object deck in addition to the procedure itself and the format of the object deck depend on the particular loader to be used. We will discuss two different types of loaders: an *absolute loader* and a *relative loader*.

6.1 An Absolute Loader

An absolute loader is the simplest type of loader. When using an absolute loader, the final memory locations of all the text of all the procedures that are loaded must have been assigned by the assembler. This means that the base address of each procedure (the address of the first instruction in the procedure) must be specified at the time the procedure is assembled. The addresses assigned by the assembler are included in the object deck. The absolute loader simply loads the procedure's text into the memory locations specified in the object deck. Note that the assembler discussed in Chapter 4 does not assemble procedures that can be loaded by an absolute loader, since it assembles every procedure with a base address of zero. In the object deck generated by that assembler, all addresses are relative to the base of the procedure. An absolute loader requires object decks in which all of the addresses are absolute, that is, they are the final, true memory addresses.

What information do we require in an object deck that is to be loaded by an absolute loader? We need the procedure's *text*, that is, the instructions and constants in the procedure. We also need the memory addresses for each word

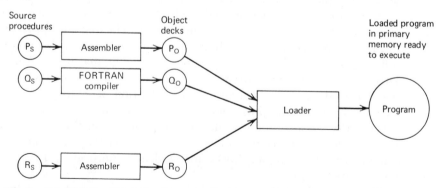

Figure 6.1 Relationship of source and object procedures to a loaded, executable program.

of this text. We must have some way to identify the last procedure in the program, that is, the end of the collection of object decks being loaded. Finally, we need to know the address at which to begin execution of the loaded program.

To make this concrete, we will briefly look at an absolute loader for Our 360. An object deck will consist of cards, each of which contains the following information:

1. One or more bytes of the procedure's text.
2. The memory address of the first byte of text on the card, which is called the *loading address*. By convention all bytes on the card are assigned consecutive memory addresses, beginning with the loading address.
3. A count of the total number of bytes of text on the card.
4. A flag to indicate that the card is part of an object deck and not a special card that stops loading and begins execution of the program.

The object decks for the procedures in the program are combined in any sequence and terminated with a special card that contains the address at which to begin execution of the program.

Our absolute loader expects the cards in the object deck to have the format shown in Figure 6.2. If a card contains procedure text, *byte count* is the number of bytes of text on the card. The *loading address* is the memory address of the first byte of text on the card. Each successive byte of text has a memory address one greater than that of the immediately preceding byte. The text is punched one byte per card column. If a card contains "E" in its first column, loading is terminated. Execution of the loaded program is then started at the *starting address* (in columns 3 to 5 of the card).

A simple absolute loader that loads cards having this format is shown in Figure 6.3. It is basically a single loop, each time around the loop it loads the text from one card. The program end card terminates loading. The *starting address* from the program end card is returned to the system as the value of the loader's only argument. If the card read is a text card, the specified number of bytes of text are copied from the card input area S into the memory locations specified by the *loading address*.

The advantages of an absolute loader are that it is simple, fast, and small. The principal disadvantage is that the layout of the program in memory must be completely preplanned. Each separate assembly language procedure must contain the absolute address of its base. In addition, it must contain the absolute address of the entry point of every other procedure that it calls and the

Card columns	Contents
1	*card type* = blank if card contains text = 'E' if card is program end card
2	*byte count*
3–5	*loading address* (*starting address* if program end card)
6–72	1 to 67 bytes of text
73–80	card identification

Figure 6.2 Object deck card format for the absolute loader.

```
procedure loader (start);
    while forever do
        Read next card from IN into 80 characters S[1] ... S[80];
        if S[1] = 'E'
            then begin start : = address in S[3] ... S[5];
                      return;
                  end;
            else begin n : = S[2];
                        a : = S[3] ... S[5];
                        Copy S[6] ... S[6+n−1] into memory locations
                            a ... a+n−1;
                  end;
        end;
    end.
```

Figure 6.3 Definition of the absolute loader.

absolute address of the base of each data region outside the procedure that it references. This requirement has significant implications.

As modifications are made to the program in the process of development and checkout, the sizes and locations of the procedures and data regions in the program will change. These changes propagate throughout the program, necessitating secondary changes of absolute addresses in procedures that were not themselves changed. Every procedure in which an address has been changed must be assembled again before it can be used in the modified program. In the past, programmers have tried to make their programs less sensitive to changes by leaving an area of unused space between each procedure and data region into which each may be expanded. This solution is rarely effective. Expansion is never uniform, so there is always some expansion area that is not large enough. Making them different sizes does not help, since it is difficult to predict where the expansion will occur. In addition, many of the expansion areas will not be used, resulting in a considerable amount of unused memory.

A further disadvantage of the requirement that absolute addresses be assigned at assembly time is that it is difficult, if not impossible, to use subroutine libraries effectively. A rich library will contain many more subroutines than can fit together in memory at the same time. Efficient use of a library demands that they be assembled before they are put in the library. What absolute addresses are to be used when they are assembled? Whatever choice is made will be wrong for some subset of subroutines that some user will wish to use, that is, there will be two or more subroutines that some user wants to use that have been assigned the same memory addresses.

All of the problems that arise in the use of absolute loaders stem from the fact that the requirement that absolute addresses must be specified at assembly time negates, in a sense, one of the principal reasons that assembly language was used in the first place. Assembly language evolved so that the programmer need not use machine addresses in his program. We do not really achieve this goal until the need to specify any machine address at all is eliminated. We can achieve this goal if we can design a loader that is able to load a set of procedures into memory and connect them together without requiring any absolute addresses in any of the object decks. This goal is reasonable because we do not really care where the procedures and data regions are located in memory, as long as the program executes properly and gives the correct results.

The basic idea involved in such a loader is to postpone assignment of all absolute addresses until the program is loaded, letting the system and loader decide where each procedure and data region is located in memory. In order to make such a loader function, the object deck form of a procedure must be such that the procedure can be put into memory at a more or less arbitrary location and fixed up so that it will execute properly at that location. This requires that the object deck does not contain any absolute addresses, hence, such object decks are called *relative object decks,* and a loader that loads them is called a *relative loader.* Relative loaders similar to the one that we will study are used in almost every modern operating system.

6.2 Basic Concepts of Relative Loaders

There are some basic concepts that are important in understanding the problems of relative loaders. The most fundamental concept is that of a *segment.* A segment is a set of contiguous words (bytes in Our 360) of information. One or more symbolic names are associated with a segment. Each symbolic name is further associated with some specific location within the segment. A segment is the smallest unit of procedure or data that translators and loaders recognize. Whenever a source language procedure is translated, the assembler or compiler translates it into a single segment. Segments are classified as *procedure* or *data segments.* A procedure segment contains a procedure (and possibly some constants) and is the result of a single translation. A data segment contains only data.

A procedure, in a procedure segment, can make two types of reference: *intrasegment,* which refers to a location in the same segment, and *intersegment,* which refers to a location in another segment. The symbols in an object language procedure are classified as *local* or *global symbols,* depending on whether they are used for intrasegment or intersegment references, respectively. The corresponding references will be called *local* (or *internal*) and *global* (or *external*) *references.* Since all of the words in a segment will be contiguous in memory, a translator can translate all local symbols into *relative addresses,* that is, a number that is a location within the segment, expressed as an offset relative to the base of the segment.

The assembler discussed in Chapter 4 permitted only local symbols except for the entry point name in the START instruction, which is a global symbol, since other procedures can refer to it. Recall that this assembler translated a procedure as if its base address was zero, thus, all symbols referring to instructions or data were assigned addresses relative to the base of the assembled procedure segment. It is easy to convert a relative address into an absolute address when a procedure segment is loaded. Once the absolute address of the base of the segment is known, the absolute address corresponding to a relative address is simply the sum of the relative address and the absolute address of the segment base. We call this conversion of a relative address to an absolute address *relocation.*

The situation with respect to global symbols is not quite so simple. A global symbol refers to a location in another segment, and the absolute address of the base of that segment is not known until the program has been loaded. Thus, there is really very little that a translator can do. We see then that the object deck will have to contain unresolved symbolic references for all of the global symbols that appeared in the source language procedure. The final conversion of these unresolved global references to absolute addresses we call *linking*.

Both relocation and linking are special cases of the basic concept of *binding*. Binding is the resolution of variability. When a symbol is converted into an absolute address, it is *bound*. Conversion of a symbol into a relative address is only partial binding. Thus, for local symbols, the assembler does part of the binding and the loader finishes the binding. The loader does all of the binding for global symbols.

The assembly language in the preceding chapters has a third class of symbols whose values are specified by EQU pseudo instructions and do not refer to either instructions or data in any segment. For example, in the assembly language program in Figure 3.7, the symbols "S" and "K" were defined to have values 2 and 3, respectively. The symbols were then used to reference registers 2 and 3. These symbols are completely bound by the assembler. This class of symbols is usually called *absolute symbols* because their values do not depend on the location of the program in memory. In this specific example, their values are completely bound by the assembler when it processes the EQU pseudo instructions that define them.

A global symbol used to reference another segment is only half the story. In order to bind a global symbol, the loader must have a definition for it. This definition is a generalization of the entry point name that appears in the label field of a START instruction. When we defined a segment, we said that one or more symbolic names were associated with it. These names are global symbols, each of which corresponds to some location within the segment. These symbols can be used in other procedure segments to reference the corresponding locations in this segment. We will call this class of global symbols *insymbols*. The global symbols used in a segment to reference locations in other segments will be called *outsymbols*. Figure 6.4 illustrates this distinction. Notice that the classification of a global symbol as an insymbol or an outsymbol depends on the context, that is, a global symbol is an insymbol relative to one segment and an outsymbol relative to all other segments. The occurrence of a global symbol as an insymbol constitutes a definition of the symbol. For example, in Figure 6.4, Y is defined as location 400 relative to the base of segment X by occurring as an insymbol of that segment. When Y is used by segments A and B to reference location 400 in segment X, Y is classified as an outsymbol. Thus, the object deck for a segment will have to contain definition information for each insymbol associated with the segment.

6.3 Functions of a Relative Loader

We are now in a position to identify the major functions of a loader. The object deck for a procedure segment must contain at least the following information.

1. The text of the procedure and the relative location within the segment of each word of text.
2. The total size of the segment, including any reserved space (by a DS pseudo instruction, for example) and any fill required for alignment.

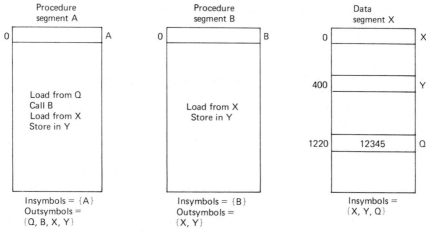

Figure 6.4 Examples of insymbols and outsymbols.

3. A list of all places in the segment text where the absolute address that results from binding an outsymbol is needed and which symbol corresponds to that address.

4. A list of all the insymbols that are defined in this segment and their locations relative to the base of the segment.

The object deck for a data segment contains the same information, except that the text does not normally include any instructions.

The loader is an ideal place for the interface between the user and the system subroutine library. The loader takes a set of procedure and data segments supplied by the user, relocates them, and links them together to form an executable program. There is no need for the user to make copies of the library subroutines that he needs and include them along with his procedures as input to the loader. It is easier and more efficient for the loader to do this for him.

The relative loader that we will study in the following sections must perform five major functions.

1. Determine where in memory each segment is to be located. We call this assignment of memory locations *memory allocation.*

2. Put each segment into its assigned locations in memory. This should properly be called *loading,* although the term loading is usually used to refer to all of the functions of a loader.

3. Relocate all relative addresses (local references) in each segment.

4. Obtain from the system library the procedure segments that contain the library subroutines that the user requires. By convention, all global symbols that are not defined in any of the segments that the user supplied are assumed to refer to library subroutines. The processing implied in functions 1 to 3 is carried out for each segment obtained from the library.

5. Link all of the segments together, that is, bind all of the outsymbols.

The result of relocation and linking is that all relative addresses and all global references are replaced by absolute addresses. The user's program is now ready for execution. The loader must recognize the last object deck supplied by the user and must obtain the starting address for execution. A special end card satisfies both these requirements. However, since the loader is assigning absolute addresses, the starting address supplied by the user on the end card will have to be a global symbol.

6.4 Assembly Language Extensions Required by the Loader

In order to include in the object deck the information required by a loader, the assembler will have to know which symbols are global. In addition it will have to know which of these are insymbols and which are outsymbols. Using the assembler of Chapter 4 as an example, two additional pseudo instructions, INSYMB and OUTSYMB, will satisfy these requirements. The operand field for each of these two pseudo instructions consists of a list of symbols separated by commas.

The INSYMB pseudo instruction identifies each of the symbols in its operand field as being an insymbol, that is, a global symbol that is defined by this segment. Every symbol appearing in an INSYMB pseudo instruction must be defined in the sense of Chapter 4, that is, the symbol must also appear in the label field of a symbolic machine instruction, a DS or DC pseudo instruction, or an EQU pseudo instruction. The OUTSYMB pseudo instruction identifies each of the symbols in its operand field as being an outsymbol, that is, a global symbol used to reference another segment.

The assembly language source procedure for segment A in Figure 6.4 would be:

```
A     START                        define segment A
      . . .
      INSYMB    A                   define insymbol
      OUTSYMB  B,Q,X,Y              define outsymbols
      . . .
      END
```

The source procedure for segment B would be:

```
B     START                        define segment B
      . . .
      INSYMB    B                   define insymbol
      OUTSYMB  X, Y                 define outsymbols
      . . .
      END
```

The source for the data segment X would be:

```
X     START                        define segment X
      INSYMB    X,Y,Q              define insymbols
      DS        F                   reserve space (a full
                                      word) for X
      . . .
Y     DS        F                   define symbol Y and
                                      reserve space
      . . .
Q     DC        F'12345'            define symbol Q, reserve
                                      space, and set initial
                                      value of that word
      . . .
      END
```

It is clear that the occurrence of a symbol in the label field of the START instruction implies that the symbol is an insymbol. Thus, it is not necessary that the symbol also appear in the operand field of an INSYMB instruction. We have done so only for emphasis.

Now that we have a means of identifying global symbols and, in particular, outsymbols, we must consider how they are used in an assembly language procedure. Recalling our discussion in Chapters 3 and 4 with respect to addressing, we see that absolute addresses will not fit into the D field of machine language instructions. We saw that this was true because the D field, being 12 bits in size, can have a maximum value of only 4095, and the resident system occupies more than the first 4096 words of primary memory. Therefore, whenever an instruction references memory, it is necessary to use a base register. The contents of this register will be some address, often the address of the base of a segment. We saw that the system standard calling sequence loaded G_{15} with the base address of the called procedure. This convention along with the USING pseudo instruction allows the programmer to use local symbols in the operand fields of symbolic machine instructions, even though the final absolute address (after relocation) corresponding to the local symbol will not fit into the D field of a machine language instruction.

Since the number of different outsymbols, and thus the number of different addresses, is not limited, we cannot have conventions governing the use of base registers for global references. The programmer must write instructions to load the address corresponding to an outsymbol into a base register and explicitly indicate that register in all instructions that reference the memory location associated with the outsymbol. The programmer may use either literals or DC pseudo instructions to refer to the required address.

For example, in Figure 6.4, procedure A may refer to segments B and X in the following way.

A	START		define segment A
	. . .		
	INSYMB	A	define insymbol
	OUTSYMB	B,Q,X,Y	define outsymbols
	. . .		
	L	2, = A'Q'	load address of Q
	L	0,(,2)	load contents of Q; G_0 now contains the number 12345
	. . .		
	L	15,ADDB	load address of B
	BALR	14,15	branch to B
	. . .		
	L	2,ADDX	load address of X
	L	0,(,2)	load contents of X
	. . .		
	L	2, = A'Y'	load address of Y
	ST	0,(,2)	store into Y
	. . .		
ADDB	DC	A'B'	define constant = address of B
ADDX	DC	A'X'	define constant = address of X
	. . .		
	END		

Both methods of referencing addresses for the purpose of loading them into a base register are illustrated. They are equivalent and use of one method as opposed to the other is the personal preference of the programmer.

The address that is referenced by the notation A 'expression' in either a literal or a DC pseudo instruction is frequently called an *address constant*, not because its value is constant, but because it is defined using literals and DC pseudo instructions. The value of an address constant in a procedure is fixed during execution of the user's program in any job containing that procedure. However, if the same procedure is used in another job, the values of its address constants may be different, since the loader may have loaded the segments of the program into different memory locations.

The programmer may also need to refer to unnamed locations in another segment. He does this by using a nonzero value in the D field. For example, if Y in Figure 6.4 is an array of full words, that is, Y is defined by:

$$\text{Y} \quad \text{DS} \quad 40\text{F} \quad \text{define Y}_0 \dots \text{Y}_{39}$$

then to load the second word in the array the programmer writes:

$$\text{L} \quad 2, = \text{A}'\text{Y}' \qquad \text{load address of Y}$$
$$\text{L} \quad 0,4(,2) \qquad \text{load contents of Y}_1$$

It may be necessary to have the address of Y_1 for use in indexing or as an argument to another procedure. The programmer can reference the appropriate address constant by:

$$\text{L} \quad 2, = \text{A}'\text{Y} + 4' \qquad \text{load address of Y}_1$$

The address loaded into G_2 by this instruction will be four larger than the address corresponding to the global symbol Y.

6.5 A Relative Loader

As an example of a relative loader, we will discuss a loader that will load object decks generated by the assembler described in Chapter 4. We assume that the assembler has been modified to accept the assembly language extensions described in the preceding section. We also assume that this loader will function as the loader command program in the system discussed in Chapter 2. Loaders very similar to this simple loader are found in almost all systems today.

In designing this loader we wish to satisfy the following objectives.

1. The loader will load any number of data and procedure segments.
2. The address constants corresponding to global symbols may refer to any location in another segment.
3. Address constants corresponding to local symbols may refer to any location within the segment in which they occur.
4. The value of an address constant may be defined by an expression that is the sum and/or difference of the values of any number of local and global symbols and a single nonnegative integer constant.
5. No restriction is placed on the structure of referencing. For example, segment A may refer to segment B, and segment B may refer to segment A.
6. Any global symbol used but not defined in any of the segments supplied by the user is assumed to refer to a library segment. The library will be searched in an attempt to find segments that define all of these undefined global symbols.

In satisfying these objectives we will place more importance on isolation of functions and clean design instead of attempting to achieve optimum efficiency. The purpose of our example is to illustrate the functions of relative loaders. If the loader were to be used in an actual system the objectives would include some constraints on the execution time and memory requirements of the loader.

Some important conclusions can be drawn from the objectives in the preceding paragraph. Objective 5 implies that linking cannot necessarily be completed in one pass over the program. Just as was the case in the assembler, the value of an address constant may be defined in terms of a global symbol that is defined in the object deck for a segment that has not yet been processed by the loader. However, the text in each segment of the program can be loaded in the first pass. Only the processing of address constants requires a second pass.

Since the expression defining the value of an address constant may involve both local and global symbols, computation of address constants will be much simpler if the treatment of local and global symbols is uniform. This is easy to achieve if every local symbol is transformed into a global symbol plus a constant. The global symbol used is the symbol appearing in the location field of the START pseudo instruction. The value of this symbol is the address of the base of the segment. Each local symbol can then be expressed as this global symbol plus the offset from the segment base of the instruction or datum associated with the local symbol. This offset is the *relative address* corresponding to the local symbol and is just the value assigned to the local symbol by the assembler, that is, it is the value of the local symbol found in the symbol table of the assembler.

The loader is designed to operate in the system described in Chapter 2. It expects to find the object decks for the segments of the program in two places: the system input device IN and the system file OBJECT. Object decks that the user included in his job are read from IN. Object decks resulting from procedures that were translated during the user's job are read from OBJECT. The loader will be called by the system with five arguments:

loader (origin, mem_length, start_add, length, status)

The first argument, **origin**, specifies where in primary memory to begin loading, and the second, **mem_length**, specifies the amount of primary memory available for the loaded program. When the loader returns to the system, the value of **status** indicates if loading was successful and, if not, what problem was encountered. Loading may not be successful because the program was larger than the available memory, global symbols not defined in the user supplied object decks were also not defined in the system library, multiple definitions occurred in the user supplied object decks for the same global symbol, or the object decks had illegal or inconsistent structure. If loading was successful, the value of **start_add** is the address at which to begin execution of the loaded program, and the value of **length** is the length of the loaded program.

6.6 Structure of the Object Deck

Before we can complete the design of the loader we must define a specific format for the object decks. An object deck has five components: the segment definition card, the text, the global symbol dictionary, the address constant definitions, and the end card. These components must occur in the order in which they were just enumerated. Only the segment definition card and end card are required, any or all of the remaining components may be omitted. The formats for the object deck cards that make up each of these five components are shown in Figures 6.5 to 6.9. A sixth type of card is shown in Figure 6.10. This follows the last object deck in the program and specifies the

Card columns	Contents
1–3	'SEG'
4–11	global symbol that is the name of the segment
12–14	length of the segment
15–72	blank
73–80	card identification

Figure 6.5 Format of segment definition card.

Card columns	Contents
1–3	'TXT'
4–6	relative address of first byte of text on this card
7–8	byte count
9–72	from 1 to 64 bytes of text
73–80	card identification

Figure 6.6 Format of text cards.

Card columns	Contents
1–3	'GSD'
4–5	index of this symbol in global symbol dictionary
6–13	global symbol
14	type of global symbol = 'T' if outsymbol = 'N' if insymbol
15–17	relative address if insymbol
18–72	blank
73–80	card identification

Figure 6.7 Format of global symbol dictionary cards.

Card columns	Contents
1–3	'ACD'
4–5	index of a global symbol dictionary entry
6	operation = '+' add value of global symbol to address constant = '−' subtract value of global symbol from address constant
7–9	relative address of first byte of address constant
10–72	blank
73–80	card identification

Figure 6.8 Format of address constant definition cards.

Card columns	Contents
1–3	'END'
4–72	blank
73–80	card identification

Figure 6.9 Format of the end card.

Card columns	Contents
1–3	'STR'
4	blank
5–12	global symbol specifying starting location for execution
13–72	blank
73–80	card identification

Figure 6.10 Format of the start card.

Relative address	Object text	Source Statement			
		A	START		define segment A
			OUTSYMB	X,Q,Z,ZN	identify outsymbols
			USING	*,15	G_{15} = base address of A
0	90 ECD00C		STM	14,12,12(13)	save all registers
4	58 2 0 F 030		L	2, = A'X'	get address of X_0
8	58 3 0 F 034		L	3, = A'NR'	get address of NR
C	1 B3 0 2 000		ST	3,(,2)	store in X_0
10	58 3 0 F 038		L	3, = A'Q + 4'	get address of Q_1
14	1 B3 0 2 004		ST	3,4(,2)	store in X_1
18	58 2 0 F 03C		L	2, = A'ZN − Z'	get length of Z
1C	1 B2 0 F 028		ST	2,HERE	store in this segment
20	98 ECD00C		LM	14,12,12(13)	restore all registers
24	07 FE		BR	14	return
26	xx x x		. . . fill		
28	xx x x x xxx	HERE	DS	F	
2C	00 000 018	NR	DC	F'24'	
30	00 000 000		. . . literal = A'X'	⎫	
34	00 00 0 02C		. . . literal = A'NR'	⎪	
38	00 000 004		. . . literal = A'Q + 4'	⎬ address constants	
3C	00 000 000		. . . literal = A'ZN − Z'	⎭	
40			END		

Figure 6.11 Assembly language source statements for procedure segment A.

Relative address	Source statement		
	X START		define segment X
0	DS	6F	reserve array of 6 full words $X_0 \ldots X_5$
18	END		

Figure 6.12 Assembly language source statements for data segment X.

Relative address	Source statement		
	Q START		define segment Q
	INSYMB	Z,ZN	identify insymbols
0	DS	4F	reserve array of 4 full words $Q_0 \ldots Q_3$
10	Z DS	20C	reserve array of 20 bytes $Z_0 \ldots Z_{19}$
24	ZN EQU	*	define ZN to be address of byte following Z_{19}
24	END		

Figure 6.13 Assembly language source statements for data segment Q.

location at which execution of the program is to begin. Figures 6.11 to 6.13 show the assembly language source statements for three segments. The object decks that are obtained by assembling these segments are shown in Figures 6.14 to 6.16. In these figures, addresses and text are written in hexadecimal. However, they will be punched on the card 1 byte per column using an appropriate code. Figure 6.17 shows how these segments are loaded into memory beginning at 10000_{hex}. The appropriate parts of these figures should be referenced during the following discussion.

The segment definition card defines the segment and has the format shown in Figure 6.5. The symbol on the segment definition card is the name of the segment. It is a global symbol (an insymbol), and its value is defined to be the base address of the segment after it is loaded. All of the cards in the object deck have a three-letter code punched in the first three columns of the card that identifies the component to which the card belongs.

The format of cards containing the procedure's text is shown in Figure 6.6. It is almost identical to the card format for the absolute loader. The principal difference is that the address on a text card for the relative loader is not an absolute loading address. It is a relative address, that is, it is the offset from the base of the segment of the first byte of text on the card. One of the jobs of the loader is to compute the absolute loading address for the text on the card. Note that there are two text cards for the procedure segment A in Figure 6.14. This is because the occurrence of a DS pseudo instruction in the source causes a gap in the text generated by the assembler. When such a gap occurs, the assembler terminates the current text card and starts a new one.

Every global symbol, either insymbol or outsymbol, used in the source segment has an entry in the global symbol dictionary. Each such entry is defined by one card in the object deck having the format shown in Figure 6.7, with one exception. Each such card identifies the type of the global symbol and the

Segment definition card

			1	1 1
1–3	4	–	1	2–4
				4
SEG	A			0

Text cards

1–3	4–6	78	9 –	4 6
		2	9ED0 52F353 F3 1 32053F3 1 32052F3 1 2 F29ED0 0F	
TXT	0	6	0C0 C800 080 04 B000800 8 B004800 CB0 088C0 C7E	
	2	1	00 0 1 000 00002 0 0000000	
TXT	C	4	00 0 8 000 0000 C0 0040000	

Global symbol dictionary cards

				1	1	1 1
1–3	45	6	–	3	4	5–7
GSD	1	X			T	0
GSD	2	Q			T	0
GSD	3	Z			T	0
GSD	4	ZN			T	0

Address constant definition cards

1–3	45	6	7–9
			3
ACD	1	+	0
			3
ACD	0	+	4
			3
ACD	2	+	8
			3
ACD	4	+	C
			3
ACD	3	–	C

End card

1–3	
END	

Figure 6.14 Object deck for Segment A in Figure 6.11.

index of its entry in the global symbol dictionary. This index will be used in the address constant definitions. The name of a segment is an insymbol for that segment, however, we do not need a global symbol dictionary entry for it, since its relative address is always zero and the symbol itself appears on the segment definition card. Therefore, we will adopt the convention that a global

Segment definition card

1-3	4	–	1	2-4
SEG	X			18

End card

1-3
END

Figure 6.15 Object deck for segment X in Figure 6.12.

Segment definition card

1-3	4	–	1	2-4
SEG	Q			24

Global symbol dictionary

1-3	45	6	–	3	4	5-7
GSD	1	Z			N	10
GSD	2	ZN			N	24

End card

1-3
END

Figure 6.16 Object deck for Segment Q in Figure 6.13.

symbol dictionary index of zero refers to the insymbol that is the name of the segment.

If the global symbol is an outsymbol its value will be defined in some other object deck. However, if the global symbol is an insymbol, its value is the absolute address of some location within the segment corresponding to this object deck. Therefore, its global symbol dictionary card contains the relative address assigned to the insymbol by the assembler (see Figure 6.16). The absolute address that is the value of the global symbol is equal to the sum of the absolute address of the base of the segment and the relative address of the global symbol. This is illustrated in Figure 6.16, where the global symbol dictionary entry for Z indicates a relative address of 10_{hex}, and Figure 16.17 where we see that the ultimate value of Z is 10068_{hex}, the base address of segment Q being 10058_{hex}.

An address constant is specified in assembly language by an expression that is composed of a sequence of symbols, local or global, and integer constants

Segment	Local symbol	Global symbol	Memory address	Contents (blanks indicate contents of that byte not loaded, only space reserved)	
↑		A	10000	90 ECD00C	
⏐			10004	58 2 0 F 030	
⏐			10008	58 3 0 F 034	
⏐			1000C	1B3 0 2 000	
⏐			10010	58 3 0 F 038	
⏐			10014	1B3 0 2 004	
⏐			10018	58 2 0 F 03C	
A			1001C	1B2 0 F 028	
⏐			10020	98 ECD00C	
⏐			10024	07 FE	
⏐	HERE		10028		
⏐	NR		1002C	00 0 0 0 018	
⏐			10030	00 0 1 0 040	= A′X′
⏐			10034	00 0 1 0 02C	= A′NR′
⏐			10038	00 0 1 0 05C	= A′Q + 4′
↓			1003C	00 0 0 0 014	= A′ZN − Z′
↑		X	10040		
⏐			10044		
X			10048		
⏐			1004C		
⏐			10050		
↓			10054		
↑		Q	10058		
⏐			1005C		
⏐			10060		
⏐			10064		
Q		Z	10068		
⏐			1006C		
⏐			10070		
⏐			10074		
↓			10078		
		ZN	1007C		

Figure 6.17 Segments A, X, and Q from Figures 6.14–6.16 after being loaded into memory beginning at 10000_{hex}.

combined with plus and minus signs. The assembler replaces each local symbol with the name of the segment plus the relative address of the local symbol. The resulting expression is then simplified as much as possible. For example, all constant terms are combined by adding or subtracting as appropriate. The result will be an expression that can be rearranged to have the form:

constant ± global symbol ± ... ± global symbol

An expression of this form cannot be reduced further by the assembler, since it does not know the values of any of the global symbols. For example, in segment A of Figure 6.11 the address constant A′NR′ is transformed into the form

$$2C_{hex} + A$$

where $2C_{hex}$ is the relative address of NR in segment A and A is the global symbol that is the name of the segment and whose value will be the base address of the segment. The address constant $A'ZN-Z'$ is transformed into the form

$$0 + ZN - Z$$

Since both ZN and Z are global symbols, the relative address part is zero.

The information in the transformed address constant definition is included in the object deck on *address constant definition cards* having the format shown in Figure 6.8. The object deck contains one address constant definition card for each global symbol in the transformed expression, and the constant is included in the text at the location assigned to the address constant. The loader computes the final value of the address constant using the location assigned to the address constant as an accumulator. Each address constant definition card corresponds to one term in the transformed expression. The sign of the term is in column 6 and the index, in the global symbol dictionary, of the global symbol is in columns 4 and 5. The loader adds (or subtracts) the value of the global symbol identified by the index to (or from) the full word value in the 4 bytes of the text beginning at the relative address specified in columns 7 to 9.

After all of the address definition cards have been processed, the final address constant will be in its proper location. For example, in Figures 6.11, 6.14, and 6.17 there is one address constant definition card for the address constant $A'NR'$, which was transformed into $2C_{hex}+A$. The constant part $2C_{hex}$ is included in the full word of text at relative location 34_{hex}. The second address constant definition card in the object deck specifies that the value of A (symbol dictionary index 0) be added to the constant at relative address 34_{hex}. The constant part of $A'ZN-Z'$, which was transformed into $0+ZN-Z$, is 0 so the contents of relative address $3C_{hex}$ is 0. Two address constant definition cards are needed to define this address constant. The fourth address constant definition card in the object deck adds the value of ZN (symbol dictionary index 4) to the contents of relative address $3C_{hex}$, and the fifth address constant definition card subtracts the value of Z.

The end card terminates an object deck for a segment. It has the format shown in Figure 6.9. The end of the object decks supplied by the user is signaled by a start card having the format shown in Figure 6.10. The loader first reads all of the object decks from OBJECT then all of the decks from IN. Thus, the start card is the last data card in the job step that called for the loader. The global symbol on the start card specifies the point where execution is to begin. The value of this symbol must be the address of the entry point of the first procedure in the program to be executed. This value is returned to the system by the loader as the address at which to start execution.

6.7 Structure of the Loader

There is a strong similarity between the loader and the assembler with respect to the definition of symbols. Both must process the entire input, collecting symbol definitions before the expressions using the symbols can be evaluated. In the loader not only the object decks supplied by the user, but also the object decks for any library segments that are needed must be processed in order to define all of the global symbols. However, the loader does not need to make a second pass over the entire program, since the only expressions that need evaluation are the address constant definitions. Thus, while the structure of the loader resembles the two passes of the assembler, it is perhaps better to call the two major modules of the loader *phases* instead of passes.

In order to define the global symbols, *phase one* of the loader must assign storage to each segment supplied by the user and each library segment that his program needs. A symbol table must be built in phase one for use in phase two. An entry in the symbol table consists of a global symbol and the absolute memory address that is assigned to it. The text of each procedure can be copied into its proper memory locations during phase one, since its memory location does not depend on the value of any symbol. Since the address constants cannot be evaluated until phase two, all information needed for their evaluation, other than the contents of the symbol table, will be saved in the temporary file SAVE for use in phase two. This information consists of all the cards in each object deck, including library object decks, except for the text cards, which are not needed after the text has been loaded.

Phase two evaluates the address constants. This is done one segment at a time. The segment definition card and the global symbol dictionary cards identify the global symbols of the segment and are used to build a global symbol dictionary for the segment. This is needed, since the address constant definition cards reference global symbols by their index in this symbol dictionary. The address constant definition cards are then processed to evaluate the address constants. An end card signals the end of the address constant definition cards for each procedure.

Figure 6.18 defines the loader ... d Figure 6.19 shows the data flow. Phase one reads all of the object decks in the program, assigning memory to each segment. It loads the text from each segment into primary memory, beginning at the address supplied by the system as the value of **origin**. The argument **mem_length** specifies the maximum amount of memory space the system will allow for the loaded program. The segment definition, global symbol dictionary, address constant definition, and end cards from each object deck are saved in SAVE. The contents of the segment definition and global symbol dictionary cards are used to build a symbol table that contains the definitions of all global symbols in the program. When phase one has finished processing all the object decks in the program, it returns with the value of **length** set equal to the length of the loaded program. If some problem was encountered, status will have a nonzero value that indicates what the problem was. In this case loading is terminated.

Phase two reads the segment definition, global symbol dictionary, and address constant definition cards from SAVE. The segment definition and global symbol

```
procedure loader (origin, mem_length, start_add, length, status);
    phase_one (origin, mem_length, length, status);
    if status = 0 then phase_two (start_add, status);
    Delete files OBJECT and SAVE;
    Close library file;
    return;
end.
```

Figure 6.18 Definition of the loader.

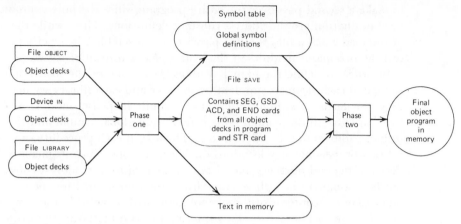

Figure 6.19 Data flow for phase one and phase two of the loader.

dictionary cards are used, along with the contents of the symbol table, to build a global symbol dictionary for each segment. The address constant definition cards are used, aided by the segment's global symbol dictionary, to derive a value for each address constant. These values are stored into the appropriate places in the segment text, which has already been loaded into the primary memory. Before phase two returns, it will set the value of **start_add** equal to the value of the global symbol on the start card. Before returning to the system, the loader deletes the temporary files OBJECT and SAVE and closes the library file.

6.8 Phase One

The major tasks of phase one are memory allocation, symbol definition, text loading, and acquisition of library procedures. The object decks for the segments in the program are processed one at a time. Any object decks in OBJECT are processed first, followed by any object decks that the user included in his job. These latter object decks will be data in the job step following the command that requested the loader. The loader will read these decks from IN.

Finally, the object decks for the library procedures needed by the program will be processed. These object decks will be read from the system's library file. The value of **read_switch** indicates from where the loader is to read the next card of the program. Its value will be 'object', 'in', or 'library'. The loader begins reading from OBJECT by setting **read_switch** equal to 'object', as shown in Figure 6.20. The procedure **read_card** is used to read the next card of the program into the array of bytes **S**. This procedure is responsible for switching the reading from OBJECT to IN and finally to the library file. It is defined in Figure 6.25 and discussed in the next section.

The segment definition card, external symbol dictionary cards, address constant definition cards, and end card for each segment are copied into SAVE for use in phase two. This file must be created and opened for writing before the loader can begin processing any object decks. Memory allocation and symbol definition are done almost exactly as in the assembler. A location counter is kept whose value is the address of the next unassigned memory location. The

```
procedure phase_one (origin, mem_length, length, status);
    Open OBJECT for reading;
    read_switch : = 'object';
    read_card;
    Create and open SAVE;
    loc_ctr : = origin;
    repeat
        Process segment definition card;
        Process text cards;
        Process global symbol dictionary cards;
        Copy address constant definition cards;
        if S ≠ END card then Error return;
        read_card;
    until S[1] ... S[3] = 'STR';
    Write S into SAVE;
    Close SAVE;
    length : = loc_ctr − origin;
    return;
end.
```

Figure 6.20 Definition of phase one of the loader.

value of the location counter is initialized to the address that is the value of the argument origin when the loader is called by the system.

The **repeat** loop processes an object deck. Each component of the object deck is processed in the order in which it appears in the deck. The first card of an object deck must be a segment definition card. The next three components, the text, the global symbol dictionary, and the address constant definitions, may each consist of an arbitrary number of cards. Processing of a component terminates as soon as a card not belonging to that component is read. If this card is not a valid card for the next component it is assumed that the next component has been omitted. After the first four components are processed, the final card of the object deck is tested to see if it is an end card. If not, the object deck is incorrectly formed, and phase one returns, indicating an error. In this case an error comment would be printed in the output for the user, and the loader would return to the system with the value of status set to indicate what problem has been encountered.

The loop to process an object deck is repeated until a start card is read. This signals the end of the object decks in the program. The start card is included in the user's job as the last data card in the loader job step. As we will see in the next section, the read_card procedure retains this card until all of the object decks required from the library have been read. Therefore, this is truly the last card in the program, and phase one is essentially finished when the start card is encountered. The start card is written into SAVE for use in phase two.

Object deck processing begins with the segment definition card, as shown in Figure 6.21. The value of loc_ctr is the address of the first unallocated byte of memory. However, this may not be a valid address for the beginning of a segment. In determining the fill needed for alignment in DS and DC pseudo

if S[1] ... S[3] $\neq$ 'SEG' **then** Error return;
Write S into SAVE;
Increment loc_ctr so that its value is evenly divisable by 4;
seg_base := loc_ctr;
Enter symbol from S[4] ... S[11] and value of loc_ctr into symbol table;
loc_ctr := loc_ctr + segment length from S[12] ... S[14];
if loc_ctr $\geq$ origin + mem_length **then** Error return;

Figure 6.21 Expansion of "Process segment definition card;" from Figure 6.20.

instructions and in literals, the assembler assumed that the base address of the segment was zero, which is evenly divisible by four. Therefore, before allocating memory to a segment, the value of loc_ctr must be adjusted so that it is evenly divisible by four.

After that, memory space is allocated to the segment beginning with the address that is the current value of the location counter. This address is the base address of the segment. It is needed later to compute the value of the segment's insymbols. Hence, it is saved in **seg_base**. The value of the symbol on the segment definition card is defined as the base address of the segment by making an entry in the symbol table. Incrementing the location counter by the length of the segment reflects the fact that that amount of memory has been allocated to the segment. If the new allocation of memory exceeds the limit specified by the system, loading is terminated.

After processing the segment definition card, the text for the segment is loaded into memory, as shown in Figure 6.22. The text cards are processed in sequence. Each card contains one or more bytes of text. The bytes are loaded into consecutive memory locations, beginning at the loading address. This address is the sum of the segment's base address and the relative loading address on the text card. The byte count indicates how many bytes of text the card contains. The loading of text continues until a card is read that does not contain "TXT" in its first three columns. Text cards are not copied into SAVE.

The next component of the object deck to be processed is the global symbol dictionary. The major function here is making entries in the symbol table, as shown in Figure 6.23. There are two types of global symbol dictionary cards, outsymbol and insymbol, distinguished by a "T" or "N", respectively, in column 14 of the card. If the card contains an insymbol, the value of that symbol is defined to be equal to the sum of the segment's base address and the symbol's relative address on the card. This definition is recorded in the symbol table.

If the card contains an outsymbol, only a reference to the symbol is indi-

read_card;
while S[1] ... S[3] = 'TXT' **do**
  n := byte count from S[7] ... S[8];
  a := seg_base + relative loading address from S[4] ... S[6];
  Copy S[9] ... S[9 + n − 1] into memory locations a ... a + n − 1;
  read_card;
end;

Figure 6.22 Expansion of "Process text cards;" from Figure 6.20.

```
while S[1] ... S[3] = 'GSD' do
    Write S into SAVE;
    if S[14] = 'T' then value := 0;
        else value := seg_base + relative address from S[15] ... S[17];
    Enter symbol from S[6] ... S[13] and value into symbol table;
    read_card;
end;
```

Figure 6.23 Expansion of "Process global symbol dictionary cards;" from Figure 6.20.

cated; the symbol is not to be defined at this time. An outsymbol may have previously been defined, will be defined in a user's object deck following the current one, or will be defined by a library segment. In order to determine what segments are needed from the library, each outsymbol must be added to the symbol table. Its value will be recorded as zero to indicate that it is undefined. If it is still undefined when all the user's object decks have been processed, the loader assumes that the symbol references a library segment.

When adding a new entry for a symbol, the loader must check to see if the symbol table already contains an entry for that symbol. If it does not, the new entry is added. However, if the symbol table already contains an entry no new entry is made. If the symbol from the global symbol dictionary card is an outsymbol, no further action is necessary. If the symbol is an insymbol, an error has occurred unless the entry already in the symbol table has a value of zero associated with the symbol indicating that the symbol has not yet been defined. If this is true, the symbol is now defined by replacing its zero value in the old entry by the contents of value. The address constant definition cards are not processed in phase one, they are simply copied into SAVE for use in phase two, as shown in Figure 6.24.

After processing all of the object decks, the symbol table contains definitions for all of the global symbols, including those that were defined by library segments. SAVE contains the information needed to compute the value of each address constant in the program, including the address constants in the library segments. This information is grouped by segment. The information for a segment begins with its segment definition card and ends with its end card. The last card in SAVE is the start card for the program.

6.9 Reading of Object Decks and Library Searching

The procedure read_card defined in Figure 6.25 reads the next card of the program. In doing so it determines when to switch reading from OBJECT to IN and finally to the library file. It also determines what library segments are

```
while S[1] ... S[3] = 'ACD' do
    Write S into SAVE;
    read_card;
end;
Write S into SAVE;
```

Figure 6.24 Expansion of "Copy address constant definition cards;" from Figure 6.20.

```
    procedure read_card;
      if read_switch = 'object' then
        begin Read next record from OBJECT into S[1] ... S[80];
          if end of file then read_switch := 'in';
            else return;
        end;
      if read_switch = 'in' then
        begin Read next record from IN into S;
          if S[1] ... S[3] ≠ 'STR' then return;
          Search symbol table and put all undefined symbols on
            undefined_list;
          if undefined_list is empty then return;
          SS := S;
          Open library file for reading;
          Read library directory and use undefined_list to determine
            identification of library segments needed by the program,
            putting these identifications on library_list;
          Sort library_list into ascending order of identifications;
          read_switch := 'library';
          S[1] ... S[3] := 'END';
          i := 0;
        end;
      if S[1] ... S[3] = 'END' then
        begin i := i + 1;
          if i ⩾ length of library_list then
            begin S := SS;
              return;
            end;
          Skip over records in library file until ready to read first record
            of library segment whose identification is in library_list [i];
        end;
      Read next record from library file into S;
      return;
    end.
```

Figure 6.25 Definition of procedure to read object decks and search the library file.

needed by the program and manages the reading of the library file so that only these segments are read. The value of **read_switch** specifies where to read the next card. Initially, cards will be read from OBJECT. Cards will be read from this file until the end of the file is reached. I/O control is used to read from files. If the **read** entry is called and all of the records have been read, it will not read a record, but will return with **status** set to indicate that the end of the file has been reached. When this occurs, **read_card** will start reading records from IN. This will read the object decks that the user supplied as data in the loader job step. Reading from IN continues until all of these decks have been read. This will be true when the start card is read, since this card must be the last data card in the loader job step.

When this occurs, all of the object decks corresponding to segments written

by the user will have been processed. The final set of object decks to be processed are for the library segments needed by the program. In order to determine which, if any, library segments are needed, the symbol table must be
searched for any global symbols that are still undefined. All of these undefined
symbols are assumed to refer to library segments. If there are no undefined
symbols, then no library segments are required. All undefined symbols are put
onto a list of undefined symbols that will be used to determine the library segments that are needed. The start card that indicated the end of user's object
decks is saved until all of the object decks for library segments have been
processed.

The library segments that are needed must now be determined. The list of
undefined symbols is not necessarily the list of library segments needed, since
a single library segment may define more than one symbol. Also, library procedure segments may reference other library segments. Thus, a library procedure segment called by a user-written procedure may reference a second library
segment that is not referenced by any user-written procedure. The name of this
second library segment will not appear in the list of undefined symbols.

The straightforward solution to the problem of finding the library segments
that are needed is very inefficient. Each undefined symbol is considered in
turn. The library file is searched from the beginning until an object deck is
found whose global symbol dictionary contains a definition for the symbol.
This object deck is then processed by phase one of the loader. Then the next
undefined symbol is considered. The library file is again searched from the
beginning. This continues until all of the undefined symbols have been considered. In the course of this action new undefined symbols will have been
introduced into the symbol table if any of the needed library segments refer
to any other library segment that none of the user's procedures references.
These new undefined symbols must also be used to search the library file. This
is continued until all symbols have been defined or no definition can be found
in the library for a symbol. In the latter case, loading is terminated with appropriate error comments.

The inefficiency of the method just described lies in the repeated searching
of the library file. We elect to use a much more efficient method at the cost of
some additional information in the library file. This improvement can be
effected if the object decks in the library file are preceded by a directory. This
directory will contain one entry for each segment in the library. The entry for
a segment contains a list of all of the symbols that are defined in the segment,
a list of all other library segments that are needed by the segment, and the
location in the library file of the object deck for the segment. The list of segments needed is complete, no matter how deep procedure calls may go. For
example, if the user calls library procedure A, and A in turn calls B and C and
so on as shown in Figure 6.26, A's list of needed segments is B, C, D, E, F, G,
and H. This list need not be the names of these segments, but merely the location of their object decks in the library file. The location of an object deck is
the sequential number of the record, in the library file, that contains the segment definition card of the object deck.

Existence of such a library directory allows very efficient acquisition of the
needed library segments. The library directory is read into memory. The entries
in the directory are then considered one at a time, in sequence. If any of the

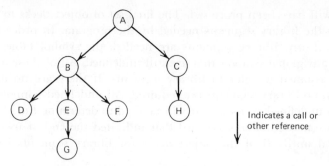

Figure 6.26 Example of library procedure segment that calls and references other library segments.

symbols that are defined in the segment corresponding to the directory entry currently being considered appear in the undefined symbol list, the location of the object deck for the segment and the locations of the object decks for all other library segments that it needs are added to library_list, which is a list of object decks to be loaded from the library file.

After the entire library directory has been scanned or all of the symbols on the undefined symbol list have been defined, whichever occurs first, library_list will contain the location of every object deck needed from the library file. This list is then sorted into ascending order by location. Object decks can then be read in the order in which they occur in the library file by scanning this list in sequence. Thus, the loader will never have to back up or start from the beginning again when reading from the library file, since the next object deck on the list is always past the one currently being read.

Since the needed object decks are normally not consecutively located in the library file, the loader will have to skip any intervening object decks. In order to do this, read_card watches for an end card. When read_card is called, and the last card read was an end card, the end of an object deck has been reached. The read routine checks to see if any object decks remain to be read. If there are, it sets up to read the next object deck by skipping all of the records in the library file between the last record read and the first record of the next object deck. The identification appearing on library_list is the record number of the first record in the next object deck. After all of the needed library segments have been read, the start card is copied back into S to signal the end of reading cards in the program.

6.10 Phase Two

The major function of phase two is to compute the final value of each address constant. The instructions for this computation are contained on the address constant definition cards, one instruction per card. The address constant definition cards are in SAVE in groups, one group for each segment in the program. Each group also contains the global symbol dictionary cards for the segment. The value of an address constant is expressed in terms of the values of entries in this global symbol dictionary. This is a local dictionary, there being a different global symbol dictionary for each procedure. The global symbol dictionary cards contain the global symbols whose values are needed in the definition of the address constants. Since a reference to a global symbol in the definition of an address constant is an index in the global symbol dictionary, there is no need to keep the actual symbol, only its value is needed, and the global symbol dictionary built for each procedure contains only symbol values.

The **while** loop in the definition of phase two in Figure 6.27 computes the address constants for one segment. The global symbol dictionary gsd is built by processing the segment definition and global symbol dictionary cards. The address constants are computed by processing the address constant definition cards. When the start card for the program is encountered, loading is complete. Before returning, phase two looks up the value of the global symbol that is specified on the start card as the starting point for execution. This value is returned to the system.

The segment definition card contains the global symbol that is the name of the segment and whose value is the base address of the segment. This symbol is always the zeroth in the global symbol dictionary. As shown in Figure 6.28, its value is put into gsd_0. The remaining entries in the global symbol dictionary are defined by the global symbol dictionary cards. Each card contains a symbol and its index in the global symbol dictionary. As shown in Figure 6.29,

```
procedure phase_two (start_add, status);
    Open SAVE for reading;
    Read next record from SAVE into S[1] ... S[80];
    while S[1] ... S[3] ≠ 'STR' do
        Process segment definition card;
        Process global symbol dictionary cards;
        Process address constant definition cards;
    end;
    start_add := value in symbol table corresponding to symbol
        in S[4] ... S[11];
    return;
end.
```

Figure 6.27 Definition of phase two of the loader.

```
gsd[0] := value in symbol table corresponding to symbol in S[4] ... S[11];
```

Figure 6.28 Expansion of "Process segment definition card;" from Figure 6.27.

```
Read next record from SAVE into S[1] ... S[80];
while S[1] ... S[3] = 'GSD' do
    i := value in S[4] ... S[5];
    gsd[i] := value in symbol table corresponding to symbol in
        S[6] ... S[13];
    Read next record from SAVE into S[1] ... S[80];
end;
```

Figure 6.29 Expansion of "Process global symbol dictionary cards;" from Figure 6.27.

the corresponding entry in **gsd** is set equal to the value of the symbol that is found in the symbol table.

The address constant definition cards contain the instructions for computing the address constants. Each card specifies the addition or subtraction of the value of a global symbol to or from a full word in the segment that contains a partially computed address constant. As shown in Figure 6.30, the symbol whose value is to be used is identified by its index in the global symbol dictionary. The address constant definition card contains this index and the relative address within the segment of the partially computed address constant that is to be modified. This relative address is transformed into an absolute address by adding it to the base address of the segment found in gsd_0.

6.11 The Evolution of Loaders

The evolution of loaders is interesting because it is an example of a trend common to many areas of both software and hardware, the trend to delay binding as long as possible. The first loaders were absolute loaders, similar to the one described in the beginning of this chapter. A program loaded by an absolute loader must be completely bound. That is, every address in the program is an absolute address. An absolute loader makes no changes to the program that it reads, it simply loads the program into memory where it is told to by the loading addresses on the cards containing the program. An absolute loader is unaware of any division into smaller units such as procedure and data segments.

The next major step in the evolution of loaders was that of relative loaders. When using a relative loader, binding of addresses is delayed until the program is loaded in preparation for execution. The concept of segment begins to play an important role with this type of loader. There are, and have been, many different relative loaders. Mostly they differ only in the way the final value of an address may be specified and what types of symbols and expressions may be used to define this value. Relative loaders also differ in the way they treat segments, some treating data segments differently than procedure segments.

A typical early relative loader permits an arbitrary number of procedure segments but only one data segment. Let us call it a *single data segment* (SDS) loader. The procedure segments are referenced symbolically. However, the data segment, called *common*, is referenced by negative relative addresses. The pro-

```
while S[1] ... S[3] = 'ACD' do
    i := value in S[4] ... S[5];
    if S[6] = '+' then delta := gsd[i];
        else delta := -gsd[i];
    a := gsd[0]+relative address from S[7] ... S[9];
    Add delta to full word located at memory address a;
    Read next record from SAVE into S[1] ... S[80];
end;
Read next record from SAVE into S[1] ... S[80];
```

Figure 6.30 Expansion of "Process address constant definition cards;" from Figure 6.27.

cedure segments are loaded from low to high memory address, beginning with the lowest address in the block of memory available for the program. The common segment is assigned in the opposite direction, from high to low memory address, beginning with the highest address in the block of memory available for the program.

Since only procedure entry points can be referenced symbolically, the SDS loader puts the values of all the global symbols that appeared in the procedure into a block at the beginning of the procedure segment. This block is called a *transfer vector*. Most of the computers using this type of loader did not have base registers. The absolute address of a memory location being referenced had to appear in the address field of the instruction, or the reference had to be indirect, that is, the address in the instruction is the absolute address of a memory location whose contents is the absolute address of the memory location being referenced. To call another procedure segment, the calling procedure transfers indirectly through one of the memory locations in the transfer vector.

All addresses in the instructions of a procedure were one of two types, reference to the procedure segment itself (including the transfer vector) or references to the common segment. These two types of address were usually distinguished by their size. Small addresses referred to the procedure segment, and large addresses referred to the common segment. The SDS loader relocated small addresses upward, adding the base address of the segment being loaded, and relocated large addresses downward, subtracting the base address of the common segment.

The SDS loader has no real advantages over the type of relative loader that we have studied in this chapter, and it does have many disadvantages. Each reference to another segment must be indirect (which is more costly in execution time than direct references), or the procedure itself must compute and store the absolute address in the instruction, using the values given in the transfer vector. The loader that we studied can be modified easily to work for computers without base registers by letting the address constant definition cards refer to the address fields of instructions in addition to full word address constants.

The limitation to only one data segment in the SDS loader is also a disadvantage. We evolved to relative loaders in the first place to avoid having to bind the location of procedures and data to specific memory addresses. Half of this advantage is lost if the locations of all data must be bound to relative addresses within a single segment.

At first glance there seem to be two advantages to the SDS loader. There is no computation of address constants. The values of global symbols are stored in the transfer vector in consecutive locations. Thus, only one pass over the program is required. The second potential advantage is that the length of the common segment need not be specified in any particular place. The length of the common segment can be deduced by the loader if each procedure declares in its object deck the smallest relative address in the common segment that it references. This feature extended to any number of data segments is easily added to the loader that we studied.

The first advantage of the SDS loader is not really true, unless the order of the object decks is restricted, which is not true for almost all existing loaders. If the order of object decks is not restricted, then the transfer vectors cannot be filled in until all of the object decks, including library procedures, have been processed. The loader then fills in the transfer vectors, so a second pass is made over the program, but in a limited way, just as in the relative loader which we studied. Computation of address constants requires more information and thus results in a slight, but not significant, increase in loading time, all other things being equal.

Loaders similar to the one that we studied are used today in most of the larger systems. All segments are treated uniformly. Expressions are used to define the values of addresses or address constants that are allowed to appear anywhere in any segment. Most loaders of this type have additional features.[1]

The latest step in the evolution of loaders is to delay the binding of some or all of the addresses until during execution of the program. This is called *dynamic loading* or *dynamic linking*. Generally, a procedure or data segment is not linked to any other segments in the program until it is actually referenced by some procedure. Usually special hardware features are available to assist this type of loader. They are not strictly necessary, but their availability results in at least an order of magnitude improvement in the efficiency of execution compared with the same computer without the special hardware features. This type of loader will be explored in more detail in Chapter 15.

EXERCISES

6.1 Modify the assembler in Chapter 4 so that it will process the INSYMB and OUTSYMB pseudo instructions described in Section 6.4 and produce an object deck, as defined in Section 6.6, that can be loaded by the relative loader.

6.2 Write IL procedures that define a *binder* (also called a *linkage editor*) that takes the object decks (as defined in Section 6.6) for two segments and binds them together into the object deck for a single, new segment. The new object deck should have the same format and be acceptable by the relative loader. The name of the new segment should be input to the binder. Link all of the global references, which become local references as a result of combining the two segments, and eliminate their GSD cards and any ACD cards that reference them.

---

[1] L. Presser and J. R. White, "Linkers and Loaders," *Computing Surveys, 4* (3) (September 1972), pp. 149–167.

7 | Compiler Languages

This chapter begins a short study of compilers. A compiler, like an assembler, is a translator from some programming language to machine language. However, a compiler differs from an assembler in respect to the character of the language that it translates. While assembly language closely resembles machine language, compiler language usually has little resemblance to machine language. In fact, most compiler languages are basically machine independent. The language that a compiler translates is often called a *compiler language*. Another term frequently used for such a language is *high-level language*, while machine language is considered a *low-level language*. The class of high-level languages is very broad and includes languages not usually considered to be compiler languages.

Actually, it is most appropriate to think of a spectrum of languages ordered according to the amount of detail with which the user must be concerned and the complexity of the basic operations in the language. Machine language of the type studied in Chapter 2 (Our 360) is at the low end of this spectrum. Compiler languages such as FORTRAN and PL/1 would be somewhere in the middle. The reader probably has not encountered any languages at the high end of the spectrum, since they tend to be quite specialized. The GPSS (General Purpose Simulation System)[1] language is one of the most common languages that is higher level than PL/1.

Machine languages properly appear at several places in the spectrum, depending on the computer. Some experimental machines have a machine language that fits near the level of PL/1 or FORTRAN. Some of the Burroughs computers have a machine language whose level is higher than Our 360, but considerably lower than FORTRAN. Some recent computers have what is called *micro programming*. Machine language for these computers is even lower in level than Our 360.

In this chapter we discuss some important concepts relating to programming languages, such as the definition and translation of programming languages. In addition, we define a simple compiler language, which we call *Instran*.

---

[1] J. E. Sammet, *Programming Languages: History and Fundamentals*, Prentice-Hall, Englewood Cliffs, N.J., 1969, contains descriptions of all languages mentioned in this chapter.

A compiler for this language will be used as a case study in the following chapters, which explore the problems of implementing a compiler.

7.1 Programming Languages

The generally accepted dictionary definition of language is that a language is, "a notational system for communication." A *programming language* is a notational system for communication of programs. This communication is most frequently from a human to a computer, however, the communication is often also between two humans. A program is basically equivalent to an algorithm. The concept of *algorithm* has been defined in many ways: a recipe specifying some computation; a set of commands that will accomplish some task when performed; or a function defined over some set of objects. A precise definition of algorithm is not required here, however, the concept of algorithm and its definition is of great interest to those who are concerned with the foundations of computer science. An algorithm is an abstract entity, and a *program*, written in some programming language, is a concrete representation of an algorithm.

When we receive a communication expressed in some language, if we are to understand the communication, we must know both the *syntax* and *semantics* of the language. The *syntax* of a programming language pertains to the form or structure of programs written in the language. In particular, it defines the set of legal programs that can be written using the language. It also defines the constituent parts of a program, for example, variables, statements, and constants. The *semantics* of a programming language pertains to the meaning of programs written in the language, that is, what algorithm a program represents. This meaning is usually defined in terms of the constituent parts, which are identified by the syntax of the language.

There are two fundamental transformations that can be applied to a program, *translation* and *interpretation*. *Translation* is the transformation of a program in one language into an equivalent program in another language. Since a program is a representation of an algorithm, translation transforms one representation of an algorithm into another representation of the same algorithm. Translation does not change meaning. *Interpretation* of a program is the performance of the algorithm that the program represents, that is, the evaluation of the function that the program represents or the performance of the commands that make up the algorithm. Interpretation of a program yields answers, while translation of a program yields another program that still needs to be interpreted in order to obtain any answers.

In order to either translate or interpret a program the translator or interpreter must know the meaning of the program. This requires a definition of the syntax and semantics of the language. A *formal definition* of a language is the specification of the syntax and semantics of the language using some formal specification language or notation. A formal definition is complete and concise. There are several important needs for a formal definition of a programming language, depending on the principal interest of the person using the definition. A systems programmer needs a complete definition of the language in order to write a compiler for it. For the computer scientist interested in the theory of programming languages, a formal definition is required if he wishes to prove any theorems about the language or even to understand programs written in that language. The existence of formal definitions for programming languages allows them to be classified according to the properties they have, such as the class of algorithms that they can represent or the complexity of their structure.

Programming linguistics is a special area of computer science that is concerned with all of the aspects of programming languages that have been mentioned so far in this chapter. This is a very broad area and covers both theoretical and practical questions. On the theoretical side programming linguistics includes the study of different methods for defining programming languages, equivalence of programs in the same or different languages, and the classification of programming languages. The practical side includes the design and implementation of translators and interpreters, particularly the application of the results of theoretical studies to the design and implementation of compilers and interpreters.

7.2 Programming Language Processors

Our principal interest is in programming language processors, that is, *translators* and *interpreters* that are the major programming language processors. Either of these kinds of processors may be a program or a hardware device. Machine language is a programming language, and the control processor of the computer is an interpreter for machine language. In this text we are concerned mostly with software processors such as the assembler, which we studied in Chapter 4, and the compiler, which we will study in the next few chapters. Both of these programs are translators. In Chapter 2 we studied an interpreter, the job control part of the operating system. The job control language is a very simple programming language. Each command specifies the execution of some command program. Thus, a sequence of commands is the representation of an algorithm. Job control in the system is an interpreter for the job control language, that is, it executes the command programs.

It should be emphasized that conceptually there is no significant difference between a software programming language processor and a hardware processor for the same language. Hardware processors have traditionally been interpreters for machine language, while most software processors have been translators. However, there are exceptions. Quite a variety of software interpreters have been written, even programs that interpret machine language. Software machine language interpreters are sometimes called *computer simulators,* and their most common use is to debug machine language programs for a new computer that is not yet completely built.

The major differences between a software interpreter and a hardware interpreter are the cost of building the interpreter and the speed at which it interprets. In the past a hardware interpreter was more expensive to build, but its interpretation speed was faster. Some of the newer computers have been designed so that software interpreters may be much faster than formerly, thus the disadvantages of software interpreters may not now be significant in many cases. Inversely, some experimental, special purpose hardware processors have been built that translate a high-level language. Such a processor would be part of a complex hardware system, since the output of the processor would still have to be interpreted by some other processor.

The most common translators are assemblers or compilers. An *assembler* usually translates a symbolic machinelike language into true machine language. A *compiler* usually translates a high-level language (often called an *algebraic* or *procedure-oriented* language) into either assembly or true machine language. The most well known compilers are translators for FORTRAN, COBOL, ALGOL, and PL/I.

Another class of translators are called *conversion programs*. These translate from one machine language to another machine language or from machine language to a higher-level language such as COBOL or PL/I. The objective of this translation is to produce a version of the program that can be executed by a different computer. It is usually not possible to make a complete translation of an arbitrary machine language program into some essentially different language. The principle reason for this is that most machine languages permit a program to modify instructions using arithmetic operations. In attempting to translate this, the conversion program may not be able to distinguish that an instruction is being modified. Direct translation of the arithmetic operations will not work, since instructions in the object language will have different formats and numeric operation codes.

Conversion programs have also been written to translate between high-level languages. MADTRAN is such a translator, translating MAD programs into FORTRAN programs. Another translator of this type is DYANA, which translates the description of a physical dynamic system into a FORTRAN program that will solve the differential equations that model the physical system.

Programs that are purely interpreters are less common than translators. A machine language simulator is usually purely an interpreter. However, interpreters more often occur as part of some larger program. This was the case with the job control language interpreter in the operating system and the macro processor in the assembler, which is really an interpreter for the macro instructions in an assembly language program. Some language processors are a combination of both translation and interpretation, the most well known being APL, BASIC, SNOBOL, and LISP. The source program is first translated into a form that is easier to interpret before interpretation begins. From the user's point of view, however, the processor is an interpreter, since he is not able to get a copy of the program in its translated form.

7.3 Syntax

There are two complementary ways to specify the syntax of a language, a set of *generation rules* or a set of *recognition rules*. One of the main reasons for formally defining the syntax of a programming language is that the syntax defines the set of all legal programs in the language. A set of *generation rules* is a set of rules for generating all of the legal programs. A set of *recognition rules* is a set of rules for determining if a given program is a legal program in the language.

In most languages, programs have some structure, that is, they are composed of some combination of elements, each element being a member of one of a small number of sets of elements called *syntactic classes*. Typical syntactic classes are *variables, contants, expressions,* and *statements.* A set of generation rules will impose this structure on the programs as they are generated, while a set of recognition rules will identify this structure as a legal program is recognized.

The simplest and most common method of defining the syntax of a language is by writing down a set of generation rules called *productions.* Unfortunately, to write a translator we need a set of recognition rules. The input to a translator is a program. Before this program can be translated it must be broken up into its constituent parts, which means identifying the structure in the program. Thus, the translator writer is usually faced with the task of transforming a set of productions that defines the syntax of a language into a set of recognition rules that will identify the structure and constituent parts of programs in the language. Application of these recognition rules is called *parsing* and will be studied in the following chapters.

A formalism called *Backus-Naur Form* (BNF), or some variant of it, is almost universally used for the formal definition of the syntax of programming languages. Backus-Naur Form is a notation for expressing productions. A set of productions that satisfies a few simple constraints defines the set of legal programs for some programming language. Each production defines a *syntatic class.* For example,

$$<identifier> ::= A|B|C$$

defines a syntactic class, called *identifier,* which contains three elements, the letters "A", "B", and "C". Thus, A is an identifier, B is an identifier, and C is an identifier.

In a production the names of syntactic classes are always enclosed in pointed brackets. Productions always have the form:

$$<name\ of\ syntactic\ class> ::= definition$$

The composite symbol "::=" can be read as, "is defined to be." The *definition* part of a production defines the syntactic class whose name appears in the left part of the production. The definition consists of one or more *alternates* separated by the vertical stroke "|". The vertical stroke means set union. Thus, the syntactic class being defined by the production consists of the union of all the elements in each alternate. An alternate consists of a concatenation of one or more syntactic class names and/or other characters.

Any character in an alternate that is not part of the name of a syntatic class, or its enclosing brackets, is called a *terminal.* The set of all terminals appearing in any production is called the *alphabet over which the language is defined,* or just the *alphabet of the language.* The name of a syntactic class along with its enclosing brackets is called a *nonterminal.* For example, the definition part of the production

$$<identifier> ::= A|B|C$$

consists of three alternates "A", "B", and "C". Each of these alternates is a terminal. The *left part* of the production is the nonterminal $<identifier>$. This part of a production must always be a single nonterminal.

When a terminal appears in an alternate, it stands for itself and can be considered as a class with one member. When a nonterminal appears in an alternate, it stands for all the elements in the syntactic class whose name appears between the pointed brackets. The set of elements in an alternate consists of all possible different combinations of the members of the constituent classes of the alternate, in the order in which they are written.

For example, given the productions:

$<letter> ::= A|B|C$

$<identifier> ::= <letter> | <letter> . <letter>$

the alphabet of this language (set of all terminals) is the set of four characters "A", "B", "C", and ".". The syntactic class *letter* has three members "A", "B", and "C". The first alternate in the second production has three elements,

$$A \quad B \quad C$$

while the second alternate has nine elements,

$$\begin{array}{ccc} A.A & B.A & C.A \\ A.B & B.B & C.B \\ A.C & B.C & C.C \end{array}$$

Therefore, the syntactic class *identifier*, which is the union of all the elements in the two alternates, has 12 members.

A syntactic class may be defined in terms of itself. Such a definition is called a *recursive definition*. It defines a syntactic class with an infinite number of members. For example,

$<term> ::= <identifier> | <term> * <identifier>$

Each element in the syntactic class *term* consists of a finite, but unbounded, number of identifiers separated by asterisks. For example, all of the following are terms:

$$\begin{array}{ll} A & A*B \\ A*A & A*B*A \\ A*A*A & A*B*A*A \\ A*A*A*A & A*B*A*A*A \end{array}$$

Any language defined by a set of productions that includes this one would be an *infinite* language, that is, there would be an infinite number of legal programs. As long as we can find an equivalent, finite set of recognition rules, a compiler will not have any serious problems in translating an infinite language.

One syntactic class is singled out as defining the set of all legal programs in the language. For example, the set of productions in Figure 7.1 defines a simple language in which a "program" is a single *assignment* statement. For example, the following are all legal programs:

$A := B;$

$A := A+A;$

$A := A*B*A+C*A*B+B+A+C*A;$

In programming language theory the nonterminal corresponding to this syntactic class is called the *sentence symbol*, and what we have been calling a program is called a *sentence*. The sentence symbol may not appear in the definition part of any production. All other nonterminals must appear in the definition part of at least one production. Each nonterminal, including the sen-

---

$<assignment> ::= <identifier> := <expression>;$
$<expression> ::= <term> | <expression> + <term>$
$<term> ::= <identifier> | <term> * <identifier>$
$<identifier> ::= A | B | C$

Figure 7.1 Set of productions for a simple language.

tence symbol, must appear in the left part of one and only one production. The language defined by a set of productions is defined to be all the members of the syntactic class named in the sentence symbol, that is, the set of all legal "programs."

Looking at the preceding examples the reader probably has an idea how sentences in the language are generated. A set of productions is used as a set of generation rules for a language in the following manner. Begin by selecting one alternate from the definition part of the production that defines the sentence symbol. In the language of Figure 7.1,

$$<assignment> ::= <identifier> := <expression>;$$

has only one alternate, so we select it,

$$<identifier> := <expression>;$$

Then replace *any* nonterminal in the selected alternate by any alternate from the definition part of the production that defines that nonterminal. In our example we can replace either $<identifier>$ or $<expression>$. We arbitrarily select the first and replace it with the third alternate from its defining production. Thus, the first two steps of our generating sequence are:

$$<identifier> := <expression>;$$
$$C \qquad := <expression>;$$

This replacement process is repeated until no nonterminals remain, and we are left with a string composed entirely of terminals. Continuing our example, we replace $<expression>$ by the second alternate in its defining production. This results in:

$$C \quad := <expression> + <term>;$$

Replacing $<expression>$ by the first alternate in its defining production, we get:

$$C \quad := <term> + <term>;$$

The second occurrence of $<term>$ is replaced by the second alternate in its defining production, yielding:

$$C \quad := <term> + <term> * <identifier>;$$

A complete generation is shown in Figure 7.2. In each step the nonterminal that is replaced is marked with an underline. The final string of terminals is a sentence in the language. The language is the set of all sentences that can be generated in this way.

$$<identifier> := <expression> ;$$

C	:=	$<expression>$;			
C	:=	$<expression>$ +	$<term>$	;	
C	:=	$<term>$ +	$<term>$	;	
C	:=	$<term>$ +	$<term>$	* $<identifier>$;	
C	:=	$<identifier>$ +	$<term>$	* $<identifier>$;	
C	:=	B +	$<term>$	* $<identifier>$;	
C	:=	B +	$<term>$	* A	;
C	:=	B + $<identifier>$ *	A	;	
C	:=	B + C *	A	;	

Figure 7.2 An example of the generation of a sentence in the language defined by the set of productions in Figure 7.1.

We have stated three constraints that a set of productions must satisfy in order to define a language.

1. Each nonterminal must appear in the left part of one and only one production.
2. The sentence symbol must not appear in the definition part of any production.
3. All nonterminals, except the sentence symbol, must appear in the definition part of at least one production.

The reader should note that none of these constraints is strictly necessary, and that programming language theory explores alternate constraints. However, the syntax of the programming languages that we will discuss can be easily defined observing these constraints. We also require another constraint on the set of productions. It should not be possible to find a sequence of replacements of nonterminals that cannot be terminated in a string consisting entirely of terminals.

To illustrate the additional constraint, consider the set of productions:

$$<sentence> ::= <stop> \mid <endless>$$
$$<stop> ::= <stop> + A \mid A$$
$$<endless> ::= <endless> + A$$

which does not satisfy this constraint. Any derivation that begins with the first alternate of $<sentence>$ can be ended in a string of terminals by replacing the nonterminal $<stop>$ with "A", which is the second alternate in its definition. For example,

$$<stop>$$
$$<stop> + A$$
$$<stop> + A + A$$
$$A + A + A$$

However, if a derivation is started with the second alternate of $<sentence>$, it can never be ended in a string consisting entirely of terminals. This is so because no alternate in the definition of $<endless>$ consists entirely of terminals, that is, the only possible replacement for $<endless>$ is:

$$<endless> + A$$

which again contains the nonterminal $<endless>$.

The productions say nothing about the meaning of a sentence or any of its components. No meaning should be assumed simply because the terminal symbols are familiar. For example, the terminal, "+", may mean integer addition, logical or, matrix addition, or something else not even related to our concept of addition. The set of productions in Figure 7.3 defines a language that is equivalent, syntactically, to the language defined in Figure 7.1. The only

$$<assignment> ::= <identifier> @ <expression> !$$
$$<expression> ::= <term> \mid <expression> \# <term>$$
$$<term> ::= <identifier> \mid <term> ? <identifier>$$
$$<identifier> ::= 1 \mid 2 \mid 3$$

Figure 7.3 Set of productions that define a language syntactically equivalent to the language defined in Figure 7.1.

difference between this new language and the previous one is the alphabet over which it is defined. The sentence

$$A := B + C * A ;$$

in the language defined in Figure 7.1 is equivalent to the sentence

$$1 @ 2 \# 3 ? 1 !$$

in the language defined in Figure 7.3.

The equivalence just mentioned is structural (syntactic), not semantic. Semantically the two languages may be as different as night and day. The semantics of the two languages can be defined so that sentences in the second language are semantically equivalent to assignment statements as we know them, while the sentences in the first language, which "look like" assignment statements, are semantically equivalent to something quite different. For example, we define a sentence in the second language to be a FORTRAN-like assignment statement by interpreting the character "@" as an assignment operator, "#" as an addition operator, and "?" as a multiplication operator. Integers are to be interpreted as symbolic variable names.

In contrast we can define a sentence in the first language to be a set of directions for drawing a pattern. The identifier on the left of ":=" indicates the shape of the basic figure: "A" standing for a square, "B" for a circle, and "C" for a lozenge. Identifiers on the right of ":=" indicate the coloring of a replication of the basic figure. "A" means dotted, "B" means black, and "C" means crosshatched. The plus sign means replication toward the right, while the asterisk means replication in a downward direction. All terms begin on the same horizontal level. Figure 7.4 shows the patterns drawn from several sentences in the language, given this semantic interpretation. It should now be clear that in order to translate programs in some language, we need a definition of the language's semantics as well as its syntax.

7.4 Semantics

Defining the semantics of a programming language is much more difficult than defining its syntax. Several different formalisms have been developed for defining semantics. All of these formalisms are more difficult to understand and use than Backus-Naur Form is. A complete formal definition of the semantics of any language like PL/1, or even FORTRAN, is rather overwhelming. Its complexity is such that its use is not practical for anyone other than experts. For this reason, informal definition of the semantics of languages is still in wide use, in contrast to the very wide use of formalisms similar to Backus-Naur Form for the definition of syntax. Unfortunately, further discussion of the formal definition of semantics is beyond the scope of this text. Whatever means is used to define the semantics of a language, one fact is

A = B + C * A   C = A * B + C * B   B = A * C + B * A * C + B

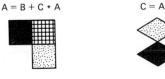 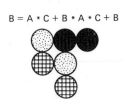

Figure 7.4 An unfamiliar interpretation of the language in Figure 7.1.

common. Not everything can be defined. Any definition will ultimately rest on a (hopefully) small number of primitive concepts that are assumed to be understood and that are undefined.

Since the class of legal sentences in most interesting programming languages is for all practical purposes infinite, we were not able to enumerate the members of the class of sentences as a definition of the syntax of the language. For the same reason it is not practical to define the semantics of a language by describing the meaning of each sentence. Our only hope is to define the meaning of a small number of basic elements in the language and describe a set of rules for deriving the meaning of any sentence from the meaning of its constituent elements.

To be effective the semantics must be coupled with the syntax. It is only through the structure imposed on a sentence by the syntax that its constituent elements can be identified. For example, the set of productions in Figure 7.5 generates the same set of sentences as did the set of productions in Figure 7.1. However, even though these productions generate the same set of sentences, the internal structure of the sentences imposed by the productions is quite different in the two cases. Suppose we wish to interpret the meaning of an assignment statement in the conventional way with the precedence of multiplication being higher than the precedence of addition, that is, multiplications are to be performed before additions. This will be easy with the productions of Figure 7.1 and impossible with the productions in Figure 7.5.

The difference in structure imposed by the two sets of productions can clearly be seen by comparing the two diagrams in Figures 7.6 and 7.7. In these diagrams, which are called *trees*, the small circles that have lines leading to and from them are called *nodes*; they represent one alternate in the definition part of some production. The lines are called *branches*. Each branch lead-

$<assignment> ::= <letter> := <letter> <beta>;$
$<letter> ::= A \mid B \mid C$
$<beta> ::= <null> \mid + <letter> <beta> \mid * <letter> <beta>$
$<null> ::=$

Figure 7.5 A different set of productions that defines the same language as the productions in Figure 7.1.

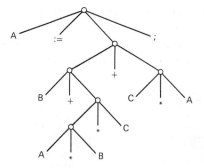

Figure 7.6 Tree representation of structure of $A := B + A * B * C + C * A$; that is imposed by the productions in Figure 7.1.

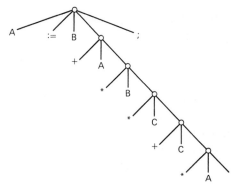

Figure 7.7 Tree representation of structure of A := B + A * B * C + C * A; that is imposed by the set of productions in Figure 7.5.

ing downward from a node represents one of the components of the alternate represented by the node. The topmost node in each diagram represents the sentence symbol $<assignment>$. Thus, the subtree in Figure 7.8a results from the definition

$$<assignment> ::= <identifier> := <expression>;$$

in Figure 7.1 and the subtree in Figure 7.8b results from the definition

$$<assignment> ::= <letter> := <letter> <beta>;$$

in Figure 7.5.

The structure displayed in Figure 7.7 does not provide any way to identify which operations should be performed first. In Figure 7.6 the structure alone indicates that first A is to be multiplied by B, then the result multiplied by C, and that result added to B. Contrast this with Figure 7.7, where the structure indicates that A is to be added to B, the result multiplied by B, and that result multiplied by C. If the conventional precedence is to be imposed on the operations in Figure 7.7, it will have to be external to the structure imposed by the productions in Figure 7.5. In fact, the end result will be the same as redefining the productions so that they impose a structure that is equivalent to that imposed by the productions in Figure 7.1.

The meaning of a compound element in a language is usually defined in terms of the meanings of the component elements. For example, an expression in languages like FORTRAN and PL/1 is an algorithm for computing a value. The rule for computing this value closely follows the syntax of an expression. Using the productions in Figure 7.1, an expression is

$$<expression> ::= <term> \mid <expression> + <term>$$

The value that an expression represents is then defined to be the value of the term if the expression is a single term; otherwise, it is the value obtained by

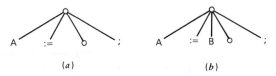

(a)   (b)

Figure 7.8 Subtrees that result from two definitions of $<assignment>$. (a) From Figure 7.6. (b) from Figure 7.7.

adding together the values of the expression and term that are the leftmost and rightmost components, respectively.

Letting $V(x)$ mean the value of x, we can define the meaning of an expression as

$$V(expression) = V(term) \mid sum \text{ of } V(expression) \text{ and } V(term)$$

This mirrors the definition of $<expression>$. It says that if a particular expression has the form specified by the first alternate, $<term>$, then its value is just the value of the corresponding term. Otherwise, if a particular expression has the form specified by the second alternate, $<expression> + <term>$, then its value is equal to the *sum* of the values of the expression and term that are its components.

To complete the definition of the semantics of an expression we must define the value of a term and an identifier.

$$V(term) = V(identifier) \mid product \text{ of } V(term) \text{ and } V(identifier)$$
$$V(identifier) = V(\text{variable named A}) \mid V(\text{variable named B}) \mid$$
$$V(\text{variable named C})$$

Thus we have defined the value of any expression and any term as a function of the values of the variables A, B, C using only the primitive functions *sum* and *product*. We mentioned earlier that not everything can be defined. Thus, we leave the primitive functions *sum* and *product* and the concept of the value of a variable undefined and assume that they are understood.

The value of a variable ranges over some set, such as the integers, the rational numbers, or the complex numbers. In most programming languages the value set for a variable is limited to a finite set, since the computer is not able to operate on arbitrarily large values. The meaning of primitive functions such as *sum* and *product* will depend on the sets of values from which the operands are taken. Thus, rules of the above type for defining the semantics of an expression will have to be supplemented with additional rules that define which specific *sum* function to use (e.g., *integer sum* or *rational sum*), depending on the value sets of its arguments.

Not all constructs in a language have a value in the above sense. The assignment statement is such a construct. The meaning of an assignment statement is usually defined as changing the value of a variable. For example, in Figure 7.1 $<assignment>$ is defined as

$$<assignment> ::= <identifier> := <expression>;$$

Its meaning is defined as

$$update \ V(identifier) \text{ to be equal to } V(expression)$$

Operationally this is equivalent to storing $V(expression)$ into the memory word that is used to store the value of the variable whose name is the identifier. Any formalism for defining the semantics of programming languages must express this assignment action in some way. This turns out to be a difficult problem, even though our informal discussion makes it seem easy.

7.5 The Instran Language

In the following chapters our discussion will be limited to compilers for languages of the FORTRAN and PL/1 type. This is not because other languages are not important, but simply because of space limitations in this text only the most common languages can be considered. All the examples in these chapters will be based on a language that is a simple subset of PL/1. The compiler for this language will be called *Instran* (for Instructional Translator), and the language will be called the *Instran Language*. In choosing this subset we have tried to include some of the most significant features of PL/1, yet keep the subset small and simple enough that the inherent problems in compiler implementation do not get buried in a mass of detail.

The complete syntax for the Instran Language is formally defined in Appendix C. The notation for the defining productions is a more compact notation than the pure Backus-Naur Form used in the preceding sections. It is explained in the appendix. There are a few restrictions on the formation of some of the elements in the language that are not easily expressible in the formalism. For example, the length of an identifier is limited to no more than 12 characters. These restrictions are stated in English as a set of notes that supplement the formal productions. Most of the Instran Language is a subset of PL/1 formed by imposing restrictions. Thus, the semantics of the Instran Language is a subset of the semantics of PL/1. An informal discussion of the semantics of the most important features of the Instran Language will be found in the following section. For a complete, detailed definition of the semantics the reader is referred to the IBM reference manual for the PL/1 language.[2]

7.6 Definition of Important Instran Features

Perhaps the easiest way to organize a discussion of the important features of any programming language is to use a checklist in the same way we used it in Chapter 3 when we discussed machine language. Our checklist for Instran-like languages will be somewhat different in detail from the computer description checklist, but will basically follow the same pattern. There are two principal differences. Instran-like languages do not usually have working registers. In addition, they usually include nonexecutable instructions, called *declarations,* as well as executable instructions.

In the Instran language the variables in a procedure correspond to the memory in a computer in that data can be saved in them and used again later. The names of variables correspond to the addresses of memory locations. Statements correspond to instructions, with the operators in an expression corresponding to the operations that a computer can perform on data. The name of a variable is used in statements to reference the value of the corresponding variable, just as the address of a memory location is used in a machine language instruction to reference the contents of the corresponding memory location. The statements in the Instran Language can be divided into groups just like the machine instructions were. In particular there are computational statements and control statements.

The memory of an Instran Language procedure consists of a collection of variables. In addition to single variables, called *scalars,* the Instran Language allows variables to be grouped into *arrays* and *structures.* A *scalar* has a single identifier associated with it that is the name of the variable. It is referenced using its name. An *array* is an ordered collection of one or more variables asso-

---

[2] *IBM System 360 PL/1 Reference Manual,* Form C28–8201, IBM Corporation, White Plains, N.Y.

ciated with a single identifier that is the name of the array. An individual variable in an array is referenced using the name of the array and the positional index of the variable in the ordered collection. The variables in an array are indexed by consecutive integers, beginning with zero. The index of a variable in an array is called its subscript. For example, the elements in an array of five variables named X are referenced as X_0, X_1, ..., X_4. In the Instran Language a subscript is written in parentheses immediately following the name of the array. A subscript may be an expression. For example, $X(I+2)$ references the element of the array X whose index is equal to the value of $I+2$ at the time the reference is made.

A *structure* is a hierarchically organized collection of variables. It consists of an arbitrary number of elements, each of which is either a scalar or another structure called a *substructure*. Thus, a structure is recursively defined. The hierarchial structure is a result of the nesting of substructures within a structure. A structure is named, and so are all of its elements. The outermost structure is sometimes called the *major structure*. All of its contained structures are called substructures. The major structure is considered to be at *level* 1 and its elements at *level* 2. In general, if a substructure is at *level* k its elements are considered to be at *level* $(k+1)$. The easiest way to visualize a structure is to view it as a tree. For example, the structure diagrammed in Figure 7.9 is named W and consists of four elements, named A, B, Z, and X, two of which are substructures, B and X. The substructure B consists of three elements X, Y, and D, all of which are scalars. The substructure X has the substructure A as one of its elements, which, in turn, has the substructure W as an element.

An element in the major structure or any of its contained substructures is referenced by a *qualified name*. A qualified name is a sequence of identifiers separated by periods. The qualified name of a (scalar) variable in the structure is composed of the names of all of the structures (major structure and sub-

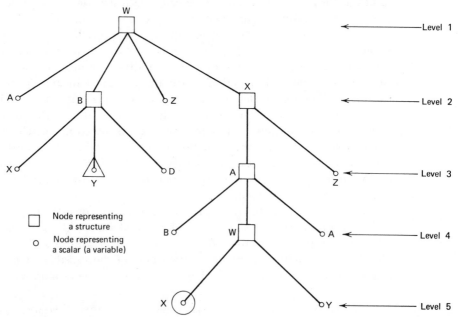

Figure 7.9 Tree diagram of a structure.

structures) that contain the scalar. For example, the qualified name of the variable in Figure 7.9, which is circled, is W.X.A.W.X, while W.B.Y is the qualified name of the variable in the triangle. A structure is said to be the *parent* of its elements. Thus, the qualified name of a variable in a structure is a list of the names of its ancestors separated by periods.

Within a structure names may be duplicated, except that the names of the elements in any particular structure must be unique. Structures may also be grouped into arrays. However, none of the elements of a structure may be an array. When referencing an element in an array of structures, the subscript follows all of the components in the qualified name. For example, A.B.C(I) refers to the variable named C that is an element of the substructure named B that is an element of the Ith structure in the array of structures named A.

In the Instran Language, declarations are used to identify arrays and structures. The only information needed for an array is the number of variables in the array. This is specified in a declaration by giving the subscript of the last variable in the array. For example,

DECLARE X(4);

declares X as an array of five variables with subscripts from 0 to 4.

The declaration for a structure is more complex, since the hierarchial relationship of the variables must be expressed. A structure declaration uses the idea of level to express this relationship. The name of each element is prefixed with its level number. Within a structure at level k each element in that structure is described in sequence. These elements are at level $(k+1)$. If one of these elements is a substructure, the entire substructure is described before the next element at level $(k+1)$ is described. For example, the structure diagrammed in Figure 7.9 would be declared in an Instran Language program with the declaration shown in Figure 7.10. Accordiing to the Instran Language syntax, the indentation and other spacing is not required, but it is permissible, and it makes the declaration easier to read.

The variables of an Instran Language procedure are divided into three classes called *storage classes*. Two of these classes are called *internal* and

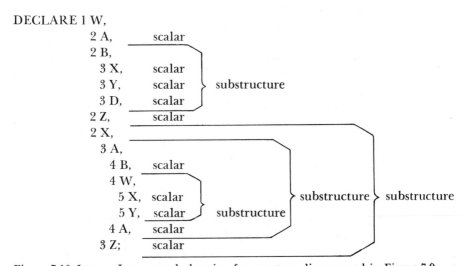

Figure 7.10 Instran Language declaration for structure diagrammed in Figure 7.9.

external and are essentially equivalent to the local and global symbols of the full assembly language that were used with the loader of Chapter 6. Variables that are internal are accessible only to the procedure in which they appear, while external variables are accessible to all of the procedures in the program. The third class of variables are those that are arguments of the procedure. If a variable is not an argument its class is indicated in the declaration for the variable using one of the key words, INTERNAL or EXTERNAL.

In discussing the computational facilities of a language the first thing we must know are the types of data that the language supports. The Instran Language supports integers (what PL/1 calls *fixed binary*), *character strings*, and *bit strings*. The set of values that a variable may assume is called its *data type*. The data type of a variable in an Instran Language procedure is specified in a declaration for the variable using one of the key words FIXED, CHARACTER, or BIT. When the data type of a variable is character string or bit string, all of its values must be the same length, that is, the same number of characters or bits in the string. This length is considered to be part of the data type and is also specified in the declaration for the variable.

A declaration for a scalar variable or an array of scalars identifies its name, its data type, and its storage class. A declaration for a structure or an array of structures specifies the name and storage class of the major structure. It also specifies the hierarchial structure, the names of all the scalars and substructures in the major structure, and the data type of all the scalars. The declaration for an array of scalars or structures specifies the number of elements in the array. The variables (scalars) in a structure may have any data type, and they do not all have to be the same type. The variables in an array must all be the same type. Thus, we say that an array is required to be *homogeneous*, while a structure need not be homogeneous.

For example,

DECLARE W(9) CHARACTER (3) INTERNAL;

declares W to be an array of 10 variables, each of whose values is from the set of all character strings that are three characters in length. Since the storage class is internal, the name W may be used to reference this array only by the procedure in which the declaration appears. If some other procedure contains a declaration for W, the variables that that procedure references when it uses W are *not* the same as the variables referenced when this procedure uses W. If two different procedures were to both contain the declaration

DECLARE W(9) CHARACTER (3) EXTERNAL;

then each would reference the same variables when they used W.

Computation is expressed in the Instran Language by expressions that are composed of variable references, constants, and operators. An expression defines a value and may appear as the right-hand part in an assignment statement, thereby defining or redefining the value of the variable referenced in the left-hand part. The variable referenced on the left in an assignment statement must have the same data type as the value of the expression on the right. Expressions may also appear as subscripts in an array reference, in which case the data type of the value of the expression must be integer. The Instran Language has the usual arithmetic operators for integers (add, subtract, . . .), relations (equals, not equals, less than, . . .), logical operators for use with bit strings, and concatenation of strings for use with bit strings and character strings.

The last item on the checklist that is applicable to the Instran Language is control. The unit of Instran Language text that is compiled at one time is a procedure. A procedure has a name and a single entry point, which is the first executable statement in the source text. The name of the procedure is implicitly declared to be an external symbol, which is defined in the procedure. Thus, other procedures may call the procedure using its name. The sequence of execution of the statements in an Instran Language procedure is the sequence in which they are written, unless a statement is encountered that explicitly changes this sequence. There are three statements that change the normal sequence of execution, the GOTO, IF, and RETURN statements. The GOTO statement sends control to the statement whose label is the identifier in the GOTO statement. The RETURN statement sends control back to the procedure that called the procedure containing the RETURN statement.

The IF statement selects one of two statements, depending on the value of an expression, and skips the other statement of the pair. The two statements that are part of the IF statement may be any statement, including IF statements. The key words DO and END are used as statement brackets. The sequence of statements that they enclose are treated as a single statement. This allows a sequence of statements to take the place of either or both of the statements that are part of the IF statement.

7.7 An Example Instran Language Procedure

As an illustration of some of the points discussed in the last section, we will briefly look at an example procedure written in the Instran Language. The procedure is defined in Figure 7.11 using our informal language. The corresponding Instran Language procedure appears in Figure 7.12. The procedure

```
procedure tablelook (key, value, type, index);
    i := 0;
    while i ⩽ st_count do
        if key field of st[i] = key then
            begin index := i;
                return;
            end;
        i := i+1;
    end;
    if i > 49 then index := −1;
        else begin st[i] := (key, value, type);
                st_count := i;
                index := i;
            end;
    return;
end.
```

Figure 7.11 Definition of a procedure to search a symbol table.

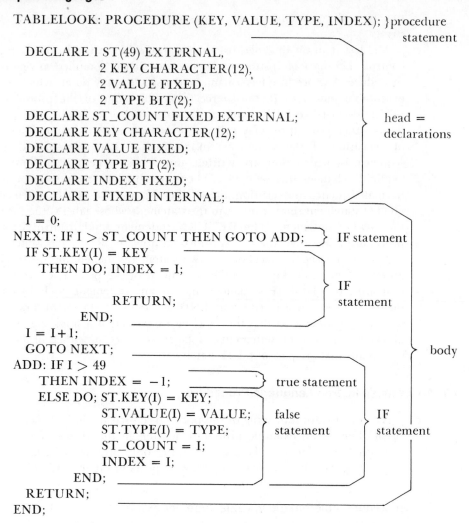

TABLELOOK: PROCEDURE (KEY, VALUE, TYPE, INDEX); } procedure statement

DECLARE 1 ST(49) EXTERNAL,
          2 KEY CHARACTER(12),
          2 VALUE FIXED,
          2 TYPE BIT(2);
DECLARE ST_COUNT FIXED EXTERNAL;
DECLARE KEY CHARACTER(12);
DECLARE VALUE FIXED;
DECLARE TYPE BIT(2);
DECLARE INDEX FIXED;
DECLARE I FIXED INTERNAL;

head = declarations

I = 0;
NEXT: IF I > ST_COUNT THEN GOTO ADD; } IF statement
  IF ST.KEY(I) = KEY
    THEN DO; INDEX = I;

IF statement

        RETURN;
      END;
  I = I+1;
  GOTO NEXT;
ADD: IF I > 49
    THEN INDEX = −1; } true statement
    ELSE DO; ST.KEY(I) = KEY;
            ST.VALUE(I) = VALUE;   false
            ST.TYPE(I) = TYPE;   statement
            ST_COUNT = I;
            INDEX = I;
        END;
    RETURN;
END;

body

IF statement

Figure 7.12 Instran Language form of procedure defined in Figure 7.11.

searches a symbol table for an entry whose key is equal to the key supplied by the calling procedure as the value of the argument **key**. If no entry has a matching key, a new entry is added to the table. The contents of this new entry are supplied by the calling procedure as values of the arguments **key**, **value**, and **type**. When the procedure returns to the calling procedure, the value of **index** is to be equal to the index in the symbol table of the entry corresponding to the key given as an argument. If the symbol table does not contain a matching entry and it is full, the value of **index** is set equal to −1. The procedure assumes that the maximum number of entries permitted in the symbol table is 50.

The procedure begins with a PROCEDURE statement that names the procedure and its arguments. Following the PROCEDURE statement are declarations for all of the variables referenced in the body of the procedure. The body of the procedure (executable statements) follows the declarations and is terminated by an END statement. The symbol table is named ST and is an array of 50 structures. We are assuming that other procedures will also reference the symbol table; therefore, ST is declared an external (global) array. Each entry in the symbol table is one of the structures in the array. Each structure consists of three elements: a symbol (KEY), its value (VALUE), and the symbol's type (TYPE). The value of the variable named ST_COUNT is the index of the last entry in the symbol table.

The data types of the arguments must be declared, but no storage class has to be indicated, since their appearance in the PROCEDURE statement identifies them as arguments. The variable I is an internal (local) variable. The body of the procedure consists of seven statements: an assignment, two IF statements, another assignment statement, a GOTO statement, another IF statement, and a RETURN statement. The first and second IF statements do not have the optional ELSE clause. The true statement in the second IF statement and the false statement in the third IF statement are groups containing two and five statements, respectively.

EXERCISES

7.1 Formally define the syntax of the assembly language described in Appendix B by writing a set of productions similar to those in Appendix C.

7.2 Represent the structure, as defined by the productions in Appendix C, of the following Instran statements, using tree diagrams similar to those in Figures 7.6 and 7.7.

    a.   IF I = 1 THEN A = B; ELSE A = C;

    b.   IF ST.KEY(I) = KEY THEN DO;
                      INDEX = I; RETURN; END;

    c.   DECLARE 1 ST(49) EXTERNAL,
                      2 KEY CHARACTER(12),
                      2 VALUE FIXED,
                      2 TYPE BIT(2);

7.3 Write Instran procedures that correspond to the following IL procedures.

    a. Figure 4.7.

    b. Figure 4.9.

    c. Figure 4.5.

8 | Instran: A Simple Compiler

How does a compiler work and how do we go about building one? There are several things we must know and do before we need to worry about how to build a compiler, or even how it works. Since a compiler translates from one language, the *source language*, to another language, the *object language*, it is clear that we must know and understand the syntax and semantics of both languages. We must derive a mapping of the source language structures into object language structures with the same meaning. That is, for each kind of statement, expression, and other structure in the source language, we must decide what statement or set of statements in the object language has an equivalent meaning. When the object language is machine language, as it is for Instran, the object language statements are machine language instructions, and they are usually referred to as *object code*.

There are several tasks in developing this mapping. We have to choose an object language representation for the data objects in the source language. We have to decide what code sequences in the object language will compute the correct result for an expression when the values used in the computation are stored in memory using the representation that was chosen. We must decide on the general layout and organization of the object procedure, and what object code sequences correspond to the control statements in the source language when the object procedure is organized in the manner that we selected.

In addition, we need a knowledge of the environment in which the object procedure will execute. Services of the host system, such as input-output, are often useful in the object code. Requirements of the host system, such as the format of the object deck required by the loader, influence the organization of the object procedure.

All of these considerations determine what the result of translating a procedure must be. This can be completely determined without any knowledge of how it will be achieved. This is because both the source and object languages are fixed and cannot be changed by the programmer who is implementing the compiler. Before exploring the correspondence between object code and source language in any depth, let us look briefly at how the translation is accomplished.

8.1 Functions of a Compiler

Basically, the compiler must recognize the structure of the source procedure, isolating each of its constituent parts. Then it must transform these into semantically equivalent fragments of object code, which are then combined to form a complete object procedure. The structure recognition we call *parsing*, and the transformation into object code we call *code generation*. There are three other major functions that play a role in translation: *input interface*, *output interface*, and *optimization*. The input interface we will call *lexical analysis*. It transforms the source procedure from its external form, which is dictated by the source language and the system which contains the compiler, into an internal form that the compiler can process more easily. The output interface is called *assembly*. This is the reverse of the input interface. It transforms an internal form of the object procedure that is more convenient for the compiler into the object deck format required by the loader. Optimization is the function of applying various transformations to the procedure in order to produce a "better" object procedure. The most common measures of better are shorter execution time and smaller memory requirements.

The five major functions identified in the preceding paragraph can be viewed as five conceptual phases in the process of translation. In the larger, more sophisticated compilers these phases are distinct; however, in many of the smaller, simpler compilers some or all of these phases are merged together. In order to simplify further discussion of each of these functions we will consider them as separate phases of a compiler. Figure 8.1 shows a diagram of such a compiler. In the following discussion we will assume that the procedure being translated is completely processed through each phase in turn. Thus, each phase transforms the procedure one step closer to the final object procedure.

There are several tables that are needed by the compiler and play an important role in the translation process. The *symbol table* is similar to the symbol table that the assembler required. It has an entry for each distinct simple variable, label, array, structure, and structure element. This entry contains all of the associated information, such as the assigned location address, data type, and storage class. Other tables include one for the key words in the source language, one for the computational operators, a table of constants, which appeared in the source procedure, and the instruction table, which contains an entry for each machine language instruction that can be generated by the compiler.

The source procedure is presented to the compiler as an unstructured string of characters. The basic elements of a source procedure are *identifiers, constants, operators*, and *punctuation symbols*. The identifiers and constants have varying lengths; for example, an identifier may be from 1 to 12 characters in length. Because of character set limitations of the computer, some of the operators are composed of two characters, for example, not equals ($"\neg ="$). It is the job of the parsing phase to recognize structure in the source procedure, but it is not interested in such fine structure as how letters and digits are combined to form an identifier. In addition, the source procedure may have an arbitrarily long string of blank spaces between the basic elements.

The lexical analysis phase replaces each basic element by a single *token* and

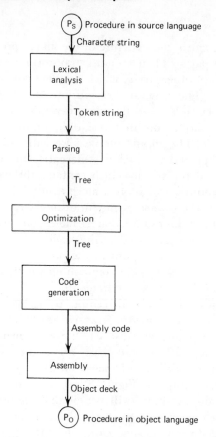

Figure 8.1 Conceptual diagram of a compiler.

deletes all of the intervening blank spaces. Thus, the original source procedure is transformed from a character string into a string of tokens, one for each basic element. All tokens have the same form and size. A token (t,i) contains two items of information: t is the type of thing it represents (identifier, operator, etc.), and i specifies which particular one of that type. The most convenient way of identifying a particular item is by the index of its entry in some table. For example, the compiler keeps a table of all identifiers that appear in the source program. Then, supposing the token $('I',k)$ corresponds to the identifier SPEED, the letter "I" specifies the identifier table and k is the index of the entry for SPEED in the identifier table.

The token string that is output from lexical analysis does not display any structure; it is just a string of tokens corresponding to the basic elements in the source procedure in the order in which they occurred there. In order to generate code we need to identify the terms, expressions, statements, and other syntactic elements that occur in the source procedure. The parsing phase transforms the token string into a form that is equivalent to the trees we used in Chapter 7 to display the syntactic structure of expressions. This tree form explicitly displays the structure of the source procedure. There are several ways in which this tree structure can be represented, and they are discussed in a later section.

While optimization transformations may occur almost anywhere in the compiler, certain global optimizations are most easily done as a separate phase operating on the tree form of the program. One such global optimization is the elimination of redundant computation of common expressions. In certain circumstances, if the same expression occurs more than once, all but one occurrence of it can be deleted. The result is that the value of the expression, instead of being computed several times, is computed only once. Its value is saved in some memory location. This memory location is referenced wherever else the expression occurs instead of recomputing the expression's value. Another global optimization is the efficient use of the working registers of the computer. The working registers are used for computation of arithmetic expressions, base registers, and index registers. How many registers are used for each of these purposes may make a considerable difference in how efficiently the object procedure executes. No fixed assignment will be most efficient all of the time, hence, the assignment of the working registers must vary, depending on the demands of each section of the object procedure.

The code generation phase takes the tree form of the procedure and generates object code from it. A standard sequence of code is selected corresponding to each of the smallest syntactic structures. These sequences are then combined, with possibly some modification, into larger structures corresponding to the larger structures of the source procedure. The object code generated by this phase is not yet in the final machine language form. Instead, its form is like assembly language. Some addresses are not yet determined, and the operation code is a pointer to an entry in the instruction table. Another task of the code generation phase is mapping the scalar variables, arrays, and structures into memory. In order to generate complete machine language instructions, relative addresses must be known for all of the internal variables in the procedure. Whenever an element of an array is referenced, code must be generated to compute its subscript and load it into an index register. All of this is part of the code generation phase.

The final phase is assembly. This phase is quite similar to some of the functions found in pass two of the assembler that we studied. Its major task is to transform the assembly code, which is output from code generation, into the standard object deck format required by the loader. The global symbol dictionary must be created with entries for each variable declared external in the source procedure. The machine language instructions are completely assembled and formatted into text cards. Finally, the address constant definition cards are generated.

We remind the reader again that this is a conceptual model of a compiler that we have been discussing. All of these functions will be found in any compiler, but in simple compilers they will not all be separate phases. Sometimes they are so simple that they are hardly recognizable. For example, simple compilers almost never have a separate optimizing phase. However, even the simplest compiler usually does some optimization. This is usually very local optimization, which can be done at the same time that the object code is generated.

8.2 Instran Structure and Tables

Our study of compilers will center around Instran, which is a compiler for the Instran Language defined in the last chapter and Appendix C. There are many different ways to build a compiler, and it is impossible to discuss them all in a text of this level. In fact, it is not possible to discuss more than one method in any detail. Instran is an example of one of the most common translation techniques. When practical we will point out other translation techniques. In this section we will briefly examine the general structure of Instran and some of the major tables it requires. Later sections in this chapter will consider translation of expressions in detail, while the following chapters will explore the complete compiler in more detail. Additional tables will be introduced in those sections as needed.

Instran is a three-pass compiler. The first pass includes lexical analysis and parsing. The second pass is code generation. The third pass is assembly. There is no global optimization in Instran, only the local optimization that can be achieved during code generation.

Parsing and lexical analysis are combined in the first pass. In fact, lexical analysis is called as a subroutine by the *parser*. Each time the *lexical analyzer* is called, it returns a single token that corresponds to the next basic item in the source procedure. There are four types of tokens that the lexical analyzer may return. They correspond to the *identifier table*, the *constant table*, the *key word table*, and the *punctuation-operator table*. The identifier table contains a list of all identifiers that occur in the source procedure, while the constant table contains a similar list of constants. Both of these tables are built up by the lexical analyzer. The key word table contains a list of all the key words in the language. The punctuation-operator table contains an entry for each punctuation and operator symbol in the language. Both of the latter two tables are constant, since their contents depend on the source language instead of on the particular procedure being translated.

As an example illustrating the action of the lexical analyzer, consider the source fragment

$$DO; \quad A = B+3;$$

Assume that the four tables mentioned in the preceding paragraph have the contents shown in Figure 8.2. The result of lexical analysis will be the token string, shown in Figure 8.3, which will be passed to the parser, one token each time the lexical analyzer is called by the parser. A token is represented as an ordered pair: a letter indicating the token type and an integer that is the index of an entry in the table corresponding to the token type.

Not all of the source procedure is transformed into the tree form by the parser. Information about a variable appearing in the declaration for the variable is put into the symbol table by the parser. The symbol table contains an entry for each simple variable, array, structure, substructure, scalar structure

Key word table (K)	Identifier table (I)	Constant table (C)	Punctuation–operator table (P)
1 PROC	1 X	1 3	1 =
2 DO	2 A	2 5	2 +
3 END	3 B		3 ;

Figure 8.2 Example of contents of tables used by the lexical analyzer. The letter enclosed in parentheses following the name of a table is the type code used in tokens that refer to the corresponding table.

Character string input to lexical analyzer	Corresponding token string output from lexical analyzer
DO	K,2
;	P,3
A	I,2
=	P,1
B	I,3
+	P,2
3	C,1
;	P,3

Figure 8.3 Example of the effect of lexical analysis based on the tables in Figure 8.2.

element, and label in the source procedure. An entry for a variable (or label) contains all of the information known to the compiler about that variable. This information is called the variable's *attributes*. The attributes for an Instran Language variable are: the variable's name, its data type, its storage class, its organization (array, structure, substructure, or scalar structure element), its length (if its data type is bit string or character string), number of elements in the array (if it is an array), and information about how it relates to other elements of the structure if it is the major structure or an element of a structure.

The parser, calling the lexical analyzer, obtains tokens for the basic elements in the source procedure and groups them in tree form according to the structure imposed by the syntax. As it is doing this, it replaces each identifier type token by a token that refers to the entry in the symbol table for the variable which the identifier names. The appearance of an identifier and a colon preceding a statement is an implicit declaration of that identifier as a label. This information is also put into the symbol table by the parser just as if the label had appeared in an explicit declaration. Tokens representing constants are not changed.

The parser also transforms all of the tokens of type punctuation-operator. Tokens of this type represent either punctuation characters such as ";" or operators such as "+" and "*". After parsing, the tokens that represent punctuation are omitted, since they were required only to determine the structure from the source procedure that was input as an unstructured sequence. In addition, the tokens that represent operators are replaced by a different type of token, *operation*. This token refers to the *operation table* and makes it easy to distinguish between different uses of the same operator symbol. For example, the equals sign "=" is used both for the assignment operator and the equals relation.

The action of the parser is illustrated by considering the assignment statement

$$A = B + 3;$$

from Figure 8.3. In this example we assume that the contents of the symbol table and operation table are as shown in Figure 8.4. The result of parsing the above assignment statement is the tree shown in Figure 8.5.

The principal task of the second pass of Instran is to generate object code.

Operation table (R)   Symbol table (S)

1	assignment (=)
2	addition (+)
3	equals (=)

1	attributes for A
2	attributes for B

Figure 8.4 Example of contents of tables used by the parser in addition to those in Figure 8.2.

Source text	Token	Transformed token
A	I, 2	S, 1
=	P, 1	R, 1
B	I, 3	S, 2
+	P, 2	R, 2
3	C, 1	Same
;	P, 3	Omitted

Tree form

Figure 8.5 Example of effect of parsing based on the tables in Figures 8.2 and 8.4.

However, the variables and constants must be mapped into memory first. The procedure that does this mapping operates only on the symbol and constant tables. All of the information that it requires is in these two tables. The assignment of memory to the constants is basically the same as for literals in the assembler. Simple internal variables are assigned consecutive memory locations. Assignment of memory to arrays and structures is somewhat more complicated and will be discussed in detail in a later chapter. Information indicating the memory assignment is then entered into both the symbol and constant tables. This information is the address of the variable or constant relative to the base of some block of memory. Conceptually the object procedure uses several blocks of memory. Two of these are the *constant block* for the constants and the *internal block* for the internal variables. An example of this memory assignment is illustrated by the symbol and constant tables in Figure 8.6. In making this assignment we have assumed that the constants and variables are all fixed binary, and that each occupies a full word in memory.

When evaluating an expression, partial results may have to be stored in memory, depending on how the registers are used. For example, suppose only two registers are allocated for use in arithmetic computation when evaluating

$$A*B+C*D$$

The first term A*B is evaluated using the two registers (recall Our 360's multiply instruction uses two registers). In order to evaluate the second term the same two registers are needed. Thus, the value of the first term will have to be stored in memory until after the second term has been evaluated. The memory used to save a partial result like the value of the first term is called a *temporary*. Temporaries needed by the object procedure are assigned by the *code generator* as the need for them arises. All temporaries are assigned in the *temporary block*, which is similar to the internal block. The *temporary table* records the relative address within the temporary block of each temporary. References to temporaries are represented by tokens with a type code indicating the temporary table.

The object code, which is output from code generation, is like assembly

Constant table (C)   Symbol table (S)

	Value	Address			Name	Address	Other attributes
1	3	0		1	A	0	fixed internal
2	5	4		2	B	4	fixed internal

Figure 8.6 Contents of the constant table from Figure 8.2 and the symbol table from Figure 8.4 after memory assignment at the beginning of pass two of Instran.

language in that instructions consist of two tokens that represent the operation code and operand. An address in the operand of most instructions is still relative to the base of some block. This "symbolic" address will be resolved during the assembly phase of the compiler. The tokens that represent the operation codes of instructions refer to the *instruction table*, which contains an entry for each machine instruction giving its numeric operation code and the operand type of the instruction.

The following example illustrates the action of the code generation phase of Instran. The object code generated here is not the most efficient. In Chapter 11 we will discuss how it can be improved. We assume the contents of the tables just discussed are as shown in Figure 8.7. We have written the symbolic operation code and instruction name to the right of each entry in the instruction table for clarity. This information is not part of the table. Figure 8.8

Temporary
table (T)   Instruction table (I)   Corresponding
   Address   Numeric code   Operand type   Our 360 instruction

	Address			Numeric code	Operand type			
1	0		1	5A300000	X		A	add
			2	58 300000	X		L	load
			3	50 300000	X		ST	store

Figure 8.7 Example contents of tables used by code generation in addition to those in Figures 8.2 and 8.4. The assembly language version of the instructions, which is shown in the rightmost column of the instruction table, is not actually stored in the computer's memory.

Tree form
input to
code generation

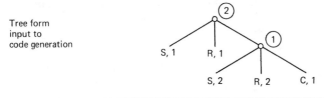

Assembly code output from code generation		Corresponding node in tree	Corresponding symbolic instruction		
I, 2	S, 2	1	L	3, B	get B
I, 1	C, 1	1	A	3, = F'3'	add 3
I, 3	T, 1	1	ST	3, temp_1	save partial result
I, 2	T, 1	2	L	3, temp_1	get B + 3
I, 3	S, 1	2	ST	3, A	A := B + 3

Figure 8.8 Example of effect of code generation based on the tables in Figures 8.6 and 8.7.

shows the assemblylike code that is output from code generation for the assignment statement whose tree is shown in Figure 8.5. In Figure 8.8 we have reproduced the tree and numbered the nodes with circled integers. The assembly code generated from each node is identified using these numbers. To help in reading the generated object code we have written the corresponding Our 360 assembly language instructions to the right of each token pair that is output from code generation.

The third pass of the compiler is the assembly phase. Its major task is to assemble the object program in the object deck format required by the loader. There are three major parts to this job: generating the global dictionary, generating the text cards, and generating the address constant definition cards. The global symbol dictionary contains one entry for each external scalar variable, array, and major structure. It also contains an entry for each procedure that is called. All of the information needed for this is in the symbol table. The text cards must contain the complete machine language instructions. The assemblylike code generated by pass two is processed to yield complete instructions. The numeric instruction code from the instruction table is combined with the operand address and the proper base register number to form a complete instruction.

Instran is to generate object code that will execute in our system. Thus, the object procedure output from Instran will observe the system standards, in particular the standard calling sequence. This means that G_{15} will contain the base address of the procedure segment when it is called. The procedure's instructions, constants, temporaries, and internal variables are all in the segment. Therefore, a machine language instruction that references any one of them will use G_{15} as a base register, and the value of the D field will be its relative address. This address must be relative to the base of the segment. However, an address in the constant table, temporary table, or symbol table is relative to the base of the constant block, temporary block, or internal block, respectively. This address is transformed into the required relative address by adding the address, relative to the base of the segment, of the base of the corresponding block of memory. These blocks are assigned sequentially in the segment in the order shown in Figure 8.9.

To finish following our example of an assignment statement through Instran, we will assume that the instruction, constant, temporary, and internal blocks

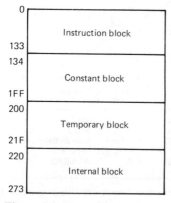

Figure 8.9 Example assignment of the blocks in a procedure segment.

have been assigned locations within the segment as shown in Figure 8.9. We have also assumed that Figures 8.6 and 8.7 show only the first few entries in the constant table, symbol table, and temporary table. Therefore, the respective blocks are larger than would be implied by Figures 8.6 and 8.7. Figure 8.10 shows the final machine code output from the assembly phase for the assignment statement

$$A = B + 3;$$

These instructions will be formatted as text cards according to the standard object deck format.

After all of the text cards have been generated for the object procedure the address constant definition cards are produced. There will be one address constant for each external symbol. The address constants are included as part of the constant block. The assembly phase must also generate the segment definition card and end card as required for the object deck. In addition, it will produce a listing in which the source procedure, a symbolic version of the object procedure, and other useful information about the object procedure are printed.

8.3 Translation of Expressions

For the remainder of this chapter we will study the translation of expressions. A very simple and general method exists for parsing expressions composed of binary operators. By treating the $"="$ character, meaning assignment, as a binary operator, we can also include assignment statements along with expressions. We consider two major functions in translating expressions, parsing and code generation. Parsing is required to locate the operands of each operator. For example, in the expression $A + B*C$, the operands of the $"*"$ operator are $"B"$ and $"C"$, however the operands of the $"+"$ operator are $"A"$ and the term $"B*C"$. Once the operands of each operator are identified the code generation is straightforward.

Before continuing with our discussion of the mechanics of translating expressions, we must decide what code is to be generated, that is, what object code sequences correspond to the various source language structures. The different source structures and the object code corresponding to them are shown in Figure 8.11. The code generated for $"+"$, $"-"$, $"\&"$ (logical **and**), and $"|"$ (logical **or**) is the same except for the second instruction. The operation code

Assembly code input to assembly	Machine instructions output from assembly	Corresponding symbolic instruction	
I,2 S,2	58 30F 224	L 3,B	get B
I,1 C,1	5A 30F 134	A 3, = F'3'	add 3
I,3 T,1	50 30F 200	ST 3,temp_1	save partial result
I,2 T,1	58 30F 200	L 3,temp_1	get B + 3
I,3 S,1	50 30F 220	ST 3,A	A: = B + 3

Figure 8.10 Example of effect of assembly based on the tables in Figures 8.6 and 8.7 and the block assignment in Figure 8.9.

Source structure	Corresponding object code
a op b	L 3,a get a OP 3,b op b ST 3,$temp$ save partial result where OP is defined as

	op	OP	
	+	A	Add
	−	S	Subtract
	&	N	Logical and
	|	O	Logical or

Source structure	Corresponding object code
$a * b$	L 3,a get a M 2,b multiply by b ST 3,$temp$ save partial result
a/b	L 2,a get a SRDA 2,32 shift a to G_3 filling G_2 with sign bit D 2,b divide by b ST 3,$temp$ save partial result
a rel b	L 3,a get a C 3,b compare with b BC m,*+10 branch if a rel b SR 3,3 set result to 0 (false) BC 15,*+8 skip next instruction L 3,=F'1' set result to 1 (true) ST 3,$temp$ save partial result where m is defined as

	rel	m
	=	8
	¬=	6
	<	4
	<=	12
	>	2
	>=	10

Source structure	Corresponding object code
$a = b$	L 3,b get b ST 3,a $a := b$

Figure 8.11 Source-object correspondence for assignment statements.

OP in this instruction depends on the operator op, as shown in the table. For example, the source expression

$$X - Q$$

causes the object code:

```
L    3,X          get X
S    3,Q          subtract Q
ST   3,temp_i     save partial result
```

to be generated.

The code sequences for multiply and divide are special because they each use a pair of registers. Recall that the resultant product of a multiply instruction is 64 bits in a pair of registers, in this case G_2 G_3. The multiplicand is originally in G_3, and the low-order part of the product (which is what we want for multiplication of integers) is also in G_3. However, the instruction addressing the multiplier references G_2. Division is the opposite. The dividend is 64 bits in a pair of registers, G_2 G_3. Since our integer values are only 32 bits, the actual value of the dividend must be put into G_3. G_2 is then filled with the sign bit of the dividend instead of zero because of the two's complement representation of negative integers. The resultant quotient is in G_3, and the instruction addressing the divisor references G_2.

The value of a relation is either 1, representing **true**, or 0, representing **false**. The code generated for all six of the relations is the same except for the third instruction. The mask m in this instruction depends on the relation rel, as shown in the table. For example, the source relation

$$R<W$$

causes the object code:

L	3,R	get R
C	3,W	compare with W
BC	4,* + 12	branch if R<W
SR	3,3	Set result to 0 (false)
BC	15,* + 8	skip next instruction
L	3, = 1	set result to 1 (true)
ST	3,temp_i	save partial result

to be generated.

8.4 Intermediate Forms

The purpose of the intermediate form is to represent explicitly the structure of the source procedure and its constituent parts. Whatever representation we choose will be equivalent to the tree diagrams that we have used earlier. There are several ways to represent such a tree structure. For example, suppose we have the tree shown in Figure 8.12. When the tree is an expression, each compound node has three branches in a downward direction. The leftmost branch is the left-hand operand, the middle branch is the operator, and the rightmost branch is the right-hand operand. The two forms of representation called *prefix* and *postfix* are similar. They differ only in the position of operators with respect to their operands. In both forms the two operands of a node are written in left-to-right order, as shown in Figure 8.13. In prefix form the operator immdiately precedes the left-hand operand, and in postfix form it

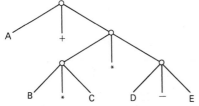

Figure 8.12 Tree form of A + B * C * (D − E).

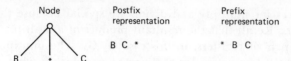

Figure 8.13 Postfix and prefix representation of a compound node.

immediately follows the right-hand operand. If an operand branch leads to a compound node then the complete description of that node is written as the operand before any following operand description or operator is written. For example, the prefix and postfix forms of the tree in Figure 8.12 are shown in Figure 8.14. The original expression is reproduced, and its operators are identified, with o_i the ith operator in the expression. The left-hand operand of operator o_i is identified by l_i and its right-hand operand by r_i.

The form that we will use is somewhat similar to the prefix form, but the structure is more explicit. Each compound node is numbered, and the node descriptions are written in sequence in a table called the *node table*. Each entry in the table describes one node. Figure 8.15 shows the tree in Figure 8.12 in

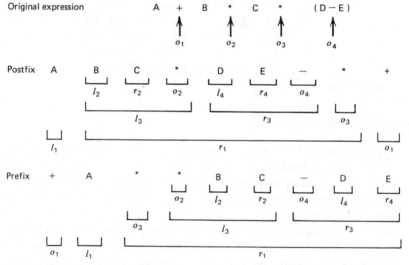

Figure 8.14 Postfix and prefix forms of example expression whose tree form is in Figure 8.12.

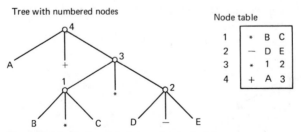

Figure 8.15 Node table form of tree in Figure 8.12.

node table form. In this form, as in the prefix form, the operator is written first, then the description of the two operands is written. However, when an operand branch leads to a compound node, only the number of that node is written. Thus, a node description consists of three elements, the operator, the name or node number of the left-hand operand, and the name or node number of the right-hand operand. An additional property that we will force this representation to have is that no node is referenced in a node description until after it is described. This is an ideal property, since it is exactly the order in which we want the generated object code to be executed.

8.5 Parsing Expressions

This section develops an algorithm that transforms a token string, representing an expression, into node table form. Figure 8.16 shows the productions that define the structure of an assignment. (We parse assignments like expressions by treating the $'='$ as an operator.) The first production says that the left operand of a $"+"$ or $"-"$ is some expression, while the right operand is some term (a-term). The second production says that the left operand of a $"*"$ or $"/"$ is some term, while the right operand is some primary (a-primary). The last production says that any expression enclosed in parentheses is a primary. The effect of these rules is to insure that an expression containing either a plus or minus sign cannot be the operand of a multiply or divide unless the plus or minus sign is enclosed in parentheses. The reason for these seemingly elaborate rules is to allow expressions to be written without a lot of parentheses. For example, the expression,

$$X = A + B^*C^*(D - E)$$

is structurally equivalent to the expression

$$(X = (A + ((B^*C)^*(D - E))))$$

The tree forms of these expressions are identical, since they contain no parentheses. The tree for these two expressions is shown in Figure 8.12.

The following four steps will transform any fully parenthesized expression (in token string form) into the node table representation of its structure as defined in Section 8.4.

1. Beginning at the left end of the expression scan from left to right until a right parenthesis is encountered. At this point we have located a group of five tokens having the form

    (left_operand   operator   right_operand)

    and we are looking at the right parenthesis.
2. Copy the three tokens between the left and right parentheses as the next entry in the node table. Let this be the *i*th entry in that list.
3. Replace the group of five tokens by a single token whose type is node table

---

assignment :: = a-variable = a-expression ;
a-expression :: = a-expression { + | − } a-term | a-term
a-term :: = a-term { * | / } a-primary | a-primary
a-primary :: = a-variable | a-constant | (a-expression)

Figure 8.16 Syntax definition of an arithmetic expression from Appendix C.

(N) and whose index part is the index of the entry just made in the node table, that is, the token ('N',i).

4. Repeat steps 1 to 3 until nothing remains but a single token.

Figure 8.17 shows the application of the preceding algorithm to the fully parenthesized expression

$$(X = (A + ((B^*C)^*(D - E))))$$

The corresponding tree and node table forms of this expression appear in Figure 8.15. Each line represents one repetition of steps 1 to 3 of the algorithm. The composition of the token string is shown both before step 2 and after step 3. The new entry, which is added to the node table in step 2, is shown in the rightmost column. A token of the form ('N',i) is represented as N_i and other tokens are represented by their identifier or operator. The small arrow indicates the point to which the scanning has progressed. Note that the scanning never has to begin again or back up. This is true because no new parentheses are introduced into the expression by the algorithm, and each repetition begins with step 1, which scans left to right until the first right parenthesis is encountered.

The problem with incompletely parenthesized expressions is knowing when to stop the left to right scan and make an entry in the node list. Consider the previous example with only the necessary parentheses:

$$X = A + B^*C^*(D - E)$$
$$\uparrow$$

We want to scan right until we reach the position indicated by the small arrow, then stop and enter the preceding three tokens in the node list. Why stop here? The syntax rules tell us that the right operand of a "*" must be a primary and that "C" is the primary immediately to the right of the "*". The effect of this syntax rule is to say that a sequence of multiplies is done in left to right order. Why not stop the scan sooner, say at "B"? If the scan was stopped at "B", then "A + B" is an expression, but the syntax rules say that the left operand of a multiply may be at most a term and not an expression. The effect of this syn-

Token string between step 1 and step 2	Token string after step 3	Node table entry made in step 2		
$(X = (A + ((B^*C)^*(D - E))))$ $\uparrow$	$(X = (A + (N_1^*(D - E))))$ $\uparrow$	1	*	B C
$(X = (A + (N_1^*(D - E))))$ $\uparrow$	$(X = (A + (N_1^*N_2)))$ $\uparrow$	2	–	D E
$(X = (A + (N_1^*N_2)))$ $\uparrow$	$(X = (A + N_3))$ $\uparrow$	3	*	N_1 N_2
$(X = (A + N_3))$ $\uparrow$	$(X = N_4)$ $\uparrow$	4	+	A N_3
$(X = N_4)$ $\uparrow$	N_5 $\uparrow$	5	=	X N_4

Figure 8.17 Parsing a fully parenthesized expression. Actual identifiers and punctuation symbols are used to represent the corresponding tokens, since they are more readable.

tax rule is to say that a multiply operator has *precedence* over an add operator, or that multiply has a higher *precedence* than add.

We make use of this concept of precedence to derive an algorithm that will parse incompletely parenthesized expressions as defined by the productions in Figure 8.16.

1. Scan from left to right until we find a pair of operator tokens (ignoring a single token between them) such that the precedence of the left operator of the pair is greater than or equal to the precedence of the right operator, or until we find a token for a right parenthesis. If the token immediately preceding the right parenthesis token is an operand token and the token immediately preceding that is a token for a left parenthesis, then the tokens for both the left and right parentheses are deleted and the scan continues. Otherwise proceed to step 2.
2. Copy the three tokens preceding the right operator of the pair or the right parenthesis into the node table as the ith entry.
3. Replace these three tokens with the token N_i.
4. Repeat steps 1 to 3 beginning with the test in step 1 until the test fails, then continue the left to right scan.

As the replacements are made, gaps begin to appear in the token string. These become a problem in actual implementation when it comes to rules such as "Copy the three tokens preceding, . . ." since an unknown number of gaps must be skipped over in order to find these three tokens. This problem is easily overcome by using a pushdown stack. Everything to the left of the current position in the left to right scan is transferred to a pushdown stack. Then when three tokens are to be replaced by a new one, the three tokens are popped off the stack, and the new token is pushed onto the stack. Thus, if there are n tokens in the token string immediately preceding the current scan position, they are always the top n items on the stack.

Figure 8.18 is the definition of an expression parser that uses the precedence of operators. The stack is implemented as an array S. The current contents of the stack is S_0 to S_j, where S_j is the top of the stack. The node table is named N and the next available position in the table is given by the value of the index i. In order that there always be at least one operator on the stack with which to compare a new operator from the token string input, it is initialized with a token corresponding to a left bracket operator $"|-"$, which has a lower precedence than any other operator in the language. The punctuation characters $"("$, $")"$, and $";"$ are also considered to have a precedence like the operators. The precedence ranking is shown in Figure 8.19.

A number of errors are detected in the algorithm. These are all due to an incorrectly formed expression. What to do in case of an error is often a difficult question to answer. Most certainly it should be flagged with an error comment. There is often a temptation on the part of the programmer to try to work out a way for the algorithm to proceed in the hope of being able to detect more errors. This is a very desirable course of action, if it is possible. The difficulty in proceeding is that additional errors may show up that are not genuine, that is, they occur as a result of trying to proceed. These errors may cause the user more confusion than the possibility of finding additional genuine errors is worth.

```
procedure parser;
  S[0] := token('|−');
  j := 0;
  i := 1;
  repeat j := j+1;
    lexical_analyzer (S[j]);
    if S[j] not a variable token, constant token, or token('(') then
      begin if S[j−2] not an operator token then Error return;
        while precedence of S[j−2] ⩾ precedence of S[j] do
          if S[j−1] and S[j−3] not both operand tokens then
            Error return;
          N[i] := ( S[j−2], S[j−3], S[j−1] );
          S[j−3] := token('N',i);
          S[j−2] := S[j];
          i := i+1;
          j := j−2;
        end;
        if S[j] = token(')') then
          begin if S[j−2] ≠ token('(') or S[j−1] not an operand token
                then Error return;
            S[j−2] := S[j−1];
            j := j−2;
          end;
      end;
  until S[j] = token (';');
  if j ≠ 2 then Error return;
  return;
end.
```

Figure 8.18 Definition of parser for arithmetic expression and assignment statements based on the precedence ranking in Figure 8.19.

```
*   /      highest precedence
+   −
=
)   ;
(
⊢             lowest precedence
```

Figure 8.19 Precedence ranking for arithmetic operators, the assignment operator, and punctuation symbols.

8.6 Generating Code To Evaluate Expressions

The final aspect of translation that we examine is code generation for arithmetic expressions. If the reader will return briefly to Section 8.3 he should be able to note the similarity between the source language structures given there and the entries in the node table that we are using to represent structure. The form $a\ op\ b$ given there corresponds to the form $op\ a\ b$ in the node table. The corresponding object code sequence for a given operator is fixed except for the addresses of *a* and *b*. The similarity between this object code pattern and a macro definition suggests that macro expansion is a promising avenue to explore in looking for a code generation algorithm.

Following this direction we will consider each entry in the node table as a macro instruction, in the sense of Chapter 5. For each different operator in the language we define a single macro. Code generation is then nothing more than macro expansion of each of the entries in the node table. We do not need the full generality of the macro facilities, which were discussed in Chapter 5. Each entry in the node table consists of a single operator and two operands. The operator determines which macro definition is to be expanded. The two operands are the parameter values to use while expanding the macro definition.

Just to make things quite simple, each macro definition that computes a result will store it in some temporary instead of leaving it in a register. Thus, there are actually three parameters that may occur in a macro definition: the left operand, the right operand, and a temporary. At most, one temporary is required for each entry in the node table. However, once its contents have been used, there is no reason that some other result cannot be stored in the same temporary. When the result stored in a temporary is used, it will be referenced with an N-type token appearing as an operand in some entry in the node table.

The macro definitions we need for the assignment statement in Figure 8.17 are add, multiply, subtract, and assignment. These macro definitions are shown in Figure 8.20. These definitions must be represented some way in Instran.

Operator	Macro definition			Internal representation
+	L	3,*left*	get left operand	1,1
	A	3,*right*	add right operand	3,2
	ST	3,*temp*	save partial result	2,0
				0,0
−	L	3,*left*	get left operand	1,1
	S	3,*right*	subtract right operand	4,2
	ST	3,*temp*	save partial result	2,0
				0,0
*	L	3,*left*	get left operand	1,1
	M	2,*right*	multiply right operand	5,2
	ST	3,*temp*	save partial result	2,0
				0,0
=	L	3,*right*	get right operand	1,1
	ST	3,*left*	store into right operand	2,2
				0,0

Figure 8.20 Macro definitions for add, subtract, multiply, and assignment operators and the internal representation of these definitions based on the instruction table in Figure 8.21.

An ordered pair of integers of the form (*op*, *par*) will be used to represent one line in the macro definition. The first integer *op* is the index in the instruction table of the machine instruction corresponding to the line in the macro definition that the pair represents. The instruction table required for our example is in Figure 8.21. The second integer *par* is the parameter that is the operand of the instruction: zero corresponding to a temporary, one to the left operand, and two to the right operand. A macro definition is terminated with a pair of zeros. This form of representation for each of the macro definitions in Figure 8.20 is written beside each macro definition. In this discussion of code generation for expressions, we are limiting ourselves to internal variables. Thus, we are able to put the base register number in with the numeric instruction code. Later we will deal with external variable references, which will require a more complicated algorithm than the one we are discussing here.

Figure 8.22 defines a simplified macro processor, which is used for code generation. The object code generated is still a list of token pairs, one for each machine instruction. The first token in the pair corresponds to the operation code and the second corresponds to the D field of the instruction. The R, X, and B fields of each instruction are included in the numeric operation code. The D field will be filled in by the assembly phase of the compiler. In this macro processor i indexes the current line in the node table, k the last machine instruction generated, and j the current line in the current macro definition. The macro definitions are stored in the array MD, the node table in the array N, and the generated machine instructions in the array C. The link between an operation token in the node table and its corresponding macro definition is accomplished via the operation table, which contains an entry for each operation. This entry is the index in MD of the first line in the corresponding macro definition.

8.7 A Complete Example

Figures 8.23 to 8.25 illustrate the parsing and code generation algorithms for arithmetic expressions, which have just been discussed. Figure 8.23 shows the contents of the major permanent tables used by Instran. These tables are never modified by the compiler.

The source language assignment statement being translated in this example is

$$X = A + B*C*(D - E)$$

	Numeric code	Operand type	Our 360 instruction	
1	58 30F000	X	L	Load
2	50 30F000	X	ST	Store
3	5A30F000	X	A	Add
4	5B30F000	X	S	Subtract
5	5C20F000	X	M	Multiply

Figure 8.21 Instruction table for example in Figure 8.17.

```
procedure code_generation;
  k := 0;
  i := 1;
  while i ⩽ length of N do
    opn := first element of N[i];
    left := second element of N[i];
    right := third element of N[i];
    j := operation_table[opn];
    while MD[j] ≠ (0,0) do
      op := first element of MD[j];
      type := second element of MD[j];
      if type = 0 then
        begin Search temporary_table for smallest n such that
                 temporary_table[n] = 0;
          opnd := ('T',n);
          temporary_table[n] := i;
        end;
        else if type = 1 then opnd := left;
              else opnd := right;
      if opnd = ('N',m) for some m then
        begin Search temporary_table for n such that
                 temporary_table[n] = m;
          opnd := ('T',n);
          temporary_table[n] := 0;
        end;
      k := k+1;
      C[k] := (('I',op), opnd);
      j := j+1;
    end;
    i := i+1;
  end;
  return;
end.
```

Figure 8.22 Definition of code generation procedure for expressions.

Operation table (R)
   Index of macro
   definition in MD   Operation

1	1	+	add
2	5	−	subtract
3	9	*	multiply
4	13	=	assignment

Macro definition table (MD)
   Internal
   form   Instruction

1	1,1	L	3,*left*
2	3,2	A	3,*right*
3	2,0	ST	3,*temp*
4	0,0		
5	1,1	L	3,*left*
6	4,2	S	3,*right*
7	2,0	ST	3,*temp*
8	0,0		
9	1,1	L	3,*left*
10	5,2	M	2,*right*
11	2,0	ST	3,*temp*
12	0,0		
13	1,2	L	3,*right*
14	2,1	ST	3,*left*
15	0,0		

Instruction table (I)

	Numeric code	Operand type	Instruction	
1	58 30F000	X	L	load
2	50 30F000	X	ST	store
3	5A30F000	X	A	add
4	5B30F000	X	S	subtract
5	5C20F000	X	M	multiply

Figure 8.23 Permanent tables used by Instran for translation of the assignment statement in Figures 8.24 and 8.25.

Symbol table (S)

1	entry for A
2	entry for B
3	entry for C
4	entry for D
5	entry for E
6	entry for X

Node table (N)

	Operation	Left operand	Right operand			
1	R,3	S,2	S,3	*	B	C
2	R,2	S,4	S,5	−	D	E
3	R,3	N,1	N,2	*	N_1	N_2
4	R,1	S,1	N,3	+	A	N_3
5	R,4	S,6	N,4	=	X	N_4

Figure 8.24 Output from the parser when the source input is the assignment statement X = A + B * C * (D − E).

Figure 8.24 shows the output from the parser defined in Figure 8.18, given this statement as input. Figure 8.25 shows the output from the code generator defined in Figure 8.22, given the node table in Figure 8.24 as input. In the rightmost column of Figure 8.25 we have written the assembly language form of the machine language instruction, which will ultimately be assembled from each entry in the code table.

Code table (C)

	Instruction	Operand	Assembly language
1	I,1	S,2	L 3,B(,15)
2	I,5	S,3	M 2,C(,15)
3	I,2	T,1	ST 3,t_1(,15)
4	I,1	S,4	L 3,D(,15)
5	I,4	S,5	S 3,E(,15)
6	I,2	T,2	ST 3,t_2(,15)
7	I,1	T,1	L 3,t_1(,15)
8	I,5	T,2	M 2,t_2(,15)
9	I,2	T,1	ST 3,t_1(,15)
10	I,1	S,1	L 3,A(,15)
11	I,3	T,1	A 3,t_1(,15)
12	I,2	T,1	ST 3,t_1(,15)
13	I,1	T,1	L 3,t_1(,15)
14	I,2	S,6	ST 3,X(,15)

Figure 8.25 Output from code generation for the assignment statement in Figure 8.24.

EXERCISES

8.1 Show the contents of the node table, the symbol table, and the constant table that result from applying the parser in Figure 8.18 to the expression:
$$W = W + 2*(Z - A*(B + 18) - 20*(C + D))*B$$
Assume the permanent tables are as shown in Figure 8.23.

8.2 Show the code generated for the example in Exercise 8.1 using the code generator in Figure 8.22.

9 | Parsing

In Chapter 8 we explored a simple parsing algorithm that is quite well suited for parsing expressions. However, it is less suitable for parsing other constructions in programming languages such as the Instran Language. In this chapter we will discuss more general parsing methods and describe in detail one that will be used in Instran.

The principal goal of parsing statements and expressions is to represent explicitly the structure of the statement or expression. The node table defined in Chapter 8 is such a representation of structure. The structure is explicit in this representation, since an operation and each of its operands, or pointers to them, are all recorded as part of the same entry in the node table. Statements and expressions ultimately cause the generation of executable object code, and the explicit association of each operation with its operands is needed to generate code.

Declarations are different. Their major purpose is to give information, such as the attributes of a variable, to the compiler. The main response to a declaration by a compiler is to enter the information that it contains into the appropriate tables. The parsing of declarations does not ordinarily result in entries being made in the node table.

In Chapter 7 we discussed the distinction between generation and recognition with respect to sentences in a language. The productions that define the syntax of a language are used to generate sentences, that is, to *produce* sentences. Recognition is the act of deciding if a given string is a sentence. This is accomplished by *reducing* the string to the sentence symbol. Production of a sentence begins with the sentence symbol. In sentence generation nonterminals are replaced by any alternate from the definition part of their defining productions until no nonterminals remain. Reduction is the reverse process. Beginning with a string of terminals, *phrases* are replaced by nonterminals until only the sentence symbol remains. A *phrase* is defined as any string of terminals and nonterminals that is some alternate appearing in the definition part of some production. The phrase is replaced by the nonterminal that is defined by the production in which the phrase appears as an alternate. Reduction is similar to parsing.

A *parse tree* is a representation of the structure of some legal string in a language. It illustrates both production and reduction. Figure 9.1 shows the parse tree for the assignment statement

$$X = A * B + C;$$

as defined by the Instran Language productions in Appendix C. Each node of the tree is labeled with a terminal or nonterminal. Production proceeds from top to bottom in the diagram. A production step consists of replacing the non-terminal labeling a node by the phrase indicated by all of the branches that emanate from the node in a downward direction. For example, *assignment* is replaced by

$$variable = expression;$$

and *expression* is replaced by *a-expression*. Reduction proceeds in the opposite direction, from bottom to top. A phrase is reduced by replacing it with the nonterminal from which all downward branches lead to the elements in the phrase. For example, "X" is reduced to *identifier* and

$$a\text{-}term * a\text{-}primary$$

is reduced to *a-term*.

Pure reduction does not generate an explicit representation of the structure. However, by associating some action with each reduction step, we can generate such a representation. This coupling of reduction with actions we will call *parsing*. In terms of the node table of Chapter 8, each nonterminal node in the parse tree corresponds to an entry in the node table. The action that we associate with reduction of a phrase is to combine all of the components of that phrase into a single entry in the node table. However, in a compiler we do not need the complete parse tree. For example, in Figure 9.1 we do not need an entry in the node table for any of the parse tree nodes that have a single

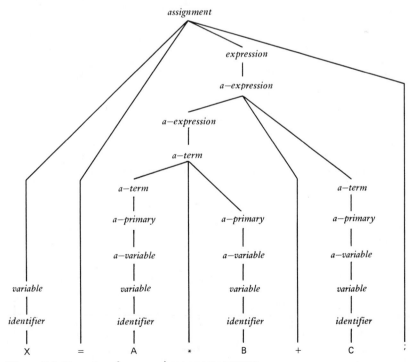

Figure 9.1 Parse tree for an assignment statement.

branch in and a single branch out such as, the nodes labeled *variable* and *a-variable*. The node list need only contain a skeleton of the parse tree, namely, only the nodes that have more than one branch emanating downward. For example, the node table entries for the example in Figure 9.1, as specified in Chapter 8, is shown in Figure 9.2 along with its skeleton parse tree.

In Instran not all of the actions associated with a reduction will make an entry in the node table. Some of the actions will enter information into other tables or perform some other required task. An example of this is the parsing of declarations. The Instran Language productions defining the declaration of a simple variable are:

$$declaration ::= \text{DECLARE } \{simple\text{-}variable \mid structure\};$$
$$simple\text{-}variable ::= identifier\ [dimension]\ type\ storage$$

Figure 9.3 shows the parse tree for

DECLARE X FIXED EXTERNAL;

based on these productions. If, with the reduction of the phrase

identifier type storage

we associate the action of making an entry in the symbol table for the variable whose name is X and whose attributes are FIXED and EXTERNAL, then the entire parse tree may be discarded. Keeping attribute information in an unstructured table like the symbol table is not adequate for more complex languages such as PL/1. At least one compiler for PL/1 keeps attribute infor-

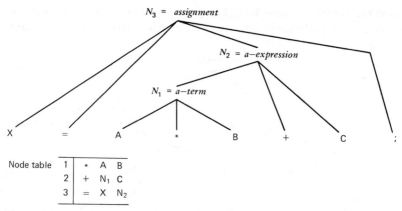

Figure 9.2 Equivalence of node table and skeleton parse tree.

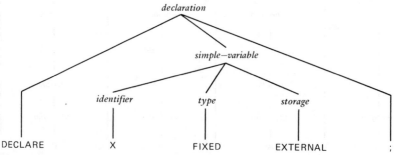

Figure 9.3 Parse tree for a declaration.

mation in a tree form.[1] However, in that case the tree form is not a parse tree, but a tree form representation of the procedure that is substantially modified from the parse tree.

9.1 Parsing Methods

There are many different methods for parsing a sentence in a programming language. These methods can be grouped into several major classes of methods. The methods in a class are all variants of the same basic method. It is not within the scope of this text even to survey all of the basic methods. Most of the basic parsing methods are based on some property of the set of productions that define the language's syntax. This set of productions is called the *grammar* for the language.

For example, the method used in Chapter 8 for parsing expressions was based on the idea of the precedence of the operators in an expression. This method requires that the grammar for expressions be an *operator precedence grammar. Operator precedence* is a property of grammars, which has nothing to do with the meaning of any of the symbols in the language. It is a property that characterizes the structure of sentences generated by the productions in the grammar. This property is quite restrictive, and many grammars do not have this property. Since the property is independent of the meaning of the symbols in the language, it can be extended to the parsing of statements and declarations as well as expressions. There are several variations of the operator precedence property, some of which are more restrictive and some of which are less restrictive. All of these variations can be parsed by essentially the same method as was used in Chapter 8.

Other more general methods exist that can be used to parse a language defined by any grammar, independent of its properties, as long as the grammar can be written as a set of productions of the form that we have been using, namely, a single nonterminal on the left and a number of alternates on the right. The syntax of many of the common programming languages cannot be defined by a set of productions of the form we are using. However, very minor changes in their grammar and a little fudging in the parser will be sufficient to allow us to use any of the more general parsing methods.

Most parsing methods can be grouped into two major classes, *top-down* and *bottom-up*. Precedence parsing is essentially a bottom-up method. A bottom-up parser looks at something in the input and says, "I have this, now what can it possibly be a part of? What can I make out of it?" In the precedence method of parsing whenever a new input token is introduced there are basically only two answers; it is not part of anything yet, so keep on going, or else it is part of a phrase that is to be reduced, and we know exactly what the phrase is. More general bottom-up methods exist and some of them are *nondeterministic*, that is, there may be several phrases that a new input token may be a part of, and there is no way of telling which is the correct one. Hence, all of the possibilities must be tried in order to determine ultimately which is the correct phrase.

Top-down parsers are *predictive*. A top-down parser has a *goal* that it is try-

---

[1] R. A. Freiburghouse, "The Multics PL/1 Compiler," *Proceedings of AFIPS 1969 FJCC, 35,* AFIPS Press, Montvale, N.J., pp. 187–199.

ing to satisfy. It looks at the input and tries to find a way to group the tokens that satisfies this goal. The goal is a prediction, and the parser is trying to confirm this prediction. The initial goal is the sentence symbol. The parser tries to find a phrase that is the first alternate in the definition part of the production defining the sentence symbol. This, in turn, may generate further subgoals, one for each nonterminal in the alternate. When a phrase is found that satisfies a goal or subgoal, it is reduced. Subgoals are generated and satisfied until the initial goal is satisfied, at which point the sentence has been parsed. Top-down parsing may also be nondeterministic, that is, the initial part of the input may satisfy the first part of several different subgoals, but the later part of the input will satisfy the later part of only one of the subgoals.

When a parser is nondeterministic, *backup* or *backtracking* is required. Since we are interested in the parse tree of a sentence being parsed, we assume that each time a reduction is made, part of the parse tree is generated. Because of the nondeterminism we may have made a sequence of reductions only to find that we can go no further. When this happens we have to back up in the sequence of reductions until we reach a point where there was more than one alternate to try and try the next alternate. All of the parse that was generated from this point until the point where we are blocked will have to be deleted, since it is not valid. If it were valid we would not have been blocked.

Figure 9.4 defines a very simple grammar used in the following examples. The figure actually shows two different grammars. However, they are equivalent, that is, they both define exactly the same language. The second grammar violates the constraint that a nonterminal appear on the left of one and only one production. As we mentioned, that is an unnecessary constraint. The derivation of the second form from the first is obvious. In the second form the definition part never contains more than one alternate. The first form is most useful for top-down parsing, while the second form is much more useful for bottom-up parsing.

Figure 9.5 illustrates top-down parsing and Figure 9.6 illustrates bottom-up parsing. Each figure shows how the parse tree for

$$A = B * A ;$$

a   Grammar for top-down

$a ::= i = e ;$
$e ::= i + i \mid t$
$t ::= i * t \mid i$
$i ::= A \mid B$

b   Equivalent grammar for bottom-up

$a ::= i = e ;$
$e ::= i + i$
$e ::= t$
$t ::= i * t$
$t ::= i$
$i ::= A$
$i ::= B$

Figure 9.4 Two equivalent grammars for use in Figures 9.5 and 9.6.

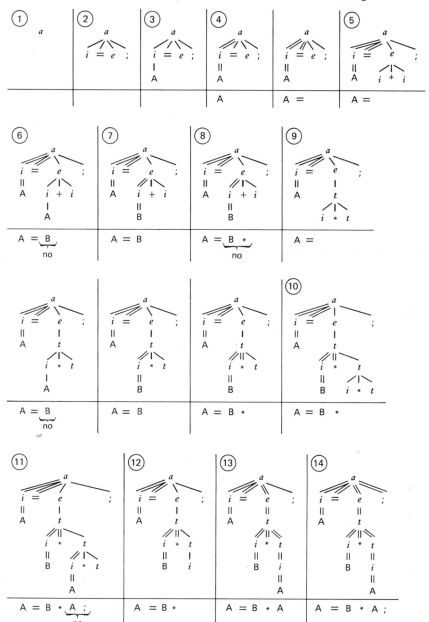

Figure 9.5 Example of top-down parsing.

which is a sentence in the language defined by the grammars in Figure 9.4, is derived by their respective methods of parsing. Selected steps in the generation of the parse tree are shown. In each of the diagrams the current, partial parse tree is shown by heavy line branches. The light line branches are possible parse tree branches not yet verified.

In top-down parsing (Figure 9.5) the initial goal is the sentence symbol a (diagram ①). Its definition is

$$a ::= i = e ;$$

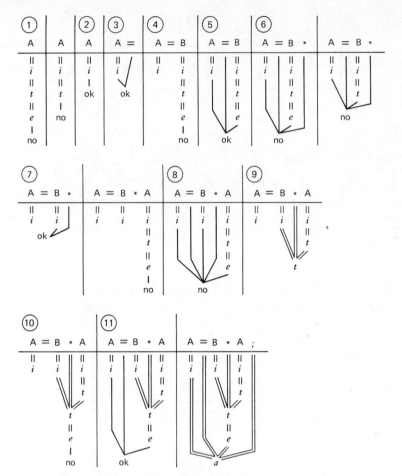

Figure 9.6 Example of bottom-up parsing.

therefore the goal a is replaced by the four subgoals i, $"="$, e, and $";"$ (diagram ②). The first of these subgoals is considered first. The definition of i is

$$i ::= A \mid B$$

therefore the first alternate $"A"$ becomes the new subgoal (diagram ③). Since this is a terminal, the new subgoal is satisfied only if it matches the next character in the input (diagram ④). The characters below the horizontal line are the input characters that have been $"used"$ so far. Since $"A"$ is the next input character, it is reduced to i.

The subgoal e is defined as

$$e ::= i + i \mid t$$

so the first alternate is tried (diagram ⑤). In diagram ⑥ the first alternate in the definition of i is $"A"$, which does not match the next input character. This failure causes backup. The backup implied by diagram ⑥ is to try the second alternate in the definition of i (diagram ⑦). Thus, the first element i, of the first alternate in the definition of e, is matched.

However, the second element does not match (diagram ⑧). Therefore, backup again occurs and the second alternate in the definition of e must be tried. This is just t (diagram ⑨), which is defined

$$t ::= i * t \mid i$$

After matching the first two elements in the first alternate, we again need to try for a t (diagram ⑩). The first alternate of t is tried, but the second element fails to match (diagram ⑪), so we back up and try the second alternate (diagram ⑫). Then "A" is reduced to i and i to t. The term

$$i * t$$

is then reduced to t, which in turn is reduced to e. The first three elements in the definition of a are now matched (diagram ⑬). Finally, the fourth element is matched, and

$$i = e ;$$

is reduced to a and we are finished (diagram ⑭).

In bottom-up parsing (Figure 9.6) we try to build up to the sentence symbol. Our strategy is always to reduce as far as possible. Then if that does not work, back up. When attempting to reduce, we search the definition parts of the productions in Figure 9.4b, looking for one that matches the partially reduced input. If we get a match, the phrase is reduced, and we try again for another reduction.

We begin by reading the first input character. In diagram ① we are able to reduce the letter "A" all the way to e, which cannot be reduced further. Here, the input characters that have been "used" are shown above the horizontal line. Since we are unable to reduce e, we check to see if e is the first element in some definition. If not we back up, undoing reductions until we reach a state where the nonterminal is the first element of a definition (diagram ②). Then we read the next input character.

In diagram ③ the partially reduced input

$$i =$$

matches the first two elements in the definition

$$a ::= i = e ;$$

so we read the next input character. This is then reduced as far as possible (diagram ④). Since it is not the first symbol in the partially reduced input, not only do we search to see if e is the first element in a definition, we also try to match

$$= e$$

with the first two elements of some definition and, that failing, succeed in matching

$$i = e$$

with the first three elements of the definition

$$a ::= i = e ;$$

(diagram ⑤).

However, when we read the next input character "*", none of

$$e *$$
$$= e *$$
$$i = e *$$

match the initial part of any definition (diagram ⑥). Therefore, we back up. This backup returns e to i (diagram ⑦), since none of

$$t$$
$$= t$$
$$i = t$$

match the first part of any definition.

In diagram ⑧ nothing can be matched to the initial part of any definition. When e is backed up to t, the phrase

$$i * t$$

can be reduced to t (diagram ⑨). This t is in turn reduced to e (diagram ⑩). In diagram ⑪ we find that

$$i = e$$

matches the initial part of the definition

$$a ::= i = e ;$$

and so the next input character is read. Now

$$i = e ;$$

can be reduced to a, and we are finished.

We have offered the two preceding examples to give the reader the flavor of the two major classes of parsing algorithms. Either class of algorithm can be made to work for an extremely large class of languages, but there are a number of nontrivial problems that we have not exposed. A discussion of these problems is not appropriate for this text.[2] In the next section we will explore in detail a deterministic, bottom-up algorithm for parsing, which is used in Instran. This algorithm uses a set of *reduction rules* that are derived from the grammar instead of using the actual productions. When the actual productions in the grammar are used, as in the preceding two examples, we say that the parser is *syntax driven*. Such parsers play an important role in compiler technology, but further discussion is beyond the scope of this text.

9.2 Parsing by Reduction Rules

Parsing in Instran is done using a set of *reduction rules*. The parsing method that uses reduction rules is a bottom-up method that is deterministic, that is, no backup is needed. A reduction rule is basically an inverted production, which we can write in the following form:

phrase → nonterminal

Application of a reduction rule consists of replacing the phrase by the non-terminal. Use of reduction rules for parsing is bottom-up, since a phrase in the partially reduced input is reduced to a single nonterminal. The basis of parsing using reduction rules is to reduce repeatedly phrases according to the reduction rules until the entire input is completely reduced to the single nonterminal that is the sentence symbol.

As in the general case of bottom-up parsing, the use of reduction rules for arbitrary grammars will be nondeterministic, that is, the parser will run into blind alleys and will have to back up and try some other sequence of reductions. However, for the languages that we are interested in, a set of reductions can be derived that enables the parser to be deterministic. In implementing parsing by reduction rules a stack will be used for the partially reduced input

---

[2] D. Gries, *Compiler Construction for Digital Computers*, Wiley, N.Y., 1971.

in order to eliminate the problems caused by gaps, in much the same way the stack was used in parsing expressions in Chapter 8.

We can view a set of reduction rules as a program, with the rules being interpreted as statements in the program. A reduction rule will have the form

$$label: phrase \rightarrow replacement \; \# \; @next\_label$$

where the number sign "#" is optional. When a reduction rule is "executed" the elements at the top of the stack are examined to see if they match *phrase*. If they do the elements matching *phrase* are popped off the stack, and the elements in *replacement* are pushed onto the stack. Then, if the reduction contains "#", the next input character is read and pushed onto the top of the stack. After this the reduction rule labeled *next_label* is executed. If the elements at the top of the stack do not match *phrase* then execution of this reduction rule immediately terminates, and the reduction rule sequentially following this one is executed.

For any given configuration of partially reduced input only a few different reductions are possible, and only these need be considered. For example, in Figure 9.6 when "*" has been read from the input in diagram ⑥, the only reduction which is possible is

$$\cdot \; * \; t \rightarrow t$$

as can be seen in the grammar in Figure 9.4*b*. Therefore, there is no reason to search for possible reductions at this point. The parser should immediately read the next input character. We use this fact and organize the reduction rules into groups. Each group is the set of all possible reductions for some particular configuration of partially reduced input. At any given point in the reduction of the input, only the set of reduction rules in the group corresponding to the current configuration of the partially reduced input are executed. If none of them match, the input is not a valid sentence in the language.

In order to illustrate how reduction rules are adapted for use in parsing in a compiler and how they are used to reduce a sentence, we will use the grammar for simple assignments from Figure 7.1, which is rewritten in Figure 9.7 in a form that is more suitable for use in deriving reduction rules. The reduction rules derived from this grammar for reducing a sentence to the sentence symbol *a* are shown in Figure 9.8. The symbol "σ" used in these reduction rules is a symbol that will match any single element on the stack.

Figure 9.9 shows how the reduction rules are used to reduce the assignment

$$A = B + C * A;$$

We begin by reading the first input character onto the stack and then execut-

$a ::= i = e \; ;$

$e ::= e + t$

$e ::= t$ $a \equiv$ assignment

$t ::= t * i$ $e \equiv$ expression

$t ::= i$ $t \equiv$ term

$i ::= A$ $i \equiv$ identifier

$i ::= B$

$i ::= C$

Figure 9.7 Grammar from Figure 7.1 rewritten for use in deriving reduction rules.

L:	$A \rightarrow i$	#	@I
L1:	$B \rightarrow i$	#	@I
L2:	$C \rightarrow i$	#	@I
L3:			@Error

I:	$i = \rightarrow$	#	@L
I1:	$t * i\,\sigma \rightarrow t\,\sigma$		@T
I2:	$i\,\sigma \rightarrow t\,\sigma$		@T

T:	$t * \rightarrow$	#	@L
T1:	$e + t\,\sigma \rightarrow e\,\sigma$		@E
T2:	$t\,\sigma \rightarrow e\,\sigma$		@E

E:	$e + \rightarrow$	#	@L
E1:	$i = e ; \rightarrow a$		@Done
E2:			@Error

Figure 9.8 Reduction rules derived from grammar in Figure 9.7.

Input	Stack	Applicable reduction
A = B + C * A ; ↑	A	L
A = B + C * A ; ↑	$i =$	I
A = B + C * A ; ↑	$i = B$	L1
A = B + C * A ; ↑	$i = i +$	I2
	$i = t +$	T2
	$i = e +$	E
A = B + C * A ; ↑	$i = e + C$	L2
A = B + C * A ; ↑	$i = e + i *$	I2
	$i = e + t *$	T
A = B + C * A ; ↑	$i = e + t * A$	L
A = B + C * A ; ↑	$i = e + t * i ;$	I1
	$i = e + t ;$	T1
	$i = e ;$	E1
	a	Done

Figure 9.9 Example of the use of reduction rules to reduce a sentence in the language defined in Figure 9.7.

ing the first reduction rule, which is labeled L. In the example the current position in the input is marked with a small arrow. Each line in the example corresponds to the reduction rule whose phrase part matched the top of the stack. The applicable reduction is identified by its label, and the complete contents of the stack at the time the match occurred is shown in each line along with the current position in the input. Reductions that were tried but did not match are omitted.

Derivation of the reduction rules for this simple grammar is fairly easy:

L	corresponds to	$i ::= A$
L1	corresponds to	$i ::= B$
L2	corresponds to	$i ::= C$
I1	corresponds to	$t ::= t * i$
I2	corresponds to	$t ::= i$
T1	corresponds to	$e ::= e + t$
T2	corresponds to	$e ::= t$
E1	corresponds to	$a ::= i = e ;$

Additional reductions are required to catch errors and to prevent premature reduction of a symbol that can be a phrase by itself, but because of its context is actually part of a larger phrase. For example, if i is followed by $"="$, then i must not be reduced to t. If i in this context is reduced to t, then eventually backup will be required in order to finally match

$$i = e ;$$

and thus effect the final reduction to a. Therefore, we add reduction rules like I, T, and E to prevent any reduction that will lead to the need for backup.

To prevent unnecessary testing of phrases for a match with the top of the stack, the reductions are grouped according to the nonterminal, which occurs in the second stack position (next to the top of the stack). For example, all of the reductions whose phrase part has the nonterminal t in the second stack position are grouped together and labeled T, T1, and T2. Then, the *next_label* part in each reduction is chosen so that control is transferred to the beginning of the group for the nonterminal known to be in the second stack position. For example, all of the reduction rules L, L1, and L2 transfer to the group beginning with I. Remember that the transfer will take place only if a rule's phrase part matched the stack. In L, L1, and L2, if the phrase matched, it is reduced to i, and then the next input character is read and pushed onto the top of the stack. Therefore, we know that i will be in the second stack position. Thus, control is transferred to I, which is the first reduction rule in the group for i in the second stack position.

Within a group the order of the rules is important, and additional rules may be needed to prevent premature reduction. If the nonterminal in the second stack position is part of a phrase that is not yet completely on the stack, this must be detected and no reduction made. For example, in reduction T, if the nonterminal t is actually part of the phrase

$$t * i$$

then a reduction cannot yet be made, since only the symbols t and $"*"$ will be on the stack. If T does not match then the nonterminal t is part of a phrase that is entirely on the stack. If there is more than one phrase to which a nonterminal could belong, the order of testing must be from most restrictive to least restrictive. For example, the phrase

$$e + t\,\sigma$$

is tested for in rule T1 before the phrase $t\,\sigma$ is tested for in rule T2. The test in reduction T2 will always succeed, since the T group was tried only when it was known that the nonterminal t was in the second stack position.

When a reduction cannot be made, as in reductions I, T, and E, the next character in the input is introduced onto the top of the stack. We know from examining the productions that the only character that may legally follow a "=", "*", or "+" is a letter. Thus, each of the rules I, T, and E transfers to the group of rules that tests for a letter on the top of the stack. If a letter is found it is reduced to the nonterminal i. If no letter is found the input is not a legal sentence, and an error is signaled.

In an actual compiler we need more than just the recognition of legal sentences. Some actions must be coupled with the reduction rules in order to transform the input into its parse tree form. In Instran this form is the *node table*. The kinds of actions that we will need in order to accomplish this transformation include adding entries to the node table, adding information to the compiler's tables, and even modifying the contents of the stack. Therefore, we extend the form of a reduction rule to include one or more *actions*. We also make invocation of the lexical analyzer an explicit action instead of the special symbol "#". The new form for a reduction rule is:

$$label:\ phrase \rightarrow replacement\ ;\ actions\ @next\_label$$

Interpretation of the new form of reduction rule is the same as for the old form, except that in addition the actions get performed *after* the replacement has been made. The phrase that was replaced is made available to the actions.

Figure 9.10 shows the reduction rules from Figure 9.8 rewritten in the new form. We have also added an initial rule S to get started. There are only two actions needed to transform an assignment into its corresponding node table form. The action **lex** is the lexical analyzer. Each time it is performed, it puts

```
S:                          ; lex
L:          A → i           ; lex    @I
L1:         B → i           ; lex    @I
L2:         C → i           ; lex    @I
L3:                                  @Error
I:          i = →           ; lex    @L
I1:   t * i σ → t σ         ; enter  @T
I2:       i σ → t σ         ;
T:          t * →           ; lex    @L
T1:   e + t σ → e σ         ; enter  @E
T2:       t σ → e σ         ;
E:          e + →           ; lex    @L
E1:   i = e ;  → a          ; enter  @Done
E2:                                  @Error
```

Figure 9.10 Reduction rules from Figure 9.8 rewritten in a form for use in building the parse tree in a compiler.

a token for the next basic element from the input onto the top of the stack.

The action **enter** makes an entry in the node table. The phrase to be entered is composed of the three tokens that *were* the second, third, and fourth items on the stack. Since the replacement is done before the actions are performed, the elements in the phrase will be lost unless they are saved before the replacement takes place. The matched phrase will be saved in an array called P. The first element of the array, P_0, contains the item from the top of the stack, the second element of the array, P_1, containis the second item from the stack, and so on for all of the items on the stack that matched the phrase part of the reduction. The action **enter** will use the contents of P_1, P_2, and P_3 to add a new entry to the node table.

For example, suppose the stack contains
$$i = t * i +$$
and reduction I1 is applicable. Then the items on the stack that match the phrase part of reduction I1 are:
$$t * i +$$
When the action **enter** is performed, which is after the replacement, the stack will contain
$$i = t +$$
and the array P will contain
$$+ i * t$$
that is, P_0 contains $"+"$, P_1 contains i, and so forth. The action **enter** then will add a new entry to the node table consisting of
$$* \quad t \quad i$$
which are found in P_2, P_3, and P_1, respectively.

The reader may have already noticed that there is a problem in what the action **enter** has done so far. In the completed node table we need the connections between the various nodes. The node table for
$$A = B + C * D ;$$
that would be generated by the **enter** action described above, is shown in Figure 9.11. It is clear that this node table is not much help when trying to generate code from it. The missing information is that the i's should point to symbol table entries and the t's and e's should point to other entries in the node table.

We saw earlier that during code generation it was not necessary to know if an operand is an e or a t; all that is required is to know which node is the operand. However, the type of node is required for matching when parsing using reduction rules. Thus, the tokens used on the stack must be triples, containing a syntactic class as well as a type (table code) and index. If the form of a token is
$$(\text{type, index, syntactic class})$$
and the second entry in the symbol table is the entry for a variable named B, then the tokens
$$('S', 2, i)$$
$$('S', 2, t)$$
$$('S', 2, e)$$
all represent the same variable name. In the first case it is syntactically an identifier, in the second case it is syntactically a term, and in the third case it

1	*	i	i
2	+	i	t
3	=	i	e

Figure 9.11 Useless node table generated by first version of **enter** action.

Input	Stack	Applicable reduction	Node table entry
A = B + C * A ;	i_A	L	
A = B + C * A ;	$i_A =$	I	
A = B + C * A ;	$i_A = i_B$	L1	
A = B + C * A ;	$i_A = i_B +$	I2	
	$i_A = t_B +$	T2	
	$i_A = e_B +$	E	
A = B + C * A ;	$i_A = e_B + i_C$	L2	
A = B + C * A ;	$i_A = e_B + i_C *$	I2	
	$i_A = e_B + t_C *$	T	
A = B + C * A ;	$i_A = e_B + t_C * i_A$	L	
A = B + C * A ;	$i_A = e_B + t_C * i_A ;$	I1	1 * C A
	$i_A = e_B + t_1 ;$	T1	2 + B N_1
	$i_A = e_2 ;$	E1	3 = A N_2
	a_3	Done	

Figure 9.12 Example from Figure 9.9 reduced using reduction rules with actions that build a parse tree in the node table form.

is syntactically an expression. This sequence of transformations in a token representing the identifier B would occur when the assignment

$$A = B;$$

is parsed using the reduction rules in Figure 9.10.

A modification is required in the interpretation of reductions as well as in the action **enter** to accommodate this new token. The matching is done considering only the syntactic class, the third element in the triple. When the reduction replacement simply changes the syntactic class, as in reductions I2 and T2, the remainder of the token is left unchanged. When a single item replaces several, as in reductions I1, T1, and E1, the replacement token contains the specified syntactic type, but the other two components of the token are left empty. The action **enter** must then fill in the empty components of the token. The token type will be set to 'N', indicating node table entry. The index will be set to the index in the node table of the new entry added by the **enter** action.

In the example that follows we will use a shorthand notation for tokens; this will make the example more readable. When a token points to a symbol table entry, the variable's name will be used as a subscript on the syntactic class. For example, the shorthand symbols for $('S',2,i)$, $('S',2,t)$, and $('S',2,e)$ are i_B, t_B, and e_B, respectively. When the token points to a node table entry, the index of that entry will be used as a subscript on the syntactic class. For example, the shorthand symbol for $('N',j,t)$ is t_j. Figure 9.12 shows the same example as Figure 9.9, but it uses the new form of reduction rule and therefore generates the node table form of the parse tree.

We will use reduction rules of the form just discussed for parsing in Instran. The rules will necessarily be somewhat more complicated than those required for the simple assignment statements, which we have used as examples in this section. However, the basic idea and mechanics will be the same.

EXERCISES

9.1 Using the reduction rules in Figure 9.10, reduce the following assignments, building their parse trees in node table form similar to the example in Figure 9.12. Assume an additional rule for each new letter.

   a.   A = B + C + D;

   b.   X = Y*Z*W;

   c.   P = Q*R + S*T*V + U;

9.2 Derive a set of reduction rules similar to those in Figure 9.10 for the grammar in Figure 8.16. Assume the sentence symbol is 'assignment'.

9.3 Using the reduction rules from Exercise 9.2, reduce the following assignments, building their parse trees in node table form similar to the example in Figure 9.12.

   a.   A = B*(C − D)/E;

   b.   X = A + (B − F*(G + H/(R + S)))*Y;

10 | Pass One of Instran

In this chapter we will explore the structure and functions of pass one of Instran in more detail. We saw in Chapter 8 that the main task of pass one is to parse the source procedure, transforming it into an internal form. The internal form that we use for Instran is the node table form. In Chapter 9 we saw how reduction rules can be used for parsing. Parsing in pass one of Instran will be based on reduction rules. So far we have considered only the translation of simple assignment statements and expressions. Now we will consider translation of the complete Instran Language. Even though this increases the complexity of parsing, the reduction rules that we have been using are still satisfactory. All of the additional complexity is reflected in a substantial increase in the number of both reduction rules and actions.

We also find a corresponding increase in the number of different kinds of entries needed in the node table. In Chapter 8 we interpreted an entry in the node table as a macro instruction, and we based object code generation on macro expansion. Therefore, the number of different macro instructions that can be used will be increased. Along with this increase we will abandon the goal of representing the complete syntactic structure of the source program in the internal representation. That is, the node list will not be a tree form of the entire source procedure, only expressions will continue to be represented in tree form. This is adequate and satisfactory for relatively simple languages such as the Instran Language. For a complex language such as full PL/1 a tree form representation of the complete syntactic structure is virtually essential. The Multics PL/1 compiler uses a tree representation for the entire source procedure.

We begin our consideration of pass one by discussing implementation of a parser based on reduction rules and summarizing the principal tables and other data needed. This is followed by a discussion of the parsing, and accompanying actions, of the major Instran Language features. These are the definition of procedures, including the PROCEDURE, RETURN, and END statements; declarations for simple variables, arrays, structures, and labels; the IF statement, including the DO group; the CALL statement; and expressions modified to include structure and array references, procedure calls, and relations for use in IF statements.

10.1 Implementation of a Parser Based on Reduction Rules

The format of the reduction rules used in Instran are basically the same as described at the end of Chapter 9. This format is:

label: *phrase* → *replacement*; *actions* @*next_label*

There are a few minor differences in the notation for writing *phrase* and *replacement*. In *phrase*, nonterminals are written in lowercase italic letters, key words are written in uppercase letters, and operators and punctuation symbols are represented by the actual characters. In order to reduce the number of rules that must be written, several auxiliary nonterminals are defined in the final section of the Instran Language grammar in Appendix C. For example, the nonterminal *any* is defined to be any terminal or nonterminal. Thus, when *any* appears in *phrase*, it will match any token whatsoever.

If *replacement* is omitted then the matched phrase is not deleted or replaced, that is, the stack is not changed. If *replacement* is the single character "#" the matched phrase is deleted and nothing replaces it. Otherwise, *replacement* consists of a sequence of P_i's. In this case the matched phrase is replaced by the sequence of tokens, which is specified by the sequence of P_i's. Each P_i corresponds to one token, namely the ith token in the matched phrase where P_0 is the token at the top of the stack. This notation should be clear if the reader recalls from Chapter 9 that the tokens on the stack that matched *phrase* in a reduction rule were stored in the array P after the match was successfully made.

The actions associated with a rule are specified by *actions*, which is a sequence of zero or more procedure names. Each name is the name of an action procedure. These procedures are executed in the sequence in which their names are written, after the specified replacement has taken place. The final part of a rule @*next_label* specifies the next rule to be executed. If this is the next rule in sequence then @*next_label* may be omitted.

There are basically two ways of implementing a parser that uses reduction rules. One way is to write, as part of the compiler, an interpreter for the reduction rule language. The other way is to translate the reduction rules into some other language, most likely the same language in which the rest of the compiler is written. The choice of which to use depends on many things, but probably the most significant issues are the amount of change expected in the reduction rules, that is, how often the reductions will be modified, replaced, or deleted, and the frequency with which the final compiler is used. Interpretation tends to be slower but more flexible with respect to changes in the reduction rules. Translation tends to be less flexible, but the resultant compiler is usually faster. Thus, if the object is to produce an efficient production compiler that will be heavily used, the reduction rules should probably be translated into the same language used to write the rest of the compiler. In other cases interpretation of the reduction rules may be more appropriate.

Even interpretation is not as flexible as it might seem at first glance. It is quite straightforward to interpret the phrase and replacement portion of a reduction rule. However, the actions portion of a reduction rule is a list of names of action procedures that are to be executed. These action procedures will have to be coded in some language. Unless we use a special language for specifying the action procedures, one for which we can write an interpreter, they will have to be written along with the rest of the compiler as ordinary procedures. Attempts have been made to design a special language for specifying the action procedures that can be efficiently interpreted. But this has generally not been very successful, mainly because some of the action procedures are rather large and complex. By the time such a special language has been developed to the point where it is reasonable to use it to describe complex action procedures, it is equivalent to high-level languages such as PL/1 or PASCAL. Therefore, we might as well use one of these languages instead of inventing a new one. An experimental program that translates a reduction rule language has been developed by the author and some of his students. This program demonstrated feasibility and shows promise for development of an efficient translator.

Except for the action procedures, the reduction rules needed for many compilers, especially Instran, are simple enough that they can easily be translated by hand. This tactic makes a great deal of sense and is a very attractive approach if the compiler itself is written in a PL/1-like language. Since the trend in writing software is generally toward using high-level languages, hand translation of reduction rules should be seriously considered. If this is done the reduction rules really are nothing more than a semiformal, systematic approach to coding a parser. The action procedures will then be written in the same language.

One method of hand translation of reduction rules is to define a set of primitive functions and procedures for matching and replacement. An example will give the flavor of this approach. Consider the reduction rule:

P2: *identifier* : PROCEDURE $\rightarrow$ P_0 ; begin_procedure lexical @P3

This rule would be translated into the following PL/1 (or Instran Language) statements:

```
P2:  IF MATCH('N',6,2) & MATCH('P',8,1) & MATCH ('K',3,0)
        THEN
            DO; CALL DELETE(3); CALL PUT(0);
                CALL BEGIN_PROCEDURE; CALL LEXICAL;
                GOTO P3;
        END;
```

Here we assume one primitive function MATCH and two primitive procedures DELETE and PUT.

The function MATCH is used to match the tokens in *phrase* from a rule with the top of the stack. Its first two arguments specify a token. The first argument is the token type. In this example 'N' indicates a nonterminal, 'P' a punctuation symbol, and 'K' a key word. The second argument is the index that identifies the particular nonterminal, punctuation symbol, or key word, respectively. The third argument specifies a position on the stack, with 0 indicating the top. The token in the specified stack position is compared with the token specified by the first two arguments. If they match, the value of the function is true, otherwise its value is false. If the two tokens match, the token in the indicated stack position is copied into the corresponding location in the array P.

The DELETE procedure pops the number of items specified by its argument off of the top of the stack. The argument of the PUT procedure specifies a location in P. A copy of the token stored in that location is pushed onto the top of the stack.

Using these primitives, we can easily translate the phrase and replacement parts of a reduction rule. The phrase part is translated into an expression that is a logical **and** (&) of invocations of the MATCH function, one for each token in *phrase*. The replacement part is translated into a call to the DELETE procedure, followed by several calls to the PUT procedure. The argument of DELETE is the number of tokens in the matched phrase. There is one call to PUT for each P_i in *replacement*, and its argument is *i*. If the replacement part is "#", then there are no calls to PUT. If the replacement part has been omitted in the reduction rule, no calls to either DELETE or PUT are made.

Following the translation of the phrase and replacement parts are calls to all of the action procedures, in the order in which they are written in *actions*. If the action is very simple, the statements that constitute its definition can simply be written in place of the CALL statement.

Translation into this form is quite simple and can easily be done by hand. Furthermore, the resulting parser is reasonably flexible. It is easy to change the reduction rules, since they very nearly retain their original form. More elaborate translation algorithms can be worked out that eliminate many of the calls to MATCH, DELETE, and PUT, replacing them with their definitions. The resulting parser will execute faster, but will lose some flexibility with respect to modification. In addition, the translation is somewhat more difficult, and hand translation may not be as practical.

We will not discuss the implementation of the reduction rules further. In implementing Instran we would use the hand translation technique just described. We would code the translated reduction rules, action procedures, and the remainder of Instran in a language such as PL/1 or PASCAL, that is, one with IF statements, flexible loop statements, and composite data structures.

10.2 Tables Used In Pass One

In this section we give a brief summary of the tables that are used in pass one of Instran. A complete description of all tables used by Instran will be found in Appendix D. The tables used in pass one are:

1. Identifier (and key word) table.
2. Punctuation-operator table.
3. Symbol table.
4. Constant table.
5. The stack.
6. Phrase table.
7. Operation table.
8. Label table.
9. Node table.

Additional auxiliary tables that are specific to the details of implementation are used, but will not be discussed here.

The *identifier table* contains all the key words in the Instran Language and the name of every variable, array, structure, structure element, and label in the source procedure being compiled. The *punctuation-operator* table contains all the punctuation symbols and operator symbols in the Instran Language. It also contains the macro operation code corresponding to each operator symbol. The *operation table* contains an entry for each macro operation that may appear in the node table form of a procedure. There are many more such macro operations than those corresponding to the operator symbols. The first part of the identifier table, which contains the key words, the punctuation-operator table, and the operation table are constant, since their contents depend only on the definition of the Instran Language. The contents of all the other tables depend on the particular source procedure being compiled.

The *symbol table* contains an entry for each *entity* declared in the source procedure. By entity we mean a scalar variable, array, structure, structure element, or label. The name of the entity is in the identifier table. All of the remaining attributes and other information concerning the entity are in its symbol table entry. The attributes for an entity include its data type, its length, its storage class, whether it is an array or not, its subscript upper bound, whether it is a structure or not, whether it is a structure element or not, and its address relative to the storage block implied by its storage class. Other information included in an entity's entry is the index of its name in the identifier table, whether the entity is undefined or multiply defined or not, and information that specifies the relationship between structures and structure elements.

Not all of the information in a symbol table entry is appropriate for all entities, and only that which is will be defined in an entity's entry. For example, the entry for a label will specify its data type as 'label' and its storage class as 'text'. The address is assumed to be relative to the text block. The entry will contain the index of the label's name in the identifier table and will indicate that the label is defined, but not multiply defined. The remaining attributes and information are not relevant for a label and will normally be equal to zero. The attributes and other information in a symbol table entry are filled in as they are discovered by the compiler. Most will be included in the declaration for the entity but, for example, the relative addresses will not be filled in until pass two.

The *constant table* contains an entry for each distinct constant that appears in the source procedure being compiled. Each entry contains the converted value of the constant, its length, its data type, and its relative address in the constant block. The *label table* contains an entry for each (internal) label generated by the compiler. These labels are needed in translating IF statements. They are generated by the parser as needed in such a way that they are unique.

The *stack* and the *phrase table* are used in parsing. They never contain anything except tokens. The phrase table contains the tokens that constitute a matched phrase. Entries in this table are referenced by P_i in the replacement part of a reduction rule. Each entry in the *node table* corresponds to one macro instruction. A macro instruction consists of up to three tokens, an R-type token for the macro operation code, followed by zero, one, or two tokens for the operands. An entry in the node table has three token positions. If the macro instruction corresponding to an entry is less than three tokens in length, only the initial token positions in the entry are used.

A *token* is an ordered pair (t,i) where t is a single letter code for one of the tables and i is the index of some entry in the table specified by t. The type codes for tokens and the tables that they reference are:

1. K—the key word part of the identifier table.
2. I —the non key word part of the identifier table.

3. P—punctuation-operator table.
4. C—constant table.
5. S —symbol table.
6. L—label table.
7. N—node table.
8. R—operation table.

Only the first seven types of token can appear in the stack, and only the last five can appear in the node table. Although a token is an ordered pair (t,i), we will usually use one of two different notations in our examples, t_i or $t_\odot$ where $\odot$ is the name of the entity in entry i of table t. For example, if entry 6 of the symbol table is the entry for the variable A, then the token ('S',6) will usually be written S_A in the examples while the token ('L',4) will usually be written L_4, since internal labels do not have any names (they do not need any). We have not made any provision in a token for the syntactic class as we did in the tokens at the end of Chapter 9. Since the macro instruction form of the procedure being compiled does not explicitly incorporate the procedure's syntactic structure and the syntactic structure of the Instran Language is quite simple, we do not need to carry along syntactic class names while we are parsing, except for expressions. The reader is cautioned that this will probably not be true for more complex languages such as the complete PL/1.

10.3 Procedure Definition

The unit of compilation in the Instran Language is a procedure. Therefore, the programmer always writes and submits to Instran a procedure definition. Figure 10.1 shows the skeleton of a procedure definition that defines a procedure named LOOK. The entry point of this procedure is the PROCEDURE statement. This procedure has three arguments. In the definition the dummy arguments are named A, B, and N. When LOOK is called, the arguments supplied by the caller will be substituted for the dummy arguments. A procedure definition may contain an arbitrary number of RETURN statements. In addition to indicating the end of the procedure definition, the END statement behaves exactly like a RETURN statement when it is executed. When either of these statements is executed, control is immediately returned to the caller.

The compilation of a procedure definition produces a single object procedure segment. This segment has the structure shown in Figure 10.2. The *text block* contains the instructions in the object procedure. The *constant block* contains the constants, including the address constants for the external variables. The *temporary block* contains the temporary storage locations needed in the computation of expressions. The *internal block* is used for all of the internal variables.

```
LOOK: PROCEDURE(A,B,N);
        declarations
        statements_1
        RETURN;
        statements_2
        END;
```

Figure 10.1 Skeleton of a procedure definition.

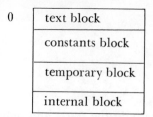

Figure 10.2 Structure of a procedure segment produced by Instran.

The instructions that are generated by the compiler for a procedure must observe the system standards. In particular, they must use the system standard calling sequence and conventions for passing arguments from the caller to the called procedure. This means that code will have to be generated for the procedure's entry point, which is the PROCEDURE statement, as well as for the RETURN and END statements. No code will get generated in pass two unless there is a corresponding macro instruction in the node table. Figure 10.3 shows the macro instructions that get put into the node table as a result of parsing the skeleton procedure definition shown in Figure 10.1. As we will see in the next chapter, the ENTER macro will be defined to be the system required code for saving the registers at the entry point. The RETURN and END macros will be defined to be the code required by the system for restoring the registers and returning to the caller. The only difference between these two macro operations is that END also indicates the end of the macro instructions in the node table.

As a result of parsing the PROCEDURE statement, an ENTER macro instruction will be entered in the node table. In addition, entries must be made in the symbol table for all of the dummy arguments. The PROCEDURE statement lists the names of all of the dummy arguments, and this information must be recorded in the symbol table. Additional attributes for these variables will be added to the symbol table later when their declarations are parsed.

The complete set of reduction rules used in Instran and their accompanying action procedures are described in Appendix E. They are organized into groups according to what part of a procedure definition they parse. The initial letters of the labels for the rules in a group were chosen to indicate the function of the group. For example, the P group parses the PROCEDURE statement, the SD group parses a structure declaration, and the S group parses a statement. The sequence of execution of the reduction rules follows the structure of a procedure definition, as shown in Figure 10.4. The first group of rules (P) parses the PROCEDURE statement. The next group (D) is repeated until all

ENTER
macro instructions for statements_1
RETURN
macro instructions for statements_2
END

Figure 10.3 Macro instructions put into the node table as a result of parsing the procedure skeleton in Figure 10.1.

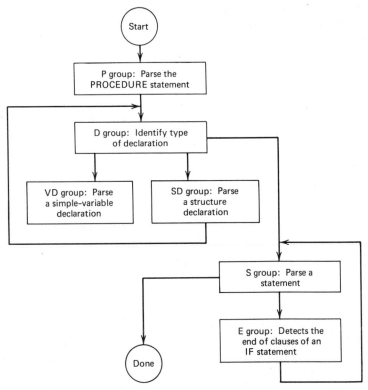

Figure 10.4 Reduction rule execution sequence.

of the declarations have been parsed. This group also identifies the type of declaration being parsed and selects the appropriate group for a simple variable declaration (VD) or a structure declaration (SD). Once all of the declarations have been parsed, the S and E groups are executed until all of the source procedure has been parsed. The S group parses a statement, and the E group is executed after each statement is parsed to detect the end of either the THEN clause or the ELSE clause of an IF statement.

Figure 10.5 illustrates the parsing of a PROCEDURE statement. The purpose of this example is simply to illustrate the interplay between matching phrases and execution of action procedures. Figure 10.5a reproduces the relevant reduction rules from Appendix E. Only the rules whose phrase part matches are indicated in Figure 10.5b. Their labels are listed in the first column. The contents of the stack at the time a match occurred is shown in the row immediately preceding the line containing the label of the matching rule. The tokens in the stack that matched the rule are underlined. The contents of the stack immediately after the replacement specified in the matching rule is shown on the same line as the rule's label. The actions associated with a rule are listed in the third column, one or more per line beginning with the line containing the rule's label. The modifications or additions to the major tables, which are the result of an action, are shown in the last column opposite the name of the responsible action. If an action modifies the contents of the stack, the modified contents are shown in the second column in the line immediately

/* The P group parses a PROCEDURE statement.
P1: → ; lexical lexical lexical
P2: *identifier* : PROCEDURE → P$_0$; begin_procedure lexical @P3
P21: → ; error
P3: (→ # ; lexical @P6
P4: ; → # ; lexical @D1
P5: → ; error
/* Parse dummy argument list.
P6: *identifier* → # ; define_dummy lexical @P7
P61: → ; error
P7: , → # ; lexical @P6
P8:) → # ; lexical @P4
P9: → ; error

Figure 10.5*a* Reduction rules for parsing a PROCEDURE statement.

following the line containing the name of the responsible action. If the reader reads each line from left to right and the lines from the top of the figure to the bottom, he will read each step in parsing a PROCEDURE statement in the sequence in which it actually takes place.

For example, the stack is initially empty, and execution of the rules begins with P1. This rule has no phrase and replacement parts; therefore, its action procedures are always executed, since its phrase part always matches. The actions specified are three calls to the lexical analyzer. Therefore, the first line of Figure 10.5*b* is:

P1   lexical lexical lexical

After three calls to lexical the stack will contain tokens for the first three basic elements in the source procedure. The second line shows the new stack contents:

$$I_{LOOK} \ P_: \ K_{PROC}$$

In this and other examples we will use obvious abbreviations for the longer key words consisting of the first three or four letters of the key word. The underline under these three tokens indicates that the phrase part of the rule named in the next line matches these three tokens.

The next rule to be executed is P2. It replaces the three matched tokens, which are stored in P$_0$, P$_1$, and P$_2$, by the single token K$_{PROC}$. After this replacement, the action begin_procedure is executed. The result is that the macro instruction

$$R_{ENTER}$$

is added to the node table. In addition, LOOK, which is the name of the procedure being compiled, is saved. Finally, the lexical analyzer is called. The contents of the stack after this is shown in the next line:

$$K_{PROC}P_($$

Again the underline beneath the token P$_($ indicates that this token is matched by the phrase part of the next rule to be executed.

Later in the example the line

P6   K$_{PROC}$ define_dummy lexical   add S$_A$(STORAGE: = 'D')

shows the result of executing rule P6, especially its accompanying action procedure define_dummy. For a complete definition of the action procedures the reader should consult Appendix E. In our examples we will show only the most

Source input

		LOOK: PROCEDURE(A,B,N);	
Rule	Stack	Actions	Table modifications
P1		lexical lexical lexical	
P2	I_{LOOK} $P_;$ K_{PROC} $\overline{K_{PROC}}$	begin_procedure	$\{R_{ENTER}\}$ entry_name: = 'LOOK'
		lexical	
P3	K_{PROC} $P_($ $\overline{K_{PROC}}$	lexical	
P6	K_{PROC} I_A $\overline{K_{PROC}}$	define_dummy lexical	add S_A(STORAGE: = 'D')
P7	K_{PROC} $P_.$ $\overline{K_{PROC}}$	lexical	
P6	K_{PROC} I_B $\overline{K_{PROC}}$	define_dummy lexical	add S_B(STORAGE: = 'D')
P7	K_{PROC} $P_,$ $\overline{K_{PROC}}$	lexical	
P6	K_{PROC} I_N $\overline{K_{PROC}}$	define_dummy lexical	add S_N(STORAGE: = 'D')
P8	K_{PROC} $P_)$ $\overline{K_{PROC}}$	lexical	
P4	K_{PROC} $P_;$ $\overline{K_{PROC}}$	lexical	

Figure 10.5*b* Parsing a PROCEDURE statement.

significant or interesting result of executing an action. In this case **define_dummy** makes an entry in the symbol table for the dummy argument A. This is indicated by writing

add S_A(STORAGE: = 'D')

When we use a token in this context, the type letter in the token identifies the table to which the new entry is being added, and the index part of the token indicates the entry. In this case, use of the identifier A suggests that the entry is for a variable whose name is A and that we do not care what its actual index is in the symbol table. The notation

(STORAGE: = 'D')

following the token S_A indicates that the storage class in the entry for A is set equal to 'D', which specifies that A is a dummy argument. An entry in the symbol table contains many components; however, the values of the components other than STORAGE are not relevant to the example in Figure 10.5, so their values are not specified in the example.

The main result of parsing a PROCEDURE statement is the addition of entries to the symbol table for each of the dummy arguments. The names of these dummy arguments appear as a list of identifiers. The parsing consists of ping-ponging between phrase matching with its accompanying replacement and execution of the action **define_dummy**. The replacement parts of rules P6 and P7 are omitted, so the matched phrase is deleted in each case. Thus, the dummy argument list in the PROCEDURE statement is consumed token by token. Whenever an identifier token is encountered, the action **define_ dummy** is executed, and it makes an entry in the symbol table for the dummy argument named by the identifier.

10.4 Declarations

The purpose of declarations is to define the identifiers, other than key words, that appear in a procedure. In order to compile a procedure, the compiler must know what entity it is that each identifier names. Most declarations are made using the DECLARE statement. The exceptions are labels, the name of the procedure, and its dummy arguments. The declaration for a label is implicit in its appearance preceding a statement. The declaration for the name of a procedure is implicit in its appearance preceding the PROCEDURE statement. The declaration for a dummy argument is split between the dummy argument list in the PROCEDURE statement and a DECLARE statement that specifies its other attributes.

All of the information associated with an entity will be collected into a single entry in the symbol table, except for its name, which will be stored in the identifier table. Since the elements of a structure may have different attributes, we need a separate entry for each. The names of entities must be unique, except for structure elements. The elements of any particular structure (or substructure) must have unique names, but they may be the same as the names of other entities (arrays, labels, elements of other structures, etc.). Thus, we may have many entities with the same name, as long as all but one of them are structure elements and no two of these are elements of the same structure. All entities with the same name will point to the same identifier in the identifier table. Storing the names separately is not a logical necessity; however, it does save space. It also makes name comparisons much simpler, since only the identifier table indexes of two names must be compared in order to tell if the names are the same.

First, let us consider all declarations except structure declarations. The basic strategy is to make an entry for an entity the first time its name is encountered. This means that the entries for the dummy arguments are made when parsing the PROCEDURE statement, as we saw in the preceding section. The attributes for each entity, except labels, must be declared in a DECLARE statement. Thus, the name of each entity other than a dummy argument or label will first be encountered when parsing a DECLARE statement. The name of a label may be encountered first, either as a label preceding a statement or in a GOTO statement. The appearance of an identifier in a GOTO statement does not define a label. However, we make an entry for it in the symbol table in anticipation of its future definition. This makes it easy to be sure that all node table entries that reference the same label will use the same token, that is, point to the same symbol table entry.

The action that accompanies the parsing of a GOTO statement is to make a symbol table entry, marked as undefined, for the referenced label, unless there is already such an entry. The action that accompanies the parsing of a label declaration is equally straightforward. A label declaration is recognized when the phrase

identifier :

preceding a statement is matched. If the symbol table does not contain an entry for the label, one is made, and it is marked as defined. If any entry for the label exists, it is simply marked as defined. In addition, the macro instruction

$$R_{EQU} S_L$$

where L is the name of the label being defined, is entered in the node table. This is required, since the address associated with the label is to be the address of the first instruction in the object code generated for the statement that follows the label declaration. The addresses of variables will be assigned at the beginning of pass two. However, the addresses of the object instructions will not be known until the instructions are generated by the code generation part of pass two. The interpretation of the EQU macro in pass two will not generate any object code. When it is encountered, the address of the next object instruction is entered into the address field of the symbol table entry for the label L.

Parsing a simple-variable declaration is straightforward. It is illustrated in Figure 10.6. The parser makes an entry in the symbol table for the identifier in the declaration. Once this entry has been made, the entity's attributes are filled in one by one as they occur in the declaration. Note that a token that points to the appropriate symbol table entry is left on the stack until the end of the DECLARE statement is reached. This identifies the entry into which each attribute is stored when it is encountered.

Whenever a new entry is to be added to the symbol table, a check is made to see if this is a multiple definition. If it is, no new entry is made, and the existing entry is marked as being multiply defined. A multiple definition occurs if an entry already exists for an entity that is not an element of a structure and that is not marked as being undefined. If an entry exists for an entity that is not an element of a structure and it is marked as being undefined, a multiple definition has not occurred unless the existing entry is not consistent with the new entry. The two entries are inconsistent if any attribute that has been set in the existing entry has a value that is different from the value of the corresponding attribute in the new entry. For example, when an entry is made for a dummy argument as a result of parsing the PROCEDURE statement, it is marked as being undefined, and its storage attribute is specified as 'D' (dummy argument). When the DECLARE statement that declares the remainder of the dummy argument's attributes is parsed, a multiple definition will not occur unless the storage is specified as external or internal.

The parsing of structure declarations is more complicated. The major problem is the requirement that the structural relationship of the elements, especially the substructures, must be recorded in the symbol table. A structure declaration has the form

DECLARE 1 *identifier* [*dimension*] *storage* {,*element*} ••• ;

where each element in the sequence is either a basic element having the form

integer identifier type

or a substructure having the form

integer identifier {, *element*} •••

The integers in the declaration are the level numbers. They specify the hierarchial relationship between the elements in the structure and the elements in the substructures.

/* The D group separates a simple variable declaration from a structure
    declaration.
D1: DECLARE → # ; lexical @D3
D2: → ; lexical @S1
D3: *identifier* → ; define lexical @VD1
D4: 1 → # ; lexical @SD1
D5: → ; error
/* The VD group parses a simple variable declaration.
VD1: (→ ; lexical lexical @VD5
VD2: *string* → # ; set_type lexical lexical lexical @VD7
/* Parse data type.
VD3: *type-word* → # ; set_type lexical lexical @VD9
VD4: → ; error
/* Parse dimension attribute.
VD5: (*integer*) → # ; set_dimension lexical @VD2
VD6: → ; error
/* Parse string attribute.
VD7: (*integer*) → # ; set_length lexical lexical @VD9
VD8: → ; error
/* Parse storage class.
VD9: *identifier storage* ; → # ; set_storage lexical @D1
VD10: *identifier* ; *any* → P_0 ; @D1
VD11: → ; error

Figure 10.6*a* Reduction rules for parsing a simple-variable declaration.

Figure 10.7 shows an example structure declaration, which we will reference in the following discussion. Figure 10.8 shows the tree diagram for this structure. It is this tree that we must record in the symbol table. Each node in the tree corresponds to a structure, substructure, or basic element. We need an entry in the symbol table corresponding to each node. Thus, the structure declaration in Figure 10.7 will require that 10 entries be added to the symbol table. There will be one for the structure X and one for each of its level 2 elements A, B, and C. Since A is a substructure, there will be one entry for each of its elements A and B (these are level 3 elements). Similar entries are needed for the elements of the other substructures.

In order to represent the relationship diagrammed in Figure 10.8, we would have to include in the symbol table entry for a structure pointers to each of the entries for its elements (its sons). This is sufficient information to record the structure, but it is rather inconvenient to work with. We need to do two things with the structure representation in the symbol table. At the beginning of pass two, when memory is assigned, we must be able to work our way through all of the elements in the structure and all of its substructures in order to compute the total amount of memory needed and to determine the relative address of the elements in the structure. This will turn out to be much easier to do if each node points to its right-hand brother. In addition, when we are generating code to access an element in the structure, we will have to be able to find the parent structure. Thus, we also want in each node a pointer to its parent node. It turns out that with these additional pointers in each node, we

Source input
DECLARE X(78) FIXED INTERNAL;

Rule	Stack	Actions	Table modifications
	$\underline{K_{DCL}}$		
D1		lexical	
	I_X		
D3	$\overline{I_X}$	define	add S_X
	S_X	lexical	
	$S_X\ P_($		
VD1	$S_X\ \overline{P_(}$	lexical lexical	
	$S_X\ P_(\ C_{78}\ P_)$		
VD5	$\overline{S_X}$	set_dimension lexical	set $S_X(UPPER:=78)$
	$S_X\ K_{FIX}$		
VD3	$\overline{S_X}$	set_type lexical lexical	set $S_X(TYPE:='F')$
	$S_X\ K_{INT}\ P_;$		
VD9	$\overline{}$	set_storage lexical	set $S_X(STORAGE:='I')$

Figure 10.6*b* Parsing a simple-variable declaration. Only the top part of the stack is shown. The token K_{PROC} is always on the stack, but is not shown here.

```
DECLARE 1 X INTERNAL,
          2 A,
            3 A FIXED,
            3 B FIXED,
          2 B FIXED,
          2 C,
            3 A,
              4 B FIXED,
              4 C FIXED,
            3 B FIXED;
```

Figure 10.7 Example structure declaration.

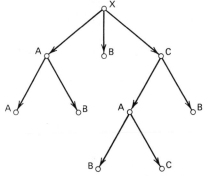

Figure 10.8 Tree diagram of structure declared in Figure 10.7.

no longer need pointers to all of the sons. A single pointer to a node's leftmost son is sufficient.

The structure is actually stored in the symbol table as a directed graph. The directed graph for the structure declared in Figure 10.7 is shown in Figure 10.9. Figure 10.10 shows the 10 symbol table entries of this structure. Each row is a single entry in the symbol table. The index of an entry is written at the left of its row. The name of each item (field) in a symbol table entry is written above the corresponding column. The portion of the identifier table that is relevant is also shown. The three pointers in an entry that were mentioned in

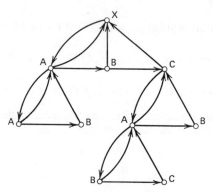

Figure 10.9 Directed graph for structure declared in Figure 10.7.

	NAME	TYPE	LENGTH	STORAGE	ARRAY	UPPER	PARENT	BROTHER	SON	DEFINED	ADDRESS	Element
11	4	S	0	I	0	0	0	0	12	D	0	X
12	5	S	0	S	0	0	11	15	13	D	0	X.A
13	5	F	0	S	0	0	12	14	0	D	0	X.A.A
14	6	F	0	S	0	0	12	0	0	D	0	X.A.B
15	6	F	0	S	0	0	11	16	0	D	0	X.B
16	7	S	0	S	0	0	11	0	17	D	0	X.C
17	5	S	0	S	0	0	16	20	18	D	0	X.C.A
18	6	F	0	S	0	0	17	19	0	D	0	X.C.A.B
19	7	F	0	S	0	0	17	0	0	D	0	X.C.A.C
20	6	F	0	S	0	0	16	0	0	D	0	X.C.B

Associated
Identifier
Table

4	X
5	A
6	B
7	C

Figure 10.10 Symbol table entries for structure in Figure 10.7.

the last paragraph are named PARENT, BROTHER, and SON. If they do not point to anything their value is zero. Otherwise their value is the index in the symbol table of the entry for the node to which they point.

The major task in parsing a structure declaration is determining the values of the three pointers in all of the symbol table entries for the structure. This is carried out by three action procedures: define_structure, which makes the symbol table entry for a major structure; set_element, which makes the symbol table entry for a structure element; and set_substructure, which does some additional bookkeeping required for a substructure. The major complication here is that in a structure declaration the elements of a substructure intervene between it and its right-hand brother. If the elements of a structure are all basic elements, the symbol table entry for an element is made immediately after the entry for its left-hand brother is made. Thus, if A and B are two consecutive elements in a structure, then the next symbol table entry to

/* The D group separates a simple variable declaration from a structure
    declaration.
D1: DECLARE → # ; lexical @D3
D2: → ; lexical @S1
D3: *identifier* → ; define lexical @VD1
D4: 1 → # ; lexical @SD1
D5: → ; error
/* The SD group parses a structure declaration.
SD1: *identifier* → ; define_structure lexical lexical @SD3
SD2: → ; error
SD3: (*any* → ; lexical lexical @SD6
SD4: *identifier storage* , → # ; set_storage lexical lexical @SD8
SD41: *identifier* , *any* → P_0 ; @SD8
SD5: → ; error
/* Parse structure dimension.
SD6: (*integer*) → # ; set_dimension lexical lexical @SD4
SD7: → ; error
/* Parse structure element.
SD8: *integer identifier* → P_0 ; set_element lexical @SD10
SD9: → ; error
/* Parse substructure element.
SD10: *commas* → ; set_substructure @SD16
/*Parse basic element.
SD11: *string* → # ; set_type lexical lexical lexical @SD14
SD12: *type-word* → # ; set_type lexical @SD16
SD13: → ; error
SD14: (*integer*) → # ; set_length @SD12
SD15: → ; error
/* Check end of structure.
SD16: *any* , → # ; lexical lexical @SD8
SD17: *any* ; → # ; lexical @D1
SD18: → ; error

Figure 10.11*a* Reduction rules for parsing a structure declaration.

Rule	Stack	Actions	Table modifications
	$\underline{K_{DCL}}$		
D1		lexical	
	$\underline{C_1}$		
D4		lexical	
	I_X		
SD1	$\underline{I_X}$	define_structure	add S_{11}(STORAGE:$=$'U') see Fig. 10.10 old_level:$=2$ parent_element:$=11$ last_element:$=11$
	S_{11}	lexical lexical	
	$S_{11}\ K_{INT}\ P_,$		
SD4		set_storage lexical lexical	set S_{11}(STORAGE:$=$'I')
	$C_2\ I_A$		
SD8	$\underline{I_A}$	set_element	add S_{12}(STORAGE:$=$'S'; TYPE:$=$'U') this_element:$=12$ new_level:$=2$ link to S_{11} last_element:$=12$
	S_{12}	lexical	
	$S_{12}\ P_,$		
SD10	$\underline{S_{12}\ \overline{P}_,}$	set_substructure	set S_{12}(TYPE:$=$'S') old_level:$=3$ parent_element:$=12$
SD16		lexical lexical	
	$C_3\ I_A$		
SD8	$\underline{I_A}$	set_element	add S_{13}(STORAGE:$=$'S'; TYPE:$=$'U') this_element:$=13$ new_level:$=3$ link to S_{12} last_element:$=13$
	S_{13}	lexical	
	$S_{13}\ K_{FIX}$		
SD12	$\underline{S_{13}}$	set_type lexical	set S_{13}(TYPE:$=$'F')
	$S_{13}\ P_,$		
SD16		lexical lexical	
	$C_3\ I_B$		
SD8	$\underline{I_B}$	set_element	add S_{14}(STORAGE:$=$'S'; TYPE:$=$'U') this_element:$=14$ new_level:$=3$ link to S_{13} and S_{12} last_element:$=14$
	S_{14}	lexical	
	$S_{14}\ K_{FIX}$		
SD12	$\underline{S_{14}}$	set_type lexical	set S_{14}(TYPE:$=$'F')
	$\underline{S_{14}\ P_,}$		

SD16		lexical lexical	
	C_2 I_B		
SD8	I_B	set_element	add S_{15}(STORAGE: $='S'$; TYPE: $='U'$)
			this_element: $=15$
			new_level: $=2$
			old_level: $=2$
			last_element: $=12$
			parent_element: $=11$
			link to S_{12} and S_{11}
			last_element: $=15$
	S_{15}	lexical	
	S_{15} K_{FIX}		
SD12	S_{15}	set_type lexical	set S_{15}(TYPE: $='F'$)
	S_{15} P,		
SD16		lexical lexical	
	$\cdots$		
	$\cdots$		
	$\cdots$		
SD12	S_{20}	set_type lexical	set S_{20}(TYPE: $='F'$)
	S_{20} P;		
SD17		lexical	

Figure 10.11*b* Parsing the structure declaration in Figure 10.7.

be made after making an entry for A is the entry for B, for example, X.A.A and X.A.B in Figure 10.7. Thus, the symbol table entries are linked as brothers in the sequence in which the entries are made. This is not true when some of the elements are substructures. For example, the symbol table entries for X.A and X.B must be linked together as brothers. However, between the time the entry for X.A is made and the time the entry for X.B is made, two other entries (for X.A.A and X.A.B) have been made.

Figure 10.11 illustrates the parsing of the structure declaration in Figure 10.7. If the reader really wants to understand the problem and our solution, he should follow through the details of the action in Figure 10.11 after reading the remainder of this section. There are three pointers in the symbol table entry for a structure or element of a structure that must be set. However, only the parent pointer can be set at the time the entry is made. The parent of an element precedes the element in the declaration; therefore, the parent's symbol table entry will already have been made when the symbol table entry for the element is made. The other two pointers point to the element's first son and right-hand brother, if it has them. Since both the first son and right-hand brother of an element follow it in the structure declaration, these pointers cannot be set when the entry for the element is made.

As an aid to determining the values of these pointers, we use five variables: last_element contains a pointer to the most recent symbol table entry; old_level contains the level number of the elements in the current structure or substructure; this_element contains a pointer to the entry currently being made; new_level contains its level number; and parent_element contains a pointer to the current parent. If the two level numbers are equal, then the entry pointed to by last_element is either its parent or its left-hand brother.

In the first case the son pointer of last_element is set equal to this_element and in the second case the brother pointer of last_element is set equal to this_element. The two cases are mutually exclusive, and the other pointer is always set to zero.

The tricky part of the algorithm occurs when the level changes, and then only when it decreases. The level increases from i to $i+1$ if and only if the last element processed at level i was a substructure. The action set_substructure takes care of this increase in level, incrementing old_level by one and resetting the value of parent_element to point to the new substructure. A decrease in level occurs at the end of a substructure. In fact, several nested substructures may end at the same place. Thus, the difference between old_level and new_level may be greater than one. In order to find the parent and left-hand brother of this_element, we must work our way up the thread of parents (as shown in Figure 10.12), that is, the parent of last_element, the parent of the parent of last_element, and so forth, until we come to the substructure that has a level number equal to new_level. This will be the left-hand brother of this_element and its parent will be the parent of this_element.

10.5 Expressions

In Chapter 8 we discussed the translation of expressions quite thoroughly, except for array and structure references. Instran uses an algorithm similar to the one described there. In this section we are concerned with how to extend it for array and structure references. The major problem in this extension is the representation of the new types of reference in the node table form of an expression. A variable reference has the form:

$$identifier\ [.\ identifier]\ \bullet\bullet\bullet\ [(a\text{-}expression)]$$

We face two problems representing a variable reference. Since we may have an

```
if new_level = old_level
  then if last_element = parent_element
          then SYMBOL.SON[parent_element] := this_element;
          else SYMBOL.BROTHER[last_element] := this_element;
     else begin
          if new_level ⩾ old_level then Error;
          if new_level ⩽ 1 then Error;
          if last_element = parent_element then Error;
          repeat
             old_level := old_level − 1;
             last_element := parent_element;
             parent_element := SYMBOL.PARENT [last_element];
          until old_level = new_level;
          SYMBOL.BROTHER[last_element] := this_element;
        end;
SYMBOL.PARENT[this_element] := parent_element;
last_element := this_element;
```

Figure 10.12 Linking the symbol table entry for the current structure element to the previous entries for the structure.

array of structures, the representations of an array reference and a structure reference must be compatible. In addition, a subscript may be an arbitrary arithmetic expression.

Looking at structures alone for a moment, the most convenient representation is simply a single token that points to the symbol table entry for the structure element that is being referenced. For example, given the declaration:

$$\text{DECLARE 1 X INTERNAL,}$$
$$\text{2 Y,}$$
$$\text{3 Z FIXED,}$$
$$\text{3 V FIXED,}$$
$$\text{2 W FIXED;}$$

the element X.Y.Z would be represented by the token S_Z and the element X.W would be represented by the token S_W. Since the symbol table entry for a structure element contains a pointer to its parent, which in turn points to its parent, and so forth, all information about the structure and its elements is accessible from the symbol table entry for any of its elements.

In representing an array reference the simplest way to cope with an arbitrary expression as a subscript is to consider subscripting as a binary operator. For example, representations for some array references are:

$$X(2) \quad \text{is represented by } \{R_{ss}\, S_X\, C_2\}$$
$$X(I) \quad \text{is represented by } \{R_{ss}\, S_X\, S_I\}$$
$$X(K+2) \text{ is represented by } \{R_{ss}\, S_X\, N_j\}$$

where N_j is a token pointing to the node table entry

$$\{R_{PLUS}\, S_K\, C_2\}$$

If the subscript is a more complicated expression the N_j token will point to the last entry for the expression in the node table.

Combining the two representations is straightforward, since a reference to a structure element is a single token, and the form for an array reference requires a single token for the array as the first operand of the subscript operator. In the combined representation it is understood that the first operand is a token pointing to the structure element, while the subscript selects one structure from the array of structures. As mentioned earlier, by following the parent thread to the top, we find the symbol table entry for the major structure, which is where the array information is stored. Figure 10.13 is an example of the combined representation.

We will not discuss the parsing of array and structure references in detail, but two points must be mentioned. Assuming that the parsing of expressions is done using a precedence-based algorithm such as the one described in Chapter 8, we can easily take care of subscripts by considering subscripting as a binary operator whose precedence is higher than any other operator. Taking this view, an array reference of the form

$$X(expression)$$

is simply a shorthand for

$$X \text{ s } (expression)$$

where "s" is the subscript operator. The simplest method of parsing is to recognize the following pair of tokens:

$$I_X\, P_(§$$

and replace them by the three tokens:

$$I_X\, P_s\, P_(§$$

Source statements
DECLARE 1 A(5) INTERNAL,
          2 B,
              3 D FIXED,
              3 E FIXED,
          2 C FIXED;
A.B.D(2) = A.B.E(I)*A.C(K+2);

Node table representation

1	R_{SS}	S_E	S_I
2	R_{PLUS}	S_K	C_2
3	R_{SS}	S_C	N_2
4	R_{TIMES}	N_1	N_3
5	R_{SS}	S_D	C_2
6	R_{ASSIGN}	N_5	N_4

Figure 10.13 Node table representation of structure references.

which makes the subscript operator explicit. Once this is done, the precedence algorithm described in Chapter 8 works without further modification.

The second point concerns the parsing of structure references. When a structure reference of the form

$$A.B.C$$

is encountered, it is not enough simply to look up C in the symbol table and construct a token pointing to its entry. The names used in structures need not be unique. The same name may be used many times, as long as no two elements with the same parent have the same name. Thus, in transforming the sequence of tokens

$$I_A \ P \ I_B \ P \ I_C$$

into a single token pointing to the symbol table entry for basic element C, the following search procedure must be used. The symbol table is searched for an entry that is a major structure whose name is A. There will only be one such entry, since all names at level 1 must be unique. If there is no such entry an error has been detected. Let S_A be a token pointing to the entry that was found. All of the sons, and only the sons, of S_A are searched for B. This is done by following the son pointer in S_A to the first son of A and then its brother pointer to the next son, and so on until the end of the brother thread is reached, which is indicated by a brother pointer that is equal to zero. There will be at most one symbol table entry on the brother thread that is named B. Let the token S_B point to this entry. The token S_A can now be discarded, as it is no longer needed. The process is then repeated using S_B, searching its sons for C. This will result in the final desired token S_C.

10.6 IF Statements

An IF statement has the form:

IF *b-expression* THEN *true-statement* [ELSE *false-statement*]

where *true-statement* and *false-statement* may be any statements, including IF statements. The compiler has to generate object code of the form:

code to evaluate *b-expression*
code to test the value of *b-expression* and branch to *label₁*
   if the value is false
code for *true-statement*
code to branch to *label₂*
label₁: code for *false-statement*
label₂: code for the statement that follows the IF statement

where the code for *false-statement* is omitted if the ELSE clause is absent. The two labels *label₁* and *label₂* are internal labels generated by the compiler. The simplest way to represent this in the node table is straightforward:

node table entries corresponding to *b-expression*
$R_{FALSEB} N_i L_n$
node table entries corresponding to *true-statement*
$R_{BRANCH} L_m$
$R_{EQU} L_n$
node table entries corresponding to *false-statement*
$R_{EQU} L_m$

where the node table entries corresponding to *false-statement* are not made if the ELSE clause is absent. The tokens L_n and L_m correspond to entries in the label table and thus function as internal labels. The token N_i points to the last node table entry in the group corresponding to *b-expression*. The FALSEB macro is expanded by pass two into code, which tests the value of N_i (*b-expression*) and branches to L_n if the value is false. The BRANCH macro expands to an unconditional branch instruction.

There are two major problems in parsing an IF statement. The special node table entries shown in the preceding paragraph must be made at three points in the parsing: when the key word "THEN" is encountered, when the key word "ELSE" is encountered, and at the end of the IF statement. A unique internal label must be generated at the first point and saved until the second point. At the second point another unique internal label must be generated and saved until the third point. The second problem is that *true-statement* and *false-statement* may be any statement, including the IF statement. Thus, IF statements may be nested to an arbitrary depth. This means that we will have to save an arbitrary number of different labels at once. Furthermore, they will have to be saved in such a way that when a key word "ELSE" or the end of an IF statement is encountered, the appropriate label is recalled. Figure 10.14 illustrates this. In the figure a new unique label is generated and used at the tail of each arrow. That label must then be saved for use again at the head of the arrow.

Our strategy for saving labels is to put them on the stack. Because of the possibility of nested IF statements, we will also need to save each occurrence of the key words "THEN" and "ELSE" on the stack until we come to the end of *true-statement* or *false-statement* in order to detect the corresponding places to insert the special node table entries. Since a label is generated when a key word "THEN" or "ELSE" is encountered in the parsing, we simply save the label on the stack immediately preceding the key word that caused it to be generated. The problem in parsing a DO group is very similar. A DO group has the form

DO; *body* END;

Source statements
IF X = Y
THEN IF X = W
    THEN A = B;
    ELSE IF Z = W
        THEN A = C;
        ELSE E = A;
ELSE P = Q;
M = J;

Node table representation

1	R_{EQUALS}	S_X	S_Y
2	R_{FALSEB}	N_1	L_1
3	R_{EQUALS}	S_X	S_W
4	R_{FALSEB}	N_3	L_2
5	R_{ASSIGN}	S_A	S_B
6	R_{BRANCH}	L_3	
7	R_{EQU}	L_2	
8	R_{EQUALS}	S_Z	S_W
9	R_{FALSEB}	N_8	L_4
10	R_{ASSIGN}	S_A	S_C
11	R_{BRANCH}	L_5	
12	R_{EQU}	L_4	
13	R_{ASSIGN}	S_E	S_A
14	R_{EQU}	L_5	
15	R_{EQU}	L_3	
16	R_{BRANCH}	L_6	
17	R_{EQU}	L_1	
18	R_{ASSIGN}	S_P	S_Q
19	R_{EQU}	L_6	
20	R_{ASSIGN}	S_M	S_J

Figure 10.14 Node table representation of nested IF statements.

The group of statements in *body* are to be treated as if they were a single state-ment. In particular, a DO group qualifies as *true-statement* or *false-statement* in an IF statement. In order to discover the correct structure of DO groups and IF statements in combination, we must save the key word "DO" on the stack until we come to its corresponding "END". Figure 10.16 illustrates this by showing the parsing of the nested IF statements and DO group in Fig-ure 10.15.

10.7 CALL Statements

A CALL statement has the form
      CALL *identifier* [(*expression* {, *expression*} • • •)];
The compiler must generate object code for a system standard call to the pro-cedure whose name is *identifier*. The principal problem here is that the stand-ard call requires an argument list. This argument list consists of the addresses of the values of each of the expressions that are the arguments. Because these

Source statements
IF X=Y
THEN IF X=W
       THEN DO; A=B;
              E=F;
         END;
      ELSE E=A;
M=J;

Node table representation

1	R_{EQUALS}	S_X	S_Y
2	R_{FALSEB}	N_1	L_1
3	R_{EQUALS}	S_X	S_W
4	R_{FALSEB}	N_3	L_2
5	R_{ASSIGN}	S_A	S_B
6	R_{ASSIGN}	S_E	S_F
7	R_{BRANCH}	L_3	
8	R_{EQU}	L_2	
9	R_{ASSIGN}	S_E	S_A
10	R_{EQU}	L_3	
11	R_{EQU}	L_1	
12	R_{ASSIGN}	S_M	S_J

Figure 10.15 Node table representation of nested IF statements and a DO group.

addresses must be absolute, the argument list cannot be generated by the compiler. So object code must be generated that will construct the argument list just before the call to the specified procedure. In addition, the compiler must generate code to evaluate each of the expressions just before the call.

The code to be generated for the CALL statement

$$\text{CALL PR } (e_1, e_2, \ldots e_n);$$

has the form:

        code to evaluate e_1
        set first argument list entry to address of value of e_1
        code to evaluate e_2
        set second argument list entry to address of value of e_2
        . . .
        code to evaluate e_n
        set nth argument list entry to address of value of e_n
        code to call PR

We can most easily represent this in the node table by using two new macros, ARG and CALL. The node table form of a CALL statement will be:

        node table entries corresponding to evaluation of e_1
        $R_{ARG} t_1$
        node table entries corresponding to evaluation of e_2
        $R_{ARG} t_2$
        . . .
        node table entries corresponding to evaluation of e_n
        $R_{ARG} t_n$
        $R_{CALL} S_{PR}$

/* The S group parses a statement.
/* Labeled statement.
S1: *identifier* : → # ; label lexical lexical
/* Assignment statement.
S2: *identifier* (→ ; compile_expression @E1
S3: *identifier* . → ; compile_expression @E1
S4: *identifier* = → ; compile_expression @E1
/* Other statements.
S5: GOTO *identifier* → P_0 ; compile_branch lexical @E1
S6: IF *any* → P_0 ; lexical compile_expression compile_if @S14
S7: RETURN ; → # ; compile_return lexical lexical @E3
S8: CALL *identifier* → P_0 ; lexical compile_expression @E1
S9: ; *any* → P_0 ; lexical @E3
S10: DO ; → P_1 ; lexical lexical @S1
/* End of group.
S11: DO END ; → # ; lexical lexical @E3
/* End of procedure.
S12: PROCEDURE END ; → ; end_procedure
S13: → ; error
/* Then clause of IF statement.
S14: THEN → ; lexical lexical @S1
S15: → ; error
/* The E group parses the end of a statement including the end of
   then and else clauses.
E1: *any* ; → # lexical lexical @E3
E2: → ; error
/* End of a then clause and the beginning of an else clause.
E3: THEN ELSE *any* → P_1 P_0 ; compile_then lexical @S1
E4: *any* THEN *any any* → P_1 P_0 ; end_if @E3
/* End of an else clause.
E5: *any* ELSE *any any* → P_1 P_0 ; end_if @E3
E6: → ; @S1

Figure 10.16a Reduction rules for parsing statements.

where t_i is a token for the ith argument. If e_i is a single identifier or constant, t_i will be S_j or C_k. If e_i is a more complex expression, t_i will be N_p, where p is the index of the last node table entry for the evaluation of e_i. Figure 10.17 shows an example of the node table entries for a CALL statement.

   In Instran we have chosen to further extend the action procedure **compile_ expression** so that it parses the CALL statement (except for the keyword "CALL"). This takes very little additional effort and will also allow function invocations to be included in expressions. Only three things need adding. We must recognize the pair of tokens

$$I_{PR} \; P_(\,$$

as the beginning of a procedure call and mark the stack accordingly, that is, the pair of tokens are replaced by

$$I_{PR} \; P_c \; P_(\,$$

However, in Section 10.5 we said that this pair of tokens would be recognized as an array reference. This apparent ambiguity is easily resolved by consulting

Rule	Stack	Actions	Table modifications
	$\underline{K_{IF}\ I_X}$		
S6	$\underline{I_X}$	lexical	
	$I_X\ \underline{P_=}$	compile_expression	$N_1 := \{R_{EQUALS}\ S_X\ S_Y\}$
	$N_1\ \underline{K_{THEN}}$	compile_if	$N_2 := \{R_{FALSEB}\ N_1\ L_1\}$
	$L_1\ \underline{K_{THEN}}$		
S14	$L_1\ \underline{K_{THEN}}$	lexical lexical	
	$L_1\ K_{THEN}\ \underline{K_{IF}\ I_X}$		
S6	$L_1\ K_{THEN}\ \underline{I_X}$	lexical	
	$L_1\ K_{THEN}\ I_X\ \underline{P_=}$	compile_expression	$N_3 := \{R_{EQUALS}\ S_X\ S_W\}$
	$L_1\ K_{THEN}\ N_3\ \underline{K_{THEN}}$	compile_if	$N_4 := \{R_{FALSEB}\ N_3\ L_2\}$
	$L_1\ K_{THEN}\ L_2\ \underline{K_{THEN}}$		
S14	$L_1\ K_{THEN}\ L_2\ \underline{K_{THEN}}$	lexical lexical	
	$L_1\ K_{THEN}\ L_2\ K_{THEN}\ \underline{K_{DO}\ P_;}$		
S10	$\dots\ \underline{K_{DO}}$	lexical lexical	
	$\dots\ K_{DO}\ \underline{I_A\ P_=}$		
S4	$\dots\ K_{DO}\ \underline{I_A\ P_=}$	compile_expression	$N_5 := \{R_{ASSIGN}\ S_A\ S_B\}$
	$\dots\ K_{DO}\ \underline{N_5\ P_;}$		
E1	$\dots\ \underline{K_{DO}}$	lexical lexical	
	$\dots\ K_{DO}\ \underline{I_E\ P_=}$		
E6	$\dots\ K_{DO}\ \underline{I_E\ P_=}$		
S4	$\dots\ K_{DO}\ \underline{I_E\ P_=}$	compile_expression	$N_6 := \{R_{ASSIGN}\ S_E\ S_F\}$
	$\dots\ K_{DO}\ \underline{N_6\ P_;}$		
E1	$\dots\ \underline{K_{DO}}$	lexical lexical	
	$\dots\ K_{DO}\ \underline{K_{END}\ P_;}$		
E6	$\dots\ K_{DO}\ \underline{K_{END}\ P_;}$		
S11	$L_1\ K_{THEN}\ L_2\ \underline{K_{THEN}}$	lexical lexical	
	$L_1\ K_{THEN}\ L_2\ K_{THEN}\ \underline{K_{ELSE}\ I_E}$		
E3	$L_1\ K_{THEN}\ L_2\ \underline{K_{ELSE}}\ I_E$	compile_then	$N_7 := \{R_{BRANCH}\ L_3\}$
			$N_8 := \{R_{EQU}\ L_2\}$
	$L_1\ K_{THEN}\ L_3\ K_{ELSE}\ \underline{I_E}$	lexical	
	$L_1\ K_{THEN}\ L_3\ K_{ELSE}\ \underline{I_E\ P_=}$		
S4	$L_1\ K_{THEN}\ L_3\ K_{ELSE}\ \underline{I_E\ P_=}$	compile_expression	$N_9 := \{R_{ASSIGN}\ S_E\ S_\Lambda\}$
	$L_1\ K_{THEN}\ L_3\ K_{ELSE}\ \underline{N_9\ P_;}$		
E1	$L_1\ K_{THEN}\ L_3\ \underline{K_{ELSE}}$	lexical lexical	
	$L_1\ K_{THEN}\ L_3\ \underline{K_{ELSE}}\ I_M\ P_=$		
E5	$L_1\ K_{THEN}\ \underline{I_M\ P_=}$	end_if	$N_{10} := \{R_{EQU}\ L_3\}$
E4	$\underline{I_M\ P_=}$	end_if	$N_{11} := \{R_{EQU}\ L_1\}$
E6	$\underline{I_M\ P_=}$		
S4	$\underline{I_M\ P_=}$	compile_expression	$N_{12} := \{R_{ASSIGN}\ S_M\ S_J\}$
	$\underline{N_{12}\ P_;}$		
E1		lexical lexical	

Figure 10.16*b* Parsing the IF statements and DO group in Figure 10.15.

Source statement

CALL LOOK(X, T(I), NW + 2*Z);

Node table representation

1	R_{ARG}	S_X	
2	R_{SS}	S_T	S_I
3	R_{ARG}	N_2	
4	$R_{MULTIPLY}$	C_2	S_Z
5	R_{ADD}	S_{NW}	N_4
6	R_{ARG}	N_5	
7	R_{CALL}	S_{LOOK}	

Figure 10.17 Node table representation of a CALL statement.

the symbol table. If the symbol table entry S_{PR} is not flagged as being an array, we will assume that PR is the name of a procedure. The second addition is to recognize a comma as signaling the end of an argument and enter an ARG macro instruction in the node table. The remaining addition is to recognize the set of tokens

$$S_{PR} \, P_c \, P_(\, t_n \, P_)$$

where t_n is the token for the nth argument, as the end of the argument list. When this is recognized, the final ARG macro instruction and the CALL macro instruction can be entered in the node table.

EXERCISES

10.1 Modify the expression parser in Figure 8.18 so that it will parse references to elements of arrays, structures, and arrays of structures, as described in Section 10.5. Be sure the modified version will serve as **compile_expression** in pass one of Instran.

10.2 Using the reduction rules in Appendix E and **compile_expression** from Exercise 10.1, parse the procedure in Figure 7.12. Show the contents, at the end of pass one, of all the relevant tables (e.g., symbol, identifier, and node).

10.3 Modify and expand the reduction rules, providing new actions if necessary, so that the following features and statements will be acceptable to pass one of Instran. Also, modify and expand the productions in Appendix C, describe and define any new macro operations needed to represent the new features in the node table, and describe any modifications and additions to the tables that are required.

    a. Arrays with a lower bound other than zero, declared as in the example,

DECLARE X (1:8)

where 1 is the lower subscript bound and 8 is the upper subscript bound.

    b. The IL **while** and **repeat** loops, defined on pages 39 and 23, adapted for Instran.

    c. The IL **case** statement, defined on page 25, adapted for Instran.

10.4 Write pass one of Instran in IL.

11 | Pass Two of Instran

In Chapter 8 we saw that the major tasks of pass two are memory assignment and code generation. In this chapter we consider how these tasks are carried out in Instran. Memory assignment for instructions and labels is done during code generation. Assignment of memory for variables and constants is done at the beginning of pass two. Instran also does some local optimization during code generation. Code is generated using the technique of macro expansion. However, the simple macros used in Chapter 8 are not adequate to accomplish the desired local optimization. For this we need a richer macro definition facility.

11.1 Memory Assignment

We are concerned with memory assignment because we must generate code in the object procedure that references its variables and constants, and even the procedure itself (labels). Thus, since all references in Our 360 are relative to a base address, we must generate object code to obtain a base address for each such reference. This means we must know where each of the different kinds of entities are stored in memory. In addition, we must know how arrays and structures are represented in memory so that we can generate object code to obtain the address of any of the individual elements.

Everything except external variables and dummy arguments is included in the procedure segment. According to the definition of the EXTERNAL storage attribute, variables having this attribute will be located in other segments. The compiler will generate address constants for all external variables, and the loader will replace the address constants by the absolute addresses of corresponding external variables. These address constants are included in the procedure segment along with the ordinary constants. In order to complete the assembly of the object deck, the compiler must have the address, relative to the base of the procedure segment, of everything in the segment. However, we saw in Chapter 8 that these relative addresses cannot all be completely determined until the end of pass two. Therefore, we split the procedure segment up into blocks, as shown in Figure 11.1, and memory assignment for an entity consists of determining its address relative to the base of the appropriate one of these blocks. We will consider memory assignment within each of these blocks.

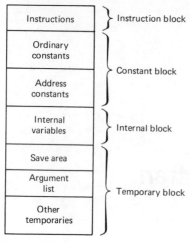

Figure 11.1 Division of a procedure segment into blocks.

The *instruction block* contains only instructions. The location addresses for instructions cannot be assigned until they are generated. Just as in the assembler, we keep a location counter. Whenever an instruction is generated during the code generation part of pass two, it is assigned to the location specified by the location counter. The location counter is then incremented by the length of the instruction. Both user labels and internal labels refer to instructions. Therefore, their values cannot be determined until their associated instruction has been assigned a memory location. As we saw in Chapter 10, an EQU macro instruction for each label was inserted into the node table immediately preceding the macro instruction with which the label is associated. During macro expansion, an EQU macro instruction defines the value of the label that is its operand to be equal to the current value of the location counter. This value is entered into the symbol table or the label table, depending on whether the label is a user label or an internal label.

The *temporary block* contains the *save area* and the *argument list,* which are part of the system standard calling sequence used to call another procedure. The save area is always 17 full words and the argument list will be N full words where N is the maximum number of arguments in any CALL statement in the procedure being compiled. The other temporaries are those needed in the evaluation of expressions. The number of this kind of temporaries that are needed will not be known until the end of pass two when all of the code has been generated. The addresses of these temporaries will be assigned at that time.

The *constant block* contains the ordinary constants that appeared in the source procedure and the address constants required for the external variables. The ordinary constants have been accumulated in the constant table. An address is assigned to each constant, observing its alignment requirement. This address assignment is exactly like the assignment of addresses for the literals at the end of pass one of the assembler in Chapter 4. Therefore, we will not describe it here. After addresses have been assigned to the ordinary constants, addresses are assigned to the address constants. An address constant is a full word, and one is needed for each different external variable. The symbol table is scanned for entries that have STORAGE = 'E'. Whenever such an entry is found, its required address constant is assigned to the next available full word in the constant block. The address of this full word, relative to the base of the constant block, is put into ADDRESS of the symbol table entry for the corresponding external variable.

In order to complete the memory assignment we must derive a value for the address field of the symbol table entry for each of the internal variables, external variables, and dummy arguments. In deciding what these values should be, we anticipate the way these entities will be referenced in the object code. As we know from previous chapters, we must use a base register to reference any entity, since its true address is not known until after the program has been loaded. Since the internal variables are part of the procedure segment, it seems reasonable to use G_{15} when referencing them because G_{15} contains the base address of the procedure segment as required by our system standards. When we use the base address of the procedure segment to reference a variable, we will need to set the displacement field of the referencing instruction equal to the variable's relative address in the procedure segment. Therefore, this is the number we want in the address field of the symbol table entry for the variable. What we actually put there is the address of the variable relative to the base of the internal block. In pass three (assembly) all such addresses will be incremented by the address, relative to the base of the procedure segment, of the base of the internal block.

References to the procedure's external variables and arguments will be made using some register other than G_{15} as a base register. Instran will have to generate code to load the proper address into this base register. The proper address for an external variable will be one of the address constants in the constant block. In order to reference this we will need its relative address, which is exactly what was put into the address field of the symbol table entry for the external variable when we did the memory assignment for the constant block. The proper address for a dummy argument will be in the argument list, which is supplied by the calling procedure. The base address of this argument list will be in G_{12} as a result of the system standard calling sequence. In order to locate the address for a particular dummy argument we need its relative address in the argument list. This relative address was put in the address field of the symbol table entry for the dummy argument when the entry was initially made during parsing of the PROCEDURE statement in pass one of Instran.

We have an additional problem with arrays and structures. When an individual element of a structure is referenced, we actually need to put the relative address of that particular element into the displacement field instead of the relative address of the base of the structure. Since there is a symbol table entry for every element of a structure, we will put the relative address of each element into the address field of its symbol table entry. For an element of an internal structure this address will be relative to the base of the internal block. In pass three it will be changed into an address relative to the base of the procedure segment. For an element of an external structure, this address will be relative to the base of the structure, since the base register used to reference it will contain the address of the base of the structure. When an element of an array is referenced, the relative address of that element within the array will have to be loaded into an index register.

In order to determine the relative address of an individual element in an array or structure we have to decide how arrays and structures will be represented in memory. We will store the elements of an array in the order of their subscripts, as shown in Figure 11.2, without any unnecessary space between them. The element with subscript 0 will have the lowest memory address. The elements of a structure will be stored in the order in which they are written in the source language declaration with no unnecessary space between them. If we are considering an array of structures, the structures in the array are stored in order of their subscripts. Then, within each structure, which is a single element of the array, the structure elements are stored in the order in which they appear in the declaration.

It may be necessary to include some unused space between the elements of an array or structure because of the alignment requirements of the different data types. An integer (FIXED) is a full word that must be stored beginning with an address that is evenly divisible by four. A character string (CHARACTER) may be stored beginning with any address one character per byte. A bit string (BIT) can be stored beginning with any bit in any byte. Thus for example, two bit strings can be stored in the same byte if their combined length is less than 9 bits. We cannot simply pack the elements of a structure together in memory with no unused space between elements, since the next element in sequence may not be able to be stored beginning exactly where the preceding element terminated. We might simply store each new element beginning with the next full word, but this could be extremely wasteful of memory for arrays of strings or structures containing strings.

We adopt a compromise and store each element beginning as soon as possible after the preceding element while satisfying its alignment constraints.

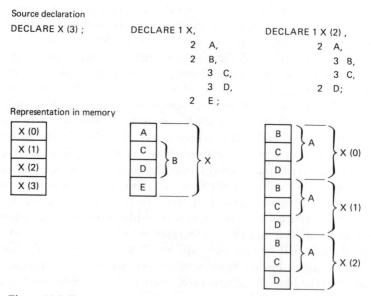

Figure 11.2 Representation of arrays and structures in memory.

This is illustrated in Figure Figure 11.3. In the table showing the relative addresses of the elements, the relative address for a bit string is shown as a pair of integers. The first integer is the relative address of the byte in which the bit string begins. The second integer is the bit number within that byte of the first bit of the string. This representation will be valid only if we insure

Source declaration

```
DECLARE 1 X INTERNAL,
          2 A BIT(3),
          2 Y,
            3 B CHARACTER(2),
            3 C FIXED,
          2 Z,
            3 D BIT(7),
            3 W,
              4 E BIT(5),
              4 F BIT(6),
              4 G CHARACTER(1),
          2 H FIXED;
```

Representation in memory

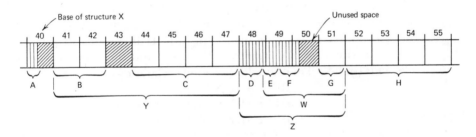

Structure element	Address of element relative to the base of the internal block	
A	(40,0)	The first integer is the relative
Y	(41,0)	address of the byte in which
B	(41,0)	the element begins, the second
C	(44,0)	integer is the bit in that
Z	(48,0)	byte where the element begins
D	(48,0)	
W	(48,7)	
E	(48,7)	
F	(49,4)	
G	(51,0)	
H	(52,0)	

Figure 11.3 Representation of a structure in memory.

that the address of the base of the structure is evenly divisible by four. This requirement also holds for arrays of integers or structures, but not for arrays of strings. The relative addresses of the elements of the structure are the values that will be put into the address field of the symbol table entries for the structure's elements.

In determining the relative addresses that will be put into the symbol table entries, we will use a location counter. Its value will be the relative address of the next available location in the internal block. The assignment of all variables except structures and arrays of structures is straightforward. The current value of the location counter, incremented as required for proper alignment, is assigned as the variable's relative address. The location counter is then incremented by the length of the variable. The length of an integer scalar is four. The length of a character string scalar is given by the length field of its symbol table entry. The length of a bit string scalar is equal to the ceiling of one eighth of its length (in bits), that is, the number of bytes required to store the bit string. For example, given the declarations:

$$\text{DECLARE A FIXED}$$
$$\text{DECLARE B CHARACTER (7)}$$
$$\text{DECLARE C BIT (11)}$$

the length of A is 4, the length of B is 7, and the length of C is 2 (= ceiling $(11/8)$ = ceiling(1.375) = 2).

The length of an array is the product of the number of elements in the array and the length of its elements. The number of elements in an array is one greater than its subscript upper bound. Since a bit string may be stored beginning at any bit, we can store the elements in an array of bits strings without any unused space. Thus, its length is the ceiling of one eighth of the product of the number of elements in the array and the length of the bit string elements (in bits). For example, given the declarations:

$$\text{DECLARE X(20) FIXED}$$
$$\text{DECLARE Y(14) CHARACTER(2)}$$
$$\text{DECLARE Z(50) BIT(3)}$$

the length of the array X is 80 bytes, the length of the array Y is 28 bytes, and the length of the array Z is 19 bytes, that is, the ceiling$((3*50)/8)$ (= ceiling $(150/8)$ = ceiling(18.75) = 19).

None of the external variables or dummy arguments need any further processing except for structures and arrays of structures. Structures, even when they are array elements, require a relative address for each of their elements. This includes external variables and dummy arguments as well as internal variables. These relative addresses are computed the same way for all structures, except that the address is relative to the base of the internal block for internal structures and relative to the base of the structure itself for structures that are external or dummy arguments. In computing the relative addresses we use an offset counter, the value of which will be the relative address in the structure of the next element of the structure. If the structure is internal, the value of the location counter gives the address of the base of the structure relative to the base of the internal block. The sum of the offset counter and the location counter is the proper relative address for the structure element and is the value that is put into the address field of the symbol table entry for the structure element. If the structure is external or a dummy argument the value of the offset counter is the proper relative address for a structure element and is the value that is put into the symbol table entry for the structure element.

The elements of a structure are assigned to memory in the order in which they appeared in the structure declaration. Since these elements may have different lengths, we must process them in this order. We can recover this order by following the son, brother, and parent pointers according to the algorithm given in Figure 11.4. The application of this algorithm to the structure in Figure 11.3 is shown in Figure 11.5. This figure shows the complete directed graph as represented by the son, brother, and parent pointers. The branches that are double lines show the sequence in which the structure's elements are assigned to memory, using the algorithm defined in Figure 11.4.

```
i: = index of symbol table entry for the structure;
while forever do
   if SYMBOL.SON[i]=0 then
      begin Assign memory to structure element corresponding
               to SYMBOL[i];
         if SYMBOL.BROTHER[i]=0 then
            repeat if SYMBOL.PARENT[i]=0 then Done;
               i:=SYMBOL.PARENT[i];
               Assign memory to structure element corresponding
                  to SYMBOL[i];
            until SYMBOL.BROTHER[i]≠0;
            else i:=SYMBOL.BROTHER[i];
      end;
      else i:=SYMBOL.SON[i];
end;
```

Figure 11.4 Algorithm to assign memory to elements of a structure in the proper order by traversing the structure's tree.

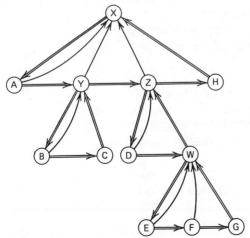

Figure 11.5 Directed graph for structure in Figure 11.3. Double lines show path traversed in memory assignment using the algorithm in Figure 11.4.

Looking at Figure 11.3, we can see that only scalar elements of a structure actually occupy any space in memory. A substructure element is a grouping imposed on the scalar elements. Nonetheless, the address field of the symbol table entry for a substructure must contain a value. This value will be identical to the value of the address field in the symbol table entry for the first scalar element in the substructure. However, in traversing the graph for a structure using the algorithm in Figure 11.4, we cannot determine this value when a substructure element is first encountered, since it depends on the alignment requirement of the first scalar in the substructure. Therefore, the address field for a substructure element is not set until that element is encountered for the second time during the traverse of the structure's graph.

It should be clear from an examination of the table of relative addresses in Figure 11.3 that an offset counter that counts bytes is not sufficient to enable bit strings to be assigned beginning at an arbitrary bit position within a byte. We actually need to count bits with the offset counter. The algorithm for incrementing and adjustment of the offset counter when assigning memory to a structure element is shown in Figure 11.6. Since the number of bits in a byte and the number of bytes in a full word are both powers of two, the offset counter (**offset** in Figure 11.6) has the format shown in Figure 11.7. The value of the function

$$\text{remainder (offset,8)}$$

is the remainder of **offset** divided by 8, that is, the first unassigned bit within the first unassigned byte. In the algorithm in Figure 11.6 we have assumed that the major structure is not internal and set the address field to the value of offset. To make the algorithm valid for internal structures as well, we would need a flag to indicate whether the major structure is internal or not. Then, everyplace in the algorithm where **offset** is stored into a symbol table entry,

```
type := SYMBOL.TYPE[i];
case type of
  'S':  SYMBOL.ADDRESS[i] := SYMBOL.ADDRESS[SYMBOL.SON[i]];
  'F':  begin align := remainder(offset,32);
        if align ≠ 0 then offset := offset + 32 − align;
        SYMBOL.ADDRESS[i] := offset;
        offset := offset + 32;
      end;
  'C':  begin align := remainder(offset,8);
        if align ≠ 0 then offset := offset + 8 − align;
        SYMBOL.ADDRESS[i] := offset;
        offset := offset + 8*SYMBOL.LENGTH[i];
      end;
  'B':  begin SYMBOL.ADDRESS[i] := offset;
        offset := offset + SYMBOL.LENGTH[i];
      end;
otherwise Error;
```

Figure 11.6 Expansion of "Assign memory to structure element corresponding to SYMBOL [i];" from Figure 11.4.

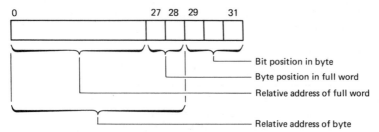

Figure 11.7 Format of offset in Figure 11.6.

we will need to test this flag and if the flag indicates an internal structure, the relative address of the base of the structure will have to be added to offset before it is stored into the symbol table entry.

11.2 Macro Definitions

Our approach to code generation will be to use macro definitions and their expansion to produce the object code in a simple assembly-like form. An overview of this approach was presented in Chapter 8. However, the simple form of macro definition and expansion that we saw there is not adequate for code generation when we are dealing with the complete Instran language in a realistic way. There are two major areas where it is deficient. The macro definition language is not rich enough to be able to generate code for accessing elements of arrays and structures. In addition, the code generated by such a simple mechanism is quite inefficient, since it generates a large number of redundant instructions. Figure 11.8 illustrates how much better the generated code could be. A more elaborate macro language and expansion facility can easily accomplish this local optimization.

Our strategy for achieving local optimization is to keep a record of what the state of the control processor will be after it executes a generated instruction. The macro definition language will provide a means of generating different code sequences, depending on the state of the control processor after execution of the preceding generated instructions. In this way the best code for that particular state can be generated. For example, the optimization shown in Figure 11.8 can be obtained by never generating a store instruction for a value until the register that contains it is needed for some other use. Then, if the next macro instruction that is expanded refers to a value that is in a register, it uses it from that register, thus eliminating two instructions (a store and a subsequent load).

A macro instruction has the form

$$OP \quad arg_1 \ldots arg_n$$

where $n \leqslant 3$, "OP" is the macro operation code, and the arg_i are tokens that stand for the arguments of the macro. A macro definition has the form:

$$DEF \quad OP \quad A1 \ldots An$$
$$body$$
$$END$$

where OP is the name of the macro being defined, and the Ai are the dummy variables of the definition for which the arguments in the macro instruction

Source statement
$$X = A + B*C*(D - E);$$

Macro instruction
(node table entry)   Chapter 8 code   Better code

1	$R_{MULTIPLY} \, S_B \, S_C$	L	3,B(,15)	L	3,B(,15)
		M	2,C(,15)	M	2,C(,15)
		ST	3,t_1(,15)	ST	3,t_1(,15)
2	$R_{MINUS} \, S_D \, S_E$	L	3,D(,15)	L	3,D(,15)
		S	3,E(,15)	S	3,E(,15)
		ST	3,t_2(,15)		
3	$R_{MULTIPLY} \, N_1 \, N_2$	L	3,t_1(,15)		
		M	2,t_2(,15)	M	2,t_1(,15)
		ST	3,t_1(,15)		
4	$R_{PLUS} \, S_A \, N_3$	L	3,A(,15)		
		A	3,t_1(,15)	A	3,A(,15)
		ST	3,t_1(,15)		
5	$R_{ASSIGN} \, S_X \, N_4$	L	3,t_1(,15)		
		ST	3,X(,15)	ST	3,X(,15)

Figure 11.8 Effect of local optimization.

will be substituted. The body of the definition is a sequence of statements each of which has the form

[*label*:] *keyword   operands*

where *keyword* identifies the type of statement and *operands* depends on the type of the statement.

The OUT statement is used to specify the generation of a single machine language instruction. Its form is

OUT   *mop   reg,adr*

The operands of this statement describe a machine instruction to be generated. In this description *mop* is the operation code for the machine instruction, *reg* specifies the R field, and *adr* specifies the remaining fields of the instruction. The value of *reg* is an integer. The operand *adr* will be an integer if the instruction has an R type operand. For example,

OUT   SR   3,3

which will generate the subtract register instruction

SR   3,3 $G_3 := G_3 - G_3$

For instructions with other operand types, *adr* may be one of several choices: a dummy variable (e.g., A2), a temporary storage reference (e.g., T), a text reference (e.g., *+4), or a literal (e.g., =F'1'). If *adr* is a dummy variable the values of the D, X, and R fields will depend on what token is supplied as the corresponding argument in the macro instruction being expanded. For example, assume that a macro definition includes the statement

OUT   L   3,A1   generate instruction to load A1

This statement will generate

L   3,380(,15)   load operand into G_3

if the token that is the first argument of the macro instruction corresponds to an internal variable whose relative address in the procedure segment is 380. On the other hand, this same definition statement will generate

$$\text{L} \quad 3,0(1,11) \qquad \text{load operand into } G_3$$

if the first argument of the macro instruction is a token corresponding to an external array. Here we are using G_{11} for the base address of the external array and G_1 for the subscript. If *adr* is a literal, temporary storage reference, or text reference, the D field in the generated instruction will contain the appropriate relative address, the X field will always be zero, and the B field will always be 15.

In addition to generation of machine instructions, we must be able to test and update the state of the control processor. All we are interested in is what values will be left in what registers as a result of executing the previously generated instructions. In the macro definitions we use the reserved symbols G0, ..., G15 to stand for the contents of the respectively numbered registers. The statement

$$\text{U} \quad \textit{symbol} \quad \textit{value}$$

is used to update the processor state. The operand *value* may be either a dummy variable of the macro definition, which refers to the token that is the corresponding argument of the macro instruction, or N*, which refers to a token pointing to the current macro. For example,

$$\text{U} \quad \text{G3} \quad \text{A2}$$

will update the processor state to indicate that as a result of executing the instructions that have just been generated, register G3 of the processor will contain whatever value the second argument of the macro instruction referenced. The statement

$$\text{U} \quad \text{G3} \quad \text{N}^*$$

updates the processor state to indicate that G_3 will contain the value computed by executing the instructions that have just been generated as a result of expanding the current macro instruction.

The state of the processor is tested by the conditional statement

$$\text{B} \quad \textit{label} \quad [\textit{condition} \ \{\text{T}|\text{F}\} \ ]$$

The specified condition is tested and, if it is true (T) or false (F), the macro expander branches to the statement with the specified label. If the condition is omitted this statement is an unconditional branch. The condition generally takes the form $Gi = Ak$, which is true if the current processor state shows that G_i will contain the value that is referenced by the kth argument of the macro instruction. For example,

$$\text{B} \quad \text{NONE} \quad \text{G3} = \text{A1} \quad \text{T}$$

will branch to the statement labeled NONE if G_3 will contain the value referenced by the first argument of the macro instruction currently being expanded.

Some examples will illustrate the use of these statements. The flowchart in Figure 11.9 displays the decisions that will generate locally optimized code for the PLUS macro. The macro definition corresponding to this code generation algorithm is shown in Figure 11.10. Figure 11.11 shows similar macro definitions for MINUS, MULTIPLY, and ASSIGN. We have adopted two conventions with respect to the control processor state and the use of the registers. First, if a register does not contain any value that is meaningful, this fact will be recorded in the processor state by a zero. If this is true the contents of the register need not be saved. This is illustrated in Figure 11.12 in the expansion

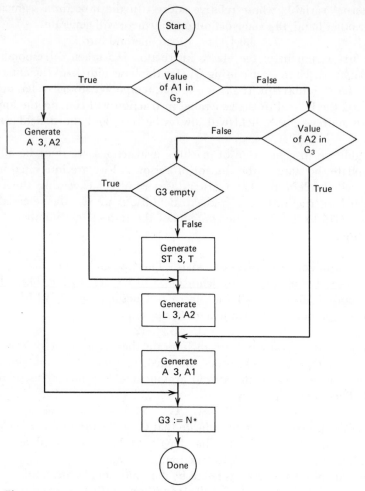

Figure 11.9 Sequence of decisions for generating optimized code for addition from the macro instruction R_{PLUS} A1   A2.

of the first macro instruction. Second, we are assuming that the only register that will contain meaningful values is G_3. Clearly, at the expense of more complex macro definitions, we could generate code, which used several of the registers. This would require more testing of the processor state, but the result would be more efficient code.

Since addition and multiplication are commutative, if either argument of these macro instructions will be in G_3, all we need do is generate an add or multiply instruction, which refers to the other operand. We cannot do this with subtraction and division, since they are not commutative. The structure of the Instran language is such that when an assignment occurs if G_3 contains any valid quantity, it must be the value of the expression on the right of the assignment operator, as in

$$A = B + C;$$

Otherwise, the macro will have come from a source statement of the form

$$A = B;$$

```
        DEF     PLUS A1 A2          define PLUS macro
        B       A1IN G3 = A1 T      branch if A1 in G₃
        B       A2IN G3 = A2 T      branch if A2 in G₃
        B       EMPTY G3 = 0 T      branch if no meaningful value in G₃
        OUT     ST   3,T            generate instruction to
                                        save contents of G₃
EMPTY:  OUT     L    3,A1           generate instruction to load A1
A1IN:   OUT     A    3,A2           generate instruction to add A2
        B       SET                 branch to SET
A2IN:   OUT     A    3,A1           generate instruction to add A1
SET:    U       G3   N*             update state indicating
                                        G₃ contains A1 + A2

        END
```

Figure 11.10 Macro definition corresponding to Figure 11.9.

```
        DEF     MINUS A1 A2         define MINUS macro
        B       A1IN G3 = A1 T      branch if A1 in G₃
        B       EMPTY G3 = 0 T      branch if no meaningful value in G₃
        OUT     ST   3,T            generate instruction to
                                        save contents of G₃
EMPTY:  OUT     L    3,A1           generate instruction to load A1
A1IN:   OUT     S    3,A2           generate instruction to subtract A2
        U       G3   N*             update state indicating
                                        G₃ contains A1 − A2

        END


        DEF     MULTIPLY A1 A2      define MULTIPLY macro
        B       A1IN G3 = A1 T      branch if A1 in G₃
        B       A2IN G3 = A2 T      branch if A2 in G₃
        B       EMPTY G3 = 0 T      branch if no meaningful value in G₃
        OUT     ST   3,T            generate instruction to
                                        save contents of G₃
EMPTY:  OUT     L    3,A1           generate instruction to load A1
A1IN:   OUT     M    2,A2           generate instruction to multiply by A2
        B       SET                 branch to SET
A2IN:   OUT     M    2,A1           generate instruction to multiply by A1
SET:    U       G3   N*             update state indicating
                                        G₃ contains A1 * A2

        END

        DEF     ASSIGN A1 A2
        B       IN   G3 = A2 T      branch if A2 in G₃
        OUT     L    3,A2           generate instruction to load A1
IN:     OUT     ST   3,A1           generate instruction to store A2 in A1
        U       G3   0              update state indicating G₃ does not
                                        contain a meaningful value

        END
```

Figure 11.11 Macro definitions for MINUS, MULTIPLY, and ASSIGN.

Node table entry	Processor state and arguments	Statements in definition that are interpreted		Assembly code generated	
1 $R_{MULTIPLY}$ S_B C_C	$G3 = 0$		DEF MULTIPLY A1 A2		
	$A1 = S_B$				
	$A2 = S_C$		B A1IN G3 = A1 T		
			B A2IN G3 = A2 T		
			B EMPTY G3 = 0 T		
		EMPTY:	OUT L 3,A1	L	$(3,0,15)S_B$
		A1IN:	OUT M 2,A2	M	$(2,0,15)S_C$
			B SET		
		SET:	U G3 N*		
	$G3 = N_1$		END		
2 R_{MINUS} S_D S_E			DEF MINUS A1 A2		
	$A1 = S_D$				
	$A2 = S_E$		B A1IN G3 = A1 T		
			B EMPTY G3 = 0 T		
			OUT ST 3,T	ST	$(3,0,15)T_1$
		EMPTY:	OUT L 3,A1	L	$(3,0,15)S_D$
		A1IN:	OUT S 3,A2	S	$(3,0,15)S_E$
			U G3 N*		
	$G3 = N_2$		END		
3 $R_{MULTIPLY}$ N_1 N_2			DEF MULTIPLY A1 A2		
	$A1 = N_1$				
	$A2 = N_2$		B A1IN G3 = A1 T		
			B A2IN G3 = A2 T		
		A2IN:	OUT M 2,A1	M	$(3,0,15)T_1$
		SET:	U G3 N*		
	$G3 = N_3$		END		
4 R_{PLUS} S_A N_3			DEF PLUS A1 A2		
	$A1 = S_A$				
	$A2 = N_3$		B A1IN G3 = A1 T		
			B A2IN G3 = A2 T		
		A2IN:	OUT A 3,41	A	$(3,0,15)S_A$
		SET:	U G3 N*		
	$G3 = N_4$		END		
5 R_{ASSIGN} S_X N_4			DEF ASSIGN A1 A2		
	$A1 = S_X$				
	$A2 = N_4$		B IN G3 = A2 T		
		IN:	OUT ST 3,A1	ST	$(3,0,15)S_X$

Figure 11.12 Macro expansion for the example in Figure 11.8.

and the value of G_3 is not meaningful. In either case, after the store instruction, the value in G_3 will no longer be needed.

We have also adopted a convention concerning the use of temporary storage. Because of the way our macro instructions are derived in pass one, a temporary is referenced only twice for a particular value. That is, initially a store instruction will put the value into a temporary and then an instruction such as load or add will use the value. After that the value will never be needed again, and that temporary location can be used to save another value. This allows us to use a very simple algorithm for the allocation of temporary locations. The temporary table is used to store the token that represents the value currently stored in the corresponding temporary location. If a temporary location is currently unused its temporary table entry will contain a zero.

A value in a temporary will originally be stored by a store instruction generated when the macro expander interprets the OUT statement

$$\text{OUT} \quad \text{ST} \quad 3,\text{T}$$

as, for example, in the fourth line of the expansion of the second node table entry in Figure 11.12. The symbol T refers to the first currently unused temporary. This is found by searching the temporary table for an entry that contains zero. If j is the index of this entry then the jth temporary is assigned. The token T_j will be used to reference this temporary and will be used in the assembly form of the generated instruction, which is put into the code table for pass three.

The assembly instruction that results from interpreting the above statement will have the form

$$\text{ST} \ (3,0,15) \ T_j$$

This format is expanded from that of Chapter 8, since we will need to use different register numbers in the R, X, and B fields, depending on the operand type (e.g., external, internal, . . .). The general format we are using is

$$i \ (R, X, B) \ t$$

where i is the index of an instruction in the instruction table, R the value of the R field, X the value of the X field, B the value of the B field, and t the token for the operand of the instruction.

When the statement

$$\text{OUT} \quad \text{ST} \quad 3,\text{T}$$

is interpreted and the jth temporary assigned, G3 in the processor state will contain a token of the form N_i. Recall that this type of token is used in a macro instruction (node table entry) to represent the value that will be computed as a result of executing the instructions that will be generated by the expansion of the macro instruction that is the ith entry in the node table. This token got into G3 as a result of the macro expander interpreting

$$\text{U} \quad \text{G3} \quad \text{N*}$$

when it expanded the macro instruction in the ith node table entry, the symbol N* being equivalent to N_i. In the example in Figure 11.12 this happens in the next to last line of the expansion of the first node table entry.

After the value N_i has been stored into temporary T_j, the token N_i will appear as an argument in a macro instruction. For example, N_1 is the first argument in the third node table entry in Figure 11.12. When this argument appears in an OUT statement, the temporary table must be searched for an entry containing N_1 in order to find the address to use in the instruction being generated. Once this entry has been found, the corresponding temporary is unallocated, so that it can be used again. For example, in Figure 11.12 the fourth line of the expansion of the third node table entry is

$$\text{RTIN:} \quad \text{OUT} \quad \text{M} \quad 2,\text{A1}$$

This will generate the instruction

$$\text{M} \ (2,0,15) \ T_1$$

since the first argument is the token N_1 and the first entry in the temporary table contains the token N_1.

The macro definitions for the EQUALS and FALSEB macros in Figure 11.13 illustrate the use of literals and text references in a definition. The EQUALS macro instruction results from the condition that appears in an IF statement, as shown in Figure 11.14. In a macro definition literals are written in the same form as they were in our assembly language, for example, $=F'1'$ and $=F'0'$. This type of reference will result in a C-type token in the assembly form, which points to the appropriate constant in the constant table, for example, the 6th and 7th generated instruction in Figure 11.14.

In a macro definition a text reference is written in the form $*+k$. A new token, which we will write as $*_k$, will be used in the assembly form. This will be transformed into a D field value that is k greater than the relative address of the instruction in which it appears. For example, in Figure 11.14 the instruction

$$BC\ (8,0,15)*_{10}$$

will be transformed into an instruction that will branch to the third instruction following, that is, to the instruction

$$L\ (3,0,15)\ C_1$$

Literal and text references will always be relative to G_{15} as a base register, and no index register will be used.

	DEF	EQUALS A1 A2	define EQUALS macro
	B	A1IN G3 = A1 T	branch if A1 in G_3
	B	A2IN G3 = A2 T	branch if A2 in G_3
	B	EMPTY G3 = 0 T	branch if no meaningful value in G_3
	OUT	ST 3,T	generate instruction to save contents of G_3
EMPTY:	OUT	L 3,A1	generate instruction to load A1
A1IN:	OUT	C 3,A2	generate instruction to compare A2
	B	SET	branch to SET
A2IN:	OUT	C 3,A1	generate instruction to compare A1
SET:	U	G3 N*	update state to indicate G_3 contains A1 = A2
	OUT	BC 8,* + 10	generate instruction to branch on equal
	OUT	SR 3,3	generate instruction to set G_3 equal to 0
	OUT	BC 15,* + 8	generate instruction to unconditionally branch
	OUT	L 3, = F'1'	generate instruction to set G_3 equal to 1
	END		
	DEF	FALSEB A1 A2	define FALSEB macro
	B	IN G3 = A1	branch if A1 in G_3
	OUT	L 3,A1	generate instruction to load A1
IN:	OUT	CL 3, = F'0'	generate instruction to compare with 0
	OUT	BC 8,A2	generate instruction to branch on equal
	U	G3 0	update state to indicate G_3 does not contain a meaningful value

Figure 11.13 Definition of the EQUALS and FALSEB macros.

Source fragment

IF A = B THEN

Node table form		Generated code	
8	$R_{EQUALS} S_A S_B$	L	$(3,0,15)S_A$
		C	$(3,0,15)S_B$
		BC	$(8,0,15)*_{10}$
		SR	$(3,3)$
		BC	$(15,0,15)*_8$
		L	$(3,0,15)C_1$
9	$R_{FALSEB} N_8 L_5$	CL	$(3,0,15)C_0$
		BC	$(8,0,15)L_5$

Figure 11.14 Example of code generated from EQUALS and FALSEB macros.

11.3 Procedure Definition and Call

We require that Instran follow the system standard with respect to the structure of procedures, procedure calling, and register usage. The system reserves four registers for special use. These registers and their use are:

G_{12}—base address of current argument list.
G_{13}—base address of current save area.
G_{14}—return point in procedure that was most recently called.
G_{15}—base address of currently executing procedure.

These are the registers that are used in the system's standard calling sequence. We used the standard sequence in Chapter 3, except that it was an abbreviated version, since the procedure that we looked at there did not call another procedure. We now must use the full standard calling sequence.

Associated with each procedure's execution is a save area. When one procedure calls another, it must provide the save area for the called procedure. This save area is used to save the contents of the registers at the time the called procedure is entered. The base address of this save area is supplied to the called procedure in G_{13}. In order to return back to the outermost procedure from a nested set of procedure calls, the save areas must be linked together. A save area consists of 17 full words, which contain the information shown in Figure 11.15 when the procedure corresponding to the save area is executing. The way save areas are linked together is shown in Figure 11.16.

The ENTER, RETURN (and END), and CALL macro instructions must generate code that conforms to the system standard calling sequence. The ENTER macro generates the code, which is executed on entry to the procedure. This code must save all of the registers and link a new save area onto the chain of save areas. Assuming that procedure Q is being called from procedure P (see Figure 11.16), the required code is:

```
STM    14,12,8(13)         save all registers
LA     2,save(,15)         get absolute address of new save area
ST     13,0(,2)            set backward link in new save area
ST     2,4(,13)            set forward link in Q's save area
```

base address +	0	base address of preceding save area
	4	base address of next save area
	8	G_{14} (return address)
	12	G_{15} (base address of current procedure)
	16	G_0
	20	G_1
		. . .
	64	G_{12} (base address of argument list for current procedure)

Figure 11.15 Contents of a save area.

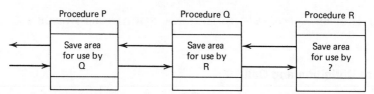

Figure 11.16 Linking of save areas.

The new save area is a block of 17 full words, beginning at relative address *save* in segment Q. Q's save area is supplied by P and is in segment P, beginning at the absolute address, which is in G_{13}.

The code to be executed when Q returns to P is

    LM   14,12,8(13)   restore all registers
    BR   14   branch to return point in P

This code is generated by both the RETURN and the END macros. There is no need to unlink explicitly the save area provided by Q. The next time Q is called, the save area will be linked onto the end of whatever chain of save areas exists at that time.

When Q calls another procedure, R, a switch must be made to the save area that Q linked onto the end of the save area chain. This will be R's save area. The code to do this switching and actually make the call will be generated by the CALL macro. The required code is:

    LA   13,save(,15)   switch to R's save area
    LA   12,arg(,15)   get address of argument list for R
    L   15,entry(,15)   get address of entry point in R
    BALR   14,15   branch to R
    L   13,0(,13)   switch back to Q's save area
    L   15,12(,13)   restore base address of segment Q
    L   12,64(,13)   restore address of Q's argument list

We are assuming that the argument list for R is constructed in Q's temporary block, as shown in Figure 11.1, beginning at relative address *arg*. The address constant needed to branch to procedure R is located in Q's constant block at relative address *entry*.

A system standard argument list consists of a list of addresses, one for each argument. If the argument is a scalar variable or constant then the argument address is the address of that variable or constant. If the argument is an element of a structure or array then the argument address is the address of that particular element of the structure or array. If the argument is the entire structure or array then the argument address is the address of the base of that structure or array. If the argument is a compound expression then the argument address is the address of a temporary in which the value of the expression is stored. The argument list may be stored anywhere. However, we will always store it in the temporary block of the procedure.

The code to construct an argument list is generated by ARG macro instructions. Each macro instruction

$$R_{ARG} \; t_k$$

generates the instructions

      code to compute the address corresponding to t_k and load it into G_2

    ST $2,4*i+arg(,15)$ store address in full word i of argument list

where t_k is a token corresponding to argument number i (counting from zero). The address of this argument is stored into full word i of the argument list. This full word is located $4*i$ bytes past the beginning of the argument list, which has a relative address of arg. For example,

$$R_{ARG} \; S_X$$

will generate the code

    LA $2,rX(,15)$

    ST $2,arg+4(,15)$

assuming that argument number 1 is X, which is an internal variable. In the first instruction we are letting rX represent the relative address of X.

11.4 Address Computation

As we know from earlier chapters it is the loader that determines where in memory each procedure is located. Thus, the compiler cannot compute any absolute addresses. All instructions that reference memory must do so relative to the contents of some base register. The system standard calling sequence requires that the calling procedure load G_{15} with the base address of the called procedure. Therefore, an instruction that references anything in the procedure segment may do so relative to G_{15}. In order to reference anything that is not in the procedure segment the object code generated by the compiler must contain instructions to load an address into some other register. Determining what address to load into a base register and generating the code to do so is one of the three major problems in address computation.

The second major problem is connected with references to array elements. In the source procedure a reference to an element of an array includes a subscript that identifies a particular array element. In the object code this requires use of an index register. The problem here is generating object code to compute the proper value and load it into an index register. The third major problem results from a reference to an element of a structure. In order to reference a structure element, the compiler must compute a value, for the D field of the instruction, that corresponds to the element's position in the structure. All three of these problems can occur in a single reference, for example, in a reference to a structure element in an array of structures that is an argument of the procedure.

When the macro expander comes to an OUT statement in a macro definition, the address processor must determine the value of the D, X, and B fields in the instruction generated by the OUT statement (unless the instruction has an R-type operand). If the *adr* operand in the OUT statement specifies a temporary storage reference (T), a text reference (*+4), or a literal (=F'1'), the referenced entity is in the procedure segment. Therefore, the B field will equal 15. Since none of these is an array reference, the X field will equal 0. The value of the D field will be the relative address of the referenced entity in the procedure segment.

If the *adr* operand is a dummy variable, the values of these fields will depend on the token that is the corresponding argument of the macro instruction being expanded. If this token has a type of L (internal label), C (constant), N (node table, which is transformed into a temporary storage reference), S (symbol table) that corresponds to a user label, or S that corresponds to an internal variable that is not an array, then the value of the B field is 15, the value of the X field is 0, and the value of the D field is the relative address of the entity in the procedure segment.

When the token corresponding to a dummy variable, which is the *adr* operand of an OUT statement, is type S corresponding to an external variable or type S corresponding to a dummy argument of the procedure being compiled, G_{15} cannot be used as a base register. In Instran we will adopt the convention that G_{11} will always be used as a base register for referencing all entities not in the procedure segment, since we cannot hope to use a separate register for each different reference. If the entity is not an array or structure element, the value of the D field and the value of the X field will both be 0, while the B field will equal 11.

In addition to determining these field values the address processor will have to generate instructions to load the proper address into G_{11}. If the reference is to an external variable the proper address is an address constant and the required instruction is

$$L \quad 11,ac(,15)$$

where *ac* is the relative address of the corresponding address constant. If the reference is to a dummy argument the proper address is in the argument list and the required instruction is

$$L \quad 11,4*i(,12)$$

when the reference is to the *i*th dummy argument.

The address processor inserts the appropriate instruction immediately preceding the instruction generated by the OUT statement. For example, consider the source language statement

$$X=Y;$$

The single macro instruction

$$R_{ASSIGN} \ S_X \ S_Y$$

results from parsing this statement. The code that results from this macro instruction depends on the attributes of X and Y. If both are internal the generated code is:

L	3,rY(,15)	get Y
ST	3,rX(,15)	store Y into X

where rX and rY are the relative addresses of X and Y. If X is external and Y is internal the generated code is:

L	3,rY(,15)	get Y
L	11,Xac(,15)	get address constant for X
ST	3,0(,11)	store Y into X

where Xac is the relative address of the address constant for X. The second instruction is inserted by the address processor when it processes the reference to X. If X is external and Y is a dummy argument the generated code is:

L	11,4*Ynr(,12)	get address of Y from argument list
L	3,0(,11)	get Y
L	11,Xac(,15)	get address constant for X
ST	3,0(,11)	store Y into X

where Ynr is the number of the dummy argument Y. The first instruction is inserted by the address processor when it is processing the reference to Y, and the third instruction is inserted when the address processor processes the reference to X.

A reference to an element of a structure requires only slightly more work by the address processor. Recall from Chapter 10 that a reference to an element of a structure is a single token, which points to the symbol table entry for that element. For example, X.Y.Z in a source statement results in a single token S_i if the ith entry in the symbol table is the entry for Z. If the structure is internal then the address field of the ith symbol table entry contains the address of the structure element Z relative to the base of the internal block. Thus, the values of the D, X, and B fields can be determined just as if Z were a nonstructure internal variable. If the structure is external or a dummy argument then the address field of the ith symbol table entry contains the address of Z relative to the base of the structure. Thus, the D field will be equal to this relative address, the X field will be equal to 0, and the B field will be equal to 11.

As before, the address processor must insert an instruction to load the corresponding address into G_{11}. However, the relative address of the address constant or the number of the dummy argument is in the address field of the symbol table entry for the major structure. Therefore, the address processor must follow the parent pointers, beginning with S_i until it reaches the symbol table entry for the major structure. In fact, the address processor will always have to follow the parent pointers to the major structure in order to find out if the structure is internal, external, or a dummy variable.

A reference to the element of an array is more complicated. The instruction that references an array element will have to specify an index register that contains the proper index value corresponding to the subscript in the source language reference. We will reserve G_1 for this index value. Pass one generates an SS macro instruction for each subscript that appears in the source language procedure. For the example in Figure 11.17, if we expand the ASSIGN macro instruction using the rules developed so far, N_1 and N_2 will produce addresses that refer to temporary storage. This is clearly not what is wanted. We want the addresses in the load and store instructions that result from the ASSIGN

Source statement	Node table entries
X(I) = Y(J);	1 R_{ss} S_X S_I
	2 R_{ss} S_Y S_J
	3 R_{ASSIGN} N_1 N_2

Figure 11.17 Node table entries for example source statement.

macro instruction to reference X(I) and Y(J), using G_1 as an index register. To accomplish this, the macro expander ignores all SS macro instructions in its normal sequencing. Then, whenever the address processor processes a token N_i, it will look at the ith entry in the node table. If this entry is not an SS macro instruction then a normal reference to a temporary is generated for N_i.

If the ith entry in the node table has the form

$$R_{ss} S_j t_k$$

then the address processor must use S_j in place of N_i and insert code to compute the index value corresponding to the subscript value represented by the token t_k and load this value into G_1. What we mean by using S_j in place of N_i is that the value of the D and B fields in the instruction generated by the OUT statement for which N_i is the operand are computed as if S_j instead of N_i were the operand.

Some examples will help in understanding the code that is inserted for an array reference and how it is merged with the code required for other reasons. In Figure 11.17 let us assume that X, Y, I, and J are all internal and that X and Y are arrays of integers. The code generated for the assignment statement will then be:

L	1,rJ(,15)	get value of subscript J
SLL	1,2	index = 4*J
L	3,rY(1,15)	get Y(J)
L	1,rI(,15)	get value of subscript I
SLL	1,2	index = 4*J
ST	3,rX(1,15)	store Y(J) into X(I)

Notice that since the elements of X and Y are integers, their length is 4 bytes, thus the required index is four times the subscript value. This multiplication can be done by a left shift of two bits.

The token t_k in an SS macro instruction is given the normal address processing. For example, if X, Y, I, and J are all external, the code generated for the assignment statement in Figure 11.17 will be:

L	11,Jac(,15)	get address constant for J
L	1,0(,11)	get subscript J
SLL	1,2	index = 4*J
L	11,Yac(,15)	get address constant for Y
L	3,0(1,11)	get Y(J)
L	11,Iac(,15)	get address constant for I
L	1,0(,11)	get subscript I
SLL	1,2	index = 4*I
L	11,Xac(,15)	get address constant for X
ST	3,0(1,11)	store Y(J) into X(I)

Even subscripts may be subscripted. The source statement

$$X = Y(I(K));$$

results in the node table entries:

1	R_{ss}	S_I S_K
2	R_{ss}	S_Y N_1
3	R_{ASSIGN}	S_X N_2

Assuming that X, Y, I, and K are all internal with data type fixed, the preceding node table entries will generate:

```
L    1,rK(,15)        get subscript K
SLL  1,2              index = 4*k
L    1,rI(1,15)       get subscript I(K)
SLL  1,2              index = 4*I(K)
L    3,rY(1,15)       get Y(I(K))
ST   3,rX(,15)        store Y(I(K)) into X
```

Subscripted subscripts to an arbitrary depth cause no problems if the address processor is coded so that it can call itself (a recursive procedure).

Since a reference to a structure element is a single token, no additional problems arise for a reference to an element of a structure that is itself an element of an array of structures. In the array of structures in Figure 11.18 the length of an element of the array X is 12. The source statement
$$Y = X.C(K);$$
will result in the node table entries

```
1  R_SS       S_C  S_K
2  R_ASSIGN   S_Y  N_1
```

which will generate the code:

```
L    1,rK(,15)        get subscript K
M    0,=F'12'         index = 12*K
L    11,Xac(,15)      get address constant for X
L    3,rC(1,11)       get X.C(K)
ST   3,rY(,15)        store X.C(K) into Y
```

The major difference between this code sequence and the previous examples is that a multiply instruction is needed instead of a left shift instruction, since 12 is not a power of 2. Also the D field of the instruction that references the structure element is the address of C relative to the base of the structure.

Source declarations
DECLARE K FIXED INTERNAL;
DECLARE Y FIXED INTERNAL;
DECLARE 1 X(20) EXTERNAL,
           2 A FIXED,
           2 B CHARACTER (2),
           2 C FIXED;
Memory representation of X

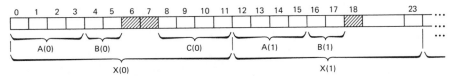

Figure 11.18 Example of an array of structures.

EXERCISES

11.1 Write an IL procedure that does the complete job of memory assignment as described in Section 11.1.

11.2 Write a code generator in IL, similar to the one in Figure 8.22, that will generate code by expanding the macros defined in Figures 11.10, 11.11, and 11.13. See Exercise 11.3 before starting this exercise.

11.3 Write an IL procedure that can be used by the code generator in Exercise 11.2 that will perform the address computation described in Section 11.4.

11.4 Using the results of Exercises 11.1 to 11.3, show the code generated for the node table that was the result of Exercise 10.2.

11.5 Using the macro definition language described in Section 11.2, define any new macros required in Exercise 10.3. Modify the memory assignment and code generation so that the features in Exercise 10.3 will be acceptable to pass two of Instran.

11.6 Write pass two of Instran in IL.

12 | Other Problems in Translation

There are several important problems in translation that were not discussed in the preceding chapters. Unfortunately, we cannot explore them in detail in this text. However, a brief exposure to the essentials of these problems will be useful in rounding out the reader's knowledge of compilers. In the following sections we examine the major problem in generating code for operations on character and bit strings. Then the extremely important idea of a recursive procedure is examined, and the problems in implementing it are explored. Finally, some of the optimization procedures that can be applied to generate better object code are pointed out.

12.1 String Operations

The Instran language includes bit and character string data. The following operations are defined for both character and bit strings.

|| (concatenation)
¬ =   (relational not equals)
=   (relational equals)
=   (assignment)

The following additional operations are defined for bit strings only.

¬   (logical complement)
&   (logical and)
| (logical or)

In order to extend Instran to translate these operations we will need macro definitions for all of the operations. We even need additional macro definitions for the ASSIGN and relation macros, since these operations for strings are different from the corresponding operations for integers. This is primarily because integers are always 4 bytes in length while strings may be any number of bytes in length. In fact, bit strings may even begin in other than the first bit of a byte.

What are these macros definitions? We will not attempt to describe them in any detail, since they are quite complex. One example will give the reader a feeling for this complexity. Suppose that A, B, and C are all elements of the structure shown in Figure 12.1. The source statement

Source declaration
DECLARE 1 X INTERNAL,
        2 P BIT(3),
        2 C BIT(9),
        2 Q BIT(25),
        2 A BIT(4),
        2 R BIT(24),
        2 B BIT(5),
        2 S BIT(26);
Memory representation

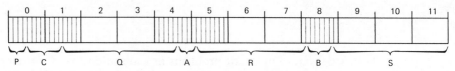

Figure 12.1 An example structure of bit strings.

$$X.C = X.A || X.B$$

results in the macro instructions

    1 R_{CAT} S_A S_B
    2 R_{ASSIGN} S_C N_1

These macro instructions must generate the code:

SR	2,2	clear G_2
L	3,rX+4(,15)	get X.A
SLL	3,5	move X.A to bit 0 of G_3
SLDL	2,4	move X.A into G_2
L	3,rX+8(,15)	get X.B
SLL	3,1	move X.B to bit 0 of G_3
SLDL	2,5	move X.B into G_2

               X.A||X.B is now in bits 23–31 of G_2

L	3,rX(,15)	get X.C
N	3,=X'1FF00000'	X.C:=0
ST	3,rX(,15)	return X.C to memory
SLL	2,20	move X.A||X.B to bit 3 of G_2
O	2,rX(,15)	store new value of X.C

It should be clear that the code required for bit string operations will be much more complex if the bit strings are longer than 32 bits, or are divided between two full words.

Character string operations are less complicated because they always consist of an integral number of bytes. In addition, computers such as the IBM 360 have special instructions for character string operations. Nonetheless, character string operations still introduce considerable complexity in code generation. In addition, if the source procedure contains a lot of string operations of either kind, a large amount of code will be generated. This may become excessive. For this reason, string operations are often implemented by generating a procedure call for each string operation. Then the system library would contain a procedure for each string operation. Each such procedure call requires generation of approximately a dozen instructions.

This number is fixed except for a variation of a few instructions, depending on the storage class of the operands and whether they are elements of arrays. This is clearly an appreciable saving in memory for many of the more complex cases of string operations; however, its cost is additional execution time. The instructions that must be executed to call and return from the procedure for a string operation would not be executed if in-line code were generated for the string operation.

When the string operation is relatively simple, using a procedure to carry out the operation incurs additional cost in both memory space and execution time compared to in-line code of a few instructions. Therefore, many compilers use a mixed strategy for implementation of string operations. The simple cases of string operations result in the generation of in-line code, while the more complex cases result in the generation of code to call a procedure to carry out the string operation. This mixed strategy introduces more complexity into the compiler, but is worthwhile if the source procedure uses string operations with any regularity.

12.2 Recursive Procedures

A recursive procedure is similar to the recursive syntax definitions that we encountered in Chapter 7 in the sense that it is defined in terms of itself. For example, we saw earlier that the address processor in pass two of Instran must generate instructions to compute and load the index corresponding to a subscript. This requires processing an address when the SS macro instruction is interpreted. Thus, the address processor is defined in terms of itself. In order to process an address it may have to process another address first. This may happen to any depth, since there is no limit to the depth of subscripting.

A recursive procedure P contains a call to itself or a call to another procedure that calls another procedure and so on until P is called again. These cases are illustrated in Figure 12.2. A recursive procedure must be able to accept calls to itself to any depth without getting mixed up or giving incorrect results. The *depth of recursion* of a recursive procedure is dynamic. At any given time, it is the number of unsatisfied calls to itself other than the initial

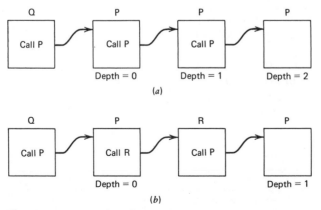

Figure 12.2 Examples of recursive calls. (*a*) Recursive call of procedure P to a depth of 2. (*b*) Hidden recursive call of procedure P to a depth of 1.

call. An unsatisfied call is one for which the corresponding return has not been executed. In Figure 12.2, if execution is in P at Depth = 2, there are three unsatisfied calls of P. In the figure, P has been drawn as if there were an additional, separate copy for each recursive call. If this were actually the case there would be no problem. However, with most recursive procedures the maximum depth of recursion is arbitrary and unknown. Therefore, there is no way to tell before execution how many copies of a recursive procedure would be needed. Even if we could tell, in most cases the memory would not be large enough to hold all of the needed copies. Therefore, we must get along with only one copy of the procedure.

The essential problem is that a single copy of a procedure segment has only one temporary block. For example, each unsatisfied recursive call requires its own, distinct argument list, but there is only one place to put an argument list, since it is in the temporary block. If we had separate copies then we would have separate places to put the argument lists. We can solve this problem by separating the temporary block from the procedure segment and making a new copy of the temporary block for every recursive call. These blocks will be linked together just as the save blocks are linked together so that the proper sequence of returns can be executed. In fact, since the save block is already in the beginning of the temporary block, the system standard calling sequence needs to be modified only slightly.

The easiest way to manage the memory required for the separate temporary blocks is put them all into a single segment. Since there is a one-to-one correspondence between these temporary blocks and unsatisfied procedure calls they need to be acquired and released in the same order as the corresponding procedure calls and returns. That is, space for temporary blocks is released in an order that is the exact inverse of the order in which it was acquired. This is precisely the way a stack behaves. Therefore, we will use a stack for the temporary blocks. We will call a temporary block in the stack a *stack frame*. The format of a stack frame is shown in Figure 12.3. The stack usage for the two cases in Figure 12.2 is shown in Figure 12.4.

In order to implement this we interpret the contents of G_{13} as the base address of the current stack frame instead of just the save area. Then all refer-

G_{13} $\rightarrow$	backward link (base address of preceding frame)
+4	forward link (base address of next frame)
+8	save registers
+68	argument list
	other temporaries

Figure 12.3 Format of stack frame.

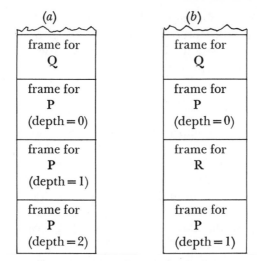

Figure 12.4 Stack usage for the recursive calls in Figure 12.2.

ences to the temporary block will be relative to G_{13} instead of relative to G_{15}. In the modified standard calling sequence, a procedure saves the contents of the registers in the save area, which is part of its caller's stack frame. This is basically the same as the old standard calling sequence. The major difference is that immediately after entry to a procedure, it must switch to a new stack frame, which will be used for its temporary storage. The forward link in this frame will also be set. Only the called procedure itself can set this link because only it knows how large the frame needs to be.

The new standard calling sequence is almost the same as the old. When a procedure is entered (the ENTRY macro instruction) the following code is executed:

```
STM    14,12,8(13)      save all registers
L      2,4(,13)         get forward link (base address of new frame)
ST     13,0(,2)         set backward link in new frame
LR     13,2             switch to new frame
LA     2,length(,13)    get address of word following new frame
ST     2,4(,13)         set forward link in new frame
```

where *length* is the length of the frame needed by this procedure. The code executed to return to the calling procedure (RETURN macro instruction) is:

```
L      13,0(,13)        switch to caller's frame
LM     14,12,8(13)      restore all registers
BR     14               branch to caller's return point
```

When a procedure calls another (CALL macro instruction), we do not switch save areas as was done in the old calling sequence, since this is done automatically when stack frames are switched. The code for calling a procedure is:

```
LA     12,arg(,13)      get base address of argument list
L      15,entry(,15)    get address of entry point
BALR   14,15            branch to called procedure
L      12,0(,13)        get backward link
L      15,12(,12)       restore base address of calling procedure
L      12,64(,12)       restore base address of its argument list
```

Since G_{13} is pointing to the stack frame for the calling procedure after return from the called procedure, in order to restore the contents of G_{15} and G_{12}, which were used in the call, we must reach back into the preceding stack frame to get the proper values.

12.3 Optimization

There are a number of ways in which the execution time or storage requirements of a procedure can be improved. In this section we will briefly point out a few of them. Generally, their implementation is quite complicated and forms a major part of the specialized study of compilers. Most optimizations can be classified as local or global. Local optimization is limited to consideration of a very short sequence of instructions, usually one statement or less. Global optimization treats the whole procedure, or a large part of it, all at the same time.

The elimination of redundant load and store instructions in pass two of Instran is a local optimization. The processor state that was used in the macro definition language reflects the action of only the instructions for one statement. There are a number of other local optimizations. The order of computing the terms in an expression can be reordered so as to minimize the number of store and load instructions that are required. Of course, this reordering must be constrained so that the resultant value of the expression is not changed. Other local optimizations attempt to utilize special instructions of the computer in order to shorten the code generated in certain special cases.

One of the most fruitful global optimizations is to move computations from inside a loop to outside the loop when the value computed does not change inside the loop. For example, suppose the source language procedure contains the statements:

```
DO I = 1 TO N;
    X(I) = A+B+W(I);
    DO J = 1 TO M;
        Z(J) = X(3*I+J);
    END;
END;
```

These statements can be compiled as if the programmer has written:

```
T_1 = A+B;
DO I = 1 TO N;
    X(I) = T_1+W(I);
    T_2 = 3*I;
    DO J = 1 TO M;
        Z(J) = X(T_2+J);
    END;
END;
```

The second set of statements computes exactly the same quantities, but executes $N-1$ fewer additions, $N(M-1)$ fewer multiplications, and $(N+1)$ additional assignments. Since addition and assignment have approximately equal execution times, they cancel out. However, in most computers the execution time for multiplication is significantly larger than for addition or assignment, so that saving approximately $N*M$ multiplications can decrease the execution time for this example by more than 30% on some computers. Studies have shown that many procedures spend most of their execution time inside loops executing a very small percentage of the total number of instructions in the procedure. Thus, a little saving in the execution time of these instructions can account for a substantial part of the execution time. The algorithm to accomplish the optimization just described is quite complicated to implement, but the payoff is well worth the effort.

Another global optimization that has significant payoff is the optimal use of the registers. We have a number of reserved registers, but there are still 10 left, 12 if we include GR_1 and GR_{11}. These registers can all be used in computing the value of expressions, for indexes of array elements, and for address constants of external variables. If all of them are used it is possible to eliminate a large number of load and store instructions in the evaluation of expressions and to eliminate load instructions when referencing array elements and external variables. The general strategy is to leave values, indexes, and address constants in these registers for future use. If the right things are left in the registers, then many load instructions can be eliminated, since the needed quantity is already in a register. The algorithm to select the right things to leave in the registers is quite complicated, but usually worth the cost.

Most of the global optimizations that have significant payoff are complicated. In addition, they need global information about the procedure and may have to reprocess the instructions generated by pass two several times. Therefore, global optimization is usually done in one or more separate, additional passes.[1]

EXERCISES

12.1 Modify Instran so that it will accept the string operations described in Section 12.1.

12.2 Modify Instran so that it will compile all procedures as recursive procedures and compile all CALL statements as calls to recursive procedures. Use the conventions described in Section 12.2.

12.3 Develop an algorithm that moves *invariant computations* from inside a loop to outside the loop. An invariant computation is one that computes a value that does not change inside the loop, as described in Section 12.3. How can you tell if the value of an expression is invariant inside a loop?

12.4 Modify the code generation in Chapter 11 so that all of the available registers (i.e., those not reserved by the system conventions) are used if their use will make the generated code more efficient in respect to the number of instructions generated.

[1] D. Gries, *Compiler Construction for Digital Computers*, Wiley, New York, 1971; A. V. Aho and J. D. Ullman, *The Theory of Parsing, Translation, and Compiling—Volume II: Compiling*, Prentice-Hall, Englewood Cliffs, N.J., 1973.

13 | Operating Systems

There are many different operating systems, at least one for each different computer, except perhaps for some of the smallest computers. In fact, several operating systems exist for many of the larger computers. Some of the larger, well-known operating systems are: OS for the IBM 360 and 370 series computers; GCOS for the Honeywell 600 and 6000 series computers; SCOPE for the CDC 6000 series computers; MCP for the Burroughs 6500 and 6700 computers; TSS for the IBM 360 model 67 computer; and Multics for the Honeywell 645 and 6180 computers. Detailed study of any one of these systems is beyond the scope of this text. Furthermore, such a study is not a realistic way to explore the basic problems with which an operating system has to cope. In the following chapters our study of the basic problems of an operating system will be mostly independent of any specific system. Whenever it is useful to explore an example, we will usually refer to the simple time-sharing system defined in the next section, just as we used a simple compiler for examples in the preceding chapters.

On the surface there seem to be many different kinds of operating systems. Some of the terms used to describe different systems are: batch processing, time sharing, multiprogramming, multiprocessing, remote batch, remote access, on-line, interactive, conversational, and sequential processing. We will see that there is more similarity among these different systems than differences. We will not attempt to explore all of the differences in detail. However, we will define a few attributes of operating systems that hopefully will illustrate the similarity of systems and show that the differences are usually in degree rather than kind.

The primary function of any operating system is to provide the best possible balance between ease of use and cost. Problems of a system concerned with ease of use we group under the heading of *the user interface*, while problems concerned with minimal cost are grouped under the heading of *resource management*. These two groups are not meant to be disjoint. Indeed, most of the problems will involve both resource management and the user interface. It is perhaps better to consider this categorization as two different viewpoints or orientations for approaching problems instead of a grouping of problems.

The three major resources of a computer system that must be managed are the control processors, the memory, and the I/O processors, with their attached devices. Our study will focus around these three resources. In this chapter we will briefly discuss the management of these three resources under the headings of *process control, memory management,* and *input-output,* which are the three major functions of an operating system. Because of their more and more significant importance, three related problems are treated under the additional heading of *sharing, privacy, and protection.* In these brief discussions we only try to expose the problems that each of these topics encompasses. In succeeding chapters we will explore each topic in more depth and discuss some solutions for the major problems. Before doing this we will describe the simple time-sharing system that we will use for examples and define some system attributes that have significant implications with respect to the problems that we will consider.

13.1 A Simple Time-Sharing System

We continue our case study approach by using a simple time-sharing system named Insys (Instructional system) to illuminate the basic concepts and problems in operating system design. In this section we describe Insys which, although simple, is similar in spirit to Multics and TSS. It contains the most significant features of these systems. However, it must be realized that many features were omitted, and those that are included have been greatly simplified in order to be able to describe the system in this text. The following paragraphs constitute a brief overview description of Insys. More details will be found in the following chapters, which explore the major functions and problems.

The hardware is basically that of Chapter 2, except that we have added another control processor, another I/O processor, and many more devices. Figure 13.1 shows the hardware configuration. All of primary memory is accessible by all of the processors, both control and I/O. Communication lines not shown in the diagram connect the control processors and the I/O processors, allowing a few very simple signals to pass between these processors. The communications control unit interfaces with a telephone exchange, thus providing a communication path between the computer system and each remote terminal through normal telephone lines. Users with a suitable device, such as a teletypewriter, may dial the computer's telephone number and get connected to the computer. Their teletypewriter then serves as an I/O device. Both system and user programs in the primary memory can be executed by either control processor without distinction. Both control processors may be executing programs at the same time and, in addition, both I/O processors may also be doing input or output.

Insys is designed so that more than one user may be using the system at the same time. This is true in several ways. More than one user program may simultaneously be in the computer's memory (in primary memory, drum, or disk). Furthermore, more than one of these programs may be partially executed, that is, execution of the program has started but not yet finished. Finally, as many as two user programs may actually be executing simultaneously, since there are two control processors. For this to be possible both of these simultaneously executing programs must be in primary memory. There may be,

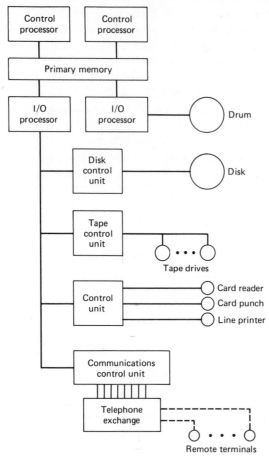

Figure 13.1 Hardware configuration for a simple time-sharing system.

and almost always are, more than two user programs in primary memory. However, the number of simultaneously executing user programs cannot be greater than the number of control processors. The I/O processors execute programs, and these must also be in primary memory. In addition to simultaneous execution of programs, all of the terminals connected to the system may be active at the same time. Thus, a user at one terminal may be typing information that is being accepted by the system or receiving typed output from the system while any or all of the other users are doing the same at other terminals.

Each user of Insys may permanently store information in the system's memory. The amount of information stored is limited only by the amount of money a user is willing to spend for the rental of storage space. The system guarantees the protection of a user's information from accidental damage and from use or modification by other users. However, any program executing on behalf of the owner of some information may access and modify that information as desired.

Insys provides facilities for cooperation between users. The owner of some information, either data or procedure, may permit selected other users to reference or modify this information. The type of access need not be the same for all of the users with whom he shares the information. An owner may share different sets of information with different sets of users. To facilitate the widest possible cooperation, Insys maintains a public library of useful procedures. Any procedure in this library may be used by any user of the system. Facilities are provided so that two or more separate, partially executed programs may communicate, that is, they may send messages to each other. These programs may belong to different users or they may all belong to the same user.

Insys is intended to be suitable for interactive, conversational use. This means that the response time for a trivial request from a terminal is relatively short, never more than a few seconds. The system must be able to be in operation continuously—24 hours a day, every day. Insys must be capable of easy expansion in both hardware and software. In the hardware it must be easy to add new devices as well as increase the number of existing devices and processors and to increase the amount of primary memory. In software it must be easy to add new procedures to the system's public library and to incorporate new subsystems and language processors.

Some significant implications follow from the requirement that an operating system have the capabilities and functions described in the preceding paragraphs. We will consider the most important of these, since they turn out to be the major problems in operating system design. Before proceeding with that task, we attempt to isolate some important attributes of operating systems in such a way that many different systems are encompassed.

13.2 Attributes That Characterize Operating Systems

If we attempt to study or compare various operating systems we soon see that they differ in many aspects. This makes comparison difficult. However, we can identify a number of different attributes that all operating systems seem to have. These attributes are similar to the attributes of a variable in the Instran language in that their values are usually descriptive words instead of numbers. (Recall that the values for the data type attribute of an Instran variable were FIXED, CHARACTER, and BIT.) Restricting our attention to the attributes defined below, we find that the difference between two operating systems is basically the difference in the values of these attributes for the two systems. The problem of comparing operating systems is confusing and poorly understood. As a result there is no method or set of attributes that is commonly accepted. The following partial list of attributes is an attempt to make the comparison of systems more rigorous. This list is not intended to be complete; instead, it contains those attributes that are important to the discussion in the following chapters.

1. *Response Ratio.* This attribute characterizes the guaranteed maximum time between a request and the response to that request. Since this response is the sum of the execution time required to process the request and the delay in this processing, response is properly measured as the ratio of elapsed time to execute time. This ratio is called the *response ratio*. The principal difference between *batch, interactive* (time-sharing), and *real-time* systems is the response ratio. In a *batch* system the response ratio is usually large. Typically it is

greater than 100, that is, the elapsed time for a job that executes for a few seconds is more than five minutes. In fact, in many batch systems the elapsed time may be an hour or more, even for jobs with short execution times. In a conversational *interactive* system the response ratio must be considerably less than 100. For example, in a good interactive system the elapsed time for a request that requires a few tenths of a second of execution would be less than five seconds. In a *real-time system* the response ratio usually needs to be less than five, that is, a request requiring one tenth of a second of execution would be serviced in less than half a second. Insys is an interactive system.

2. *Simultaneity of Use.* This attribute measures the degree to which a system can be doing things for more than one user or doing more than one thing for the same user. Numerical values for this attribute are not really useful or meaningful. This is true for all of the remaining attributes that we will discuss. A *multiprogramming* system is one in which more than one program may be in a state of partial execution, that is, several programs have started execution, but none is yet finished. However, some or even all of these programs need not actually be executing at any given time. A *sequential* system is one in which each program is executed to completion before execution of another program is begun. A *multiprocessing* system is one that has more than one control processor and is capable of simultaneously executing more than one program. A *multitasking* system is one in which a single user may have two or more interdependent programs that are in various states of execution. Effective use of the multiprocessing capability requires the multiprogramming capability. However, neither the multitasking capability nor the multiprogramming capability requires the multiprocessing capability. Insys has multiprocessing, multiprogramming, and multitasking capabilities.

Another aspect of simultaneity of use is the degree to which more than one I/O device may be simultaneously active. Most computer hardware systems are capable of having more than one device simultaneously active. This is accomplished by multiplexing a single data transfer path in an I/O processor, by use of several independent paths in the same I/O processor, or by use of more than one I/O processor. Insys allows simultaneous operation of more than one I/O processor as well as simultaneous operation of more than one device attached to a single I/O processor.

3. *Permanence of Information Storage.* This attribute characterizes the length of time a user may store information in a system's memory. This information must be retrievable on request by its owner. The storage must be reliable in the sense that the owner may expect the information to remain undamaged as long as it is left in the system memory. Insys allows user information to be stored for an *indefinite* period of time. In some systems information may be stored only for the duration of a job's execution. Many current systems allow information storage for longer than this, but not for an indefinite time, and some maximum time limit for information storage is enforced.

4. *Facilities for Cooperative Effort.* This attribute measures the degree to which users of a system may share information that is stored in the system's memory and the degree to which they can communicate with each other through the system. A system may not allow any sharing of information, may provide sharing only through a public library, or may allow sharing of all information in the system's memory. This sharing may be achieved by each

sharer making a copy of the original, or the sharing may be *direct*, each user referring directly to the single original. If some information shared by two users can be modified by both users in addition to being referenceable by both, then these two users can communicate with each other. In addition, a system may provide more direct communication between two executing programs in a multiprogramming environment. Insys allows direct sharing and modification of all information stored in the system. Direct communication between two executing programs is also provided.

5. *Privacy and Protection.* This attribute measures the degree to which a user's information is protected from unauthorized access and accidental damage. A system may protect a user's information from snooping and deliberate damage by another user. In conjunction with the facilities for sharing, a system may also control access to information on a selected basis, that is, the owner of any information may specify which other users may access it. A system also may permit the owner to specify the type of access for each user whom he has permitted access to some of his information. A system may also try to protect each user's information from accidental damage resulting from a system failure or a mistake by the user himself. Insys provides each user complete protection from other users. It also allows user control of sharing. Insys also includes some backup facilities that help to protect users from accidental damage.

6. *Expandability, Adaptability,* and *Generality.* This attribute measures the degree to which a system can encompass new hardware and software components without redesign or extensive reprogramming and the degree of limitations imposed on user programs. For example, the public library of a system must be expandable. It is often desirable to add new subroutines to the public library. A system that requires that the procedures in the library file be ordered by name may be less expandable than one that allows them to be in random order. Suppose the library file is supplemented by a dictionary; adding a new procedure to the library then only requires that a copy of the procedure be appended to the library file instead of having to rewrite the entire library file in order to get the new procedure into its required position. In like fashion a system that is able to adjust to an increase or decrease in the amount of memory or in the number of processors simply by updating the values of some table entries is more adaptable than one that requires changing instructions in the system program in order to adjust to these changes in the hardware. A system that permits user programs to be written in many different languages and imposes no maximum on their execution time is more general purpose than a system that restricts user programs to a single language and limits their execution time. One of the major objectives of Insys is that it be highly expandable, adaptable, and general purpose.

There are other important attributes of operating systems. However, our discussion will focus on the six listed above. We are interested in the values of these attributes for a proposed system, since particular values or combinations of values constrain the implementation of the system. For example, a small response ratio in combination with the capability for several users to use the system simultaneously from terminals implies that the system cannot let requests that require a long execution time complete their execution without interruption. If the system allowed a request for 10 minutes of execution to complete without interruption once it got started, the system would be prevented from satisfying the short maximum response time, as required by the small response ratio, for a one-second request made by a second user if the resquest were made more than a few seconds before completion of the first user's 10-minute request. In the following four sections we discuss the major problems of operating system design as they apply to Insys. We will also try to show how the values of the preceding attributes for Insys influence the solution of these problems.

13.3 Process Control

A major function of an operating system is resource management. We generalize the job of Chapter 2 to a conceptual entity called a *process*, which is a consumer of resources. A process needs resources to complete its execution. For example, primary memory is required for the program and its data, and a control processor is needed for execution of the program. Insys allocates and charges to some process all resources that are used.

Even though processes are not strictly resources, the system must still manage them, since they are intimately connected with the management of real resources. *Process control* is responsible for the creation and management of processes. Once a process has been created it can execute. However, it may not always be executing. There are two reasons for this. A process may not be able to execute or the system may not let it even though it is able to execute. A process cannot execute if it is waiting for something to happen. For example, it may be waiting for further information to be received from the user's terminal. Multiprogramming implies that more than one process may exist. All of the existing processes may be competing for use of the control processors. The system may not let a process execute because all of the processors are currently in use, even though the process is able to execute.

Since many processes are competing for use of the control processors, we need some policy for their allocation. This allocation is called *scheduling*. There are a variety of scheduling policies, each of which provides service of differing characteristics to the users of the system. The choice of a particular scheduling policy is partly an administrative decision. Our study will consider both the mechanics and the policy aspects of the management of control processors.

One aspect of Insys' facilities for cooperative use of the system is the need for communication between *cooperating concurrent processes*. A set of processes is concurrent if they all exist simultaneously. A group of cooperating users, or even a single user, may have a set of concurrent processes all working on a common problem. These cooperating processes have to communicate with each other. A primitive form of communication consists of stop and go signals, which allow partial synchronization of the cooperating processes. A more elaborate form permits actual messages to be passed between the processes. We will explore both forms of communication.

The final aspect of process control that we will explore is the command language and its interpreter. This is the user's principal interface to the system for managing his process or processes. A user specifies the major steps in his process using commands in the command language. For example, in Chapter 2 we saw that the user specified what translator to use for each procedure in his program by including the appropriate command in his job. Our goal of expandability and adaptability requires that the command language be expandable so that new commands can easily be added.

13.4 Memory Management

Memory is one of the major resources that needs to be allocated. A user has two types of memory needs. He needs primary memory in which to store his executing program and its data. This is a temporary, short-term need. He also needs secondary memory for permanent storage of information. Efficient utilization of the system's memory can be achieved only if the system manages both types.

There are many different algorithms for allocation of primary memory, and we explore several. One extremely important concept is that of *virtual memory*. Virtual memory provides an interface to the user that greatly simplifies his use of primary memory. Some virtual memory implementations combine the management of both primary and secondary memory to give the user the illusion of a practically unlimited primary memory. Virtual memory also facilitates the system's management of primary memory. We examine virtual memory in detail with particular attention to policies for the system's movement of user information in and out of primary memory. This movement of information is called *paging*.

Permanently stored user information is usually grouped into large units called *files*. The system must account for all files and provide facilities for creation and use of files by the users. We explore these facilities. In addition, we explore the coupling of virtual memory with permanent file storage. Because of the highly dynamic nature of the composition of a process in an interactive environment, there are significant advantages in eliminating the loader as described in Chapter 6. Instead, the loading from file storage and linking, which the loader performed, is now done dynamically during execution as needed. We explore the requirements for implementation of this dynamic loading and linking.

13.5 Input-Output

Modern computer systems frequently have a wide variety of I/O devices that have quite different operational properties, especially the length of time required to read or write information onto or from the device. These systems usually have an I/O processor, which is logically separate from the control processor and is capable of independent operation. However, overall control is still exercised by the control processor. We explore the operation and control of the I/O processor and its attached devices.

The I/O devices are resources that can be allocated to processes. Our goals of expandability and adaptability require a high degree of device independence in the way user programs reference and use devices. In addition, the use of devices is so complex that the user needs a more hospitable interface for his use. We consider the allocation and use of I/O devices. We also examine the problems of device independence and explore a simple user interface that achieves our goals.

Many systems implement a user interface that combines both permanent file storage and I/O device use. The viewpoint taken in such an implementation is that files are considered to be a kind of I/O device and are used essentially as if they were devices. This view is briefly examined to expose the differences between it and the viewpoint of file storage coupled to a virtual memory.

13.6 Sharing, Privacy, and Protection

Two of our goals are conflicting: sharing of information and privacy. We explore the problems in implementing the sharing of user information, especially direct sharing, which allows more than one user to update and use the single master copy of some information. Direct sharing of procedures presents some extra problems, which are examined. With the requirement for privacy any sharing permitted must be controlled by the owner. This requires facilities that enable him to specify who may use his information and the system must enforce his wishes. We examine the problems involved in this kind of control of access.

In order to enforce privacy both system and user information must be protected. We explore what is required of both the hardware and the software in order to achieve this protection. One of our objectives is reliable operation of the system. This also implies protection. It also implies some facilities for recovery from accidents and normal system failure. We briefly examine the aspect of protection.

EXERCISES

13.1 Describe, in terms of the attributes defined in this chapter, the operating system that you are currently using.

14 | Process Control

This chapter is a discussion of the basic problems in the management of processes and control processors. What are these problems? The major problems in the design of any system arise as a result of the requirements and objectives of the system, coupled with the limitation of the computer hardware on which it will operate. In Chapter 13 we discussed some of the possible objectives and requirements. We called them attributes of a system and gave them descriptive names. We then briefly explored the idea that any particular combination of attributes had logical implications with respect to the internal design of the system. In this chapter we limit our concern to the implications of the attributes for our example system, Insys.

One of the major problems of the system is the management of processes. Processes will be the only consumers of resources. Therefore, before we can discuss any of the problems of resource management, we must have a clear understanding of the properties and use of processes: what constitutes a process, how it is represented, who can create and destroy a process, and when and how it is created and destroyed. The *general purpose* attribute of Insys implies that no artificial limit be placed on the execution time of any command or user program, that is, they may execute as long as required to do whatever computation is required. This attribute, together with the *simultaneous use* attribute, implies multiprogramming. Since every command and user program executes as part of some process, process control must be able to manage many processes simultaneously. The nature, constitution, and representation of processes are discussed in Section 14.1. The management of processes is explored in Section 14.2.

In order to progress from start to finish, a program in a process must have the use of a control processor. However, the attributes just mentioned imply that use of the control processors must be shared among all of the processes (*processor multiplexing*). The *adaptability* attribute implies that processor multiplexing work with an arbitrary number of control processors, and that none of them are treated differently from the remainder. The *conversational interactive* attribute implies a small response ratio. In conjunction with multiprogramming, the conversational attribute has stronger implications. Even after a processor has been allocated to a process, that process cannot be permitted to use the processor for an arbitrary length of time. Process control must be able to suspend execution of any process forcibly and reallocate its processor to another process. This forcible suspension is called *preemption*. Preemptive multiplexing of processors is discussed in Section 14.3.

The Insys objective of *support for cooperating concurrent processes* implies that there must be facilities for communication between the processes in a set of cooperating concurrent processes. Not only do we need a facility for sending messages between such processes, we need a method of coordinating the execution of these processes that will prevent undesirable indeterminism. The problem of coordination is explored in Section 14.4, and facilities for interprocess communication are discussed in Section 14.5.

Since a process is a generalization of a job, the command language is a major interface between the user and the system with respect to control of the user's process. Everything that a user does, or has done for him by the system, is done as part of a process. In particular, whenever he issues a command, execution of the corresponding command program is part of a process belonging to him. The command language also functions as an interface between the user and other parts of the system. However, since it most naturally appears to him as a major way of controlling the behavior of his process, we will include the discussion of the command language under the general heading of process control. The Insys attributes of *expandability, adaptability*, and *general purpose* imply that the command language be easily expandable and not limited in the kinds of functions that a command may invoke. The command language is discussed in Section 14.6.

14.1 Processes

A *process* corresponds to what we called a job in Chapter 2. Unfortunately, the concept of job as defined there is not adequate for our purposes. We need a concept that will help in solving the problems relating to multiprogramming, processor multiplexing, and cooperating concurrent processes. In fact, the concept of process that we are about to explore resulted from the earliest attempts to solve these problems. What is wrong with the concept of a job from Chapter 2? A job was defined as a sequence of job steps. Each job step consisted of either a system or user program that was to be executed and some data that was to be used during its execution. Thus, a job was a collection of programs and data. The problem with this concept is that it is static. It does not capture the dynamic aspects of what really happens in a multiprogramming, multiprocessing system.

Resource management is a central idea in our view of operating systems. A process is a consumer of resources. Thus, it is the central entity of concern to the system. In fact, we adopt the principle that all resource usage is associated with some process, and that process is charged for use of the resource. Inversely, all resources used on behalf of a process are charged to that process. In particular, this means that whenever system procedures are executing, the resources they use are charged to the user process that requested the service, that is, a system procedure always executes as part of the user process on whose behalf it is executing. Programs and their data will be an integral part of a process, but a process will consist of much more than that. In order to complete the execution of (the programs in) a process, resources such as a control processor, primary memory, and I/O devices are needed. It is the function of resource management to allocate these needed resources to a process so that it may complete its execution.

The resources allocated to a process vary with time. This is part of what we mean by the dynamic aspects of a process. In the sequential batch system of Chapter 2 this was not an interesting aspect because all of the system's resources, except those permanently reserved for system use, are allocated to each job. This allocation was done at the beginning of a job's execution. Once this was done the job began its execution and continued uninterrupted until it was finished. The reverse is true for Insys. Many processes are competing for use of the system's resources. It is usually impossible to allocate to any process when it begins its execution all of the resources that it will need during the course of its execution. The system has to deal with the resource demands of more than one process at a time, and the resources are divided up among these processes. Since a process's resource demands vary with time, the system must be able to deal with the changing resource demands of each process. Thus, in addition to a collection of procedures and data, a process must include the resources that are currently allocated to it.

Another way of characterizing a process is to say that it is a program and its *environment*. We include in a program's environment all things that the program may reference or use that are exterior to the program. As we will see later, the system must keep a record of what is currently included in this environment. In theoretical work dealing with processes this record is called a *process state word* or *state vector*. In this text we use the common implementation-oriented term, *process control block*. This is similar to what is called a *task control block* (TCB) in OS/360.[1]

What is included in a program's environment? It certainly includes all of the procedures in the program and the data that they need. It includes all of the resources that have been allocated to the process, such as the devices that it may use. In addition, the environment includes the contents of all of the hardware registers that the procedure is able to reference, such as the general registers, the instruction counter, and the condition code. Since a process may not always actually be executing, the process control block must contain, or lead to, all the information about a process that is needed to resume its execution. It is simplest conceptually to view the process control block as containing all the information that defines a process and is needed for process management. However, in practice, this is never true, and the process control block contains mostly information that enables the system to find the process definition information.

The process control block for Insys contains at least the following information.

1. A pointer to the definition of a set of primary memory addresses.
2. Pointers to the currently allocated resources.
3. A record of resources used by the process since it was created.
4. The contents of the control processor registers.
5. Other information.

[1] IBM, "IBM Operating System/360 Concepts and Facilities (Excerpts)," *Programming Systems and Languages* (S. Rosen, editor), McGraw-Hill, New York, 1967, pp. 598–646.

The set of primary memory addresses mentioned in item 1 define the data and procedures that the process may reference. In the system in Chapter 2 this would be the integer K in Figure 2.5, since a job can reference any primary memory address that is greater than or equal to K. This will be discussed in detail in Chapter 15 and should be ignored until then. The contents of the control processor registers are stored in the process control block when the process is not executing. When the process is executing, this information will, of course, be in the actual registers. Each specific item in the process control block will be described in more detail when we encounter the need for that information.

The reader should note that nowhere in this discussion of process have we said anything that implies that a process must have a human being associated with it. In a well-designed system that deals with processes, it is not only possible but extremely useful to have processes that do not have human owners. These are system processes, and their function is to do things for the system. For example, in an interactive system there may be a process that is analogous to a telephone operator. The sole function of this *answering process* would be to answer whenever a new user tries to use a terminal and determine if the new user is to be permitted to use the system. Our discussion of process also did not make any distinction between two processes that are owned by the same person and two processes that are owned by two different persons. As long as there is no such distinction, it is straightforward to allow one user to have more than one process.

14.2 Process Management

Process control must manage all processes that currently exist. This means that it must create new processes when they are needed and destroy old ones when they have finished their execution. When creating a process the system must know what program is to be executed as part of the process and who will be the owner of the new process (the system is considered the owner of the system processes mentioned in the preceding section). When destroying an existing process the system will reclaim any resources that are still allocated to the process.

The major function in process management is to keep a record of the current *state* of all existing processes and change this state whenever necessary. The state of a process has to do with whether or not the process is executing. A process may not be able to execute continuously. There are two reasons why. There may not be an unallocated (*free*) control processor that the process can use for execution, or the process may not be able to continue its execution until some *event* occurs. A simple example of the latter case is when a process requests the use of some resource, such as an I/O device, which is currently allocated to another process. If the requesting process cannot continue its execution without the resource, then it must wait until the requested resource is free. The event in this case is the releasing of the needed resource by the other process.

When a process is waiting for an event we say that it is *blocked*. A process that is not blocked is capable of executing. However, it may not be able to do so because all the processors are currently allocated to other processes. A process in this state is said to be *ready*. A process that is actually executing is said to be *running*. A process that has been preempted is ready; that is, it is capable of executing, but does not have a processor allocated to it. A process that is in the ready state is implicitly requesting a control processor. In fact, a process never explicitly requests a processor. Because preemption occurs at unpredictable times, a user cannot write a program that would explicitly request a processor at the proper times.

In the preceding discussion we have identified three different states for a process.

1. *Running:* the process has a processor allocated to it; the process is executing.
2. *Ready:* no processor is currently allocated to the process, but the process is able to execute as soon as it has a processor.
3. *Blocked:* the process is waiting for some event to occur and cannot execute until that event occurs.

Note that in the sequential batch system of Chapter 2 a job was always in the running state. Even if it was waiting for some event, it did not release the processor. Since the system was sequential, there was no other job in the system to which the processor could be allocated. Therefore, as far as the system was concerned, a job was always running. The most important function of process control is that of effecting the transition between states for a process.

Process control needs a record of every process in the system. This record is called the *process control block* and contains all of the information (or pointers to the information) that defines a process. Control blocks are frequently used by a system to record information about the conceptual entities that the system manages, such as processes or files. A control block is nothing more than a

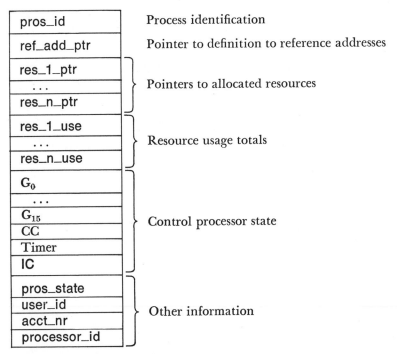

Figure 14.1 Process control block.

data structure that contains or leads to the required information. The process control blocks for all of the existing processes are kept in a table, called the *process table*. There is a one-to-one correspondence between existing processes and process control blocks in the process table; that is, a process does not exist unless there is an entry for it in the process table.

The process control block contains the state of the process (Figure 14.1). It is convenient to maintain three lists in addition to the process table, one for each of the three possible process states. The entries in these lists are simply pointers to entries in the process table. Each list contains one entry for each process whose current state is equal to the state corresponding to the list, as illustrated in Figure 14.2. For example, the ready list contains one entry for each ready process. Each of these entries points to the process control block for the corresponding process. Note that the length of the running list is equal to the number of processors, while the lengths of the other two lists must be equal to the maximum number of processes that will be allowed to exist at any one time.

In order to gain a deeper understanding of process management we will discuss the preceding problems in more detail in the remainder of this section.

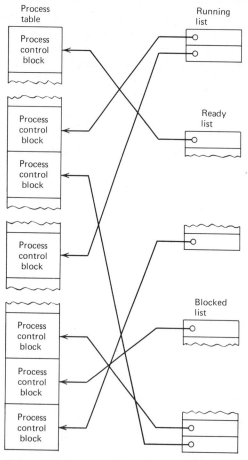

Figure 14.2 Process table and process state lists.

This discussion will be in the context of Insys and, in particular, will relate directly to the process table and state lists defined in the preceding paragraph. As in past chapters, the specific algorithms used are not the only implementation; however, any implementation will have to face the same problems and will be forced to do analogous things. We also warn the reader that this discussion is simplified and is not the complete picture. In later chapters we will see that there is interaction between process control and the other resource management modules.

The major task in creating a process is building its process control block. This is accomplished by the procedure create_process defined in Figure 14.3. A new process is created by

create_process (user_id, acct_nr, start_exec, pros_id);

The caller must supply the identification (user_id) and account number (acct_nr) of the new process's owner. Execution of the process will begin at the address specified by start_exec. When create_process returns to the caller, pros_id will contain a unique identification number for the new process. The system guarantees that no other process has, or will ever have, this identification number. This allows the system and the users to distinguish this process from all other processes in the system.

The skeleton control block, which is copied into the process table entry for the new process, has most of its items set to standard values. The pointers to allocated resources are all set equal to *null* values that do not point to anything. For example, a pointer that is pointing to an entry in some table will have a value that is the index of that entry in the table. The null value for this pointer could be a negative number, since all of our table indexes are greater than or equal to zero. The resource usage totals, the control processor state (except for the instruction counter), and processor_id are all set equal to zero.

After copying the skeleton control block into the process table, the remainder of its items must be filled in. A new unique process identification is generated. This is the new process's identification number. The reference addresses for the new process are defined, and a pointer to this definition is placed into the process control block. The definition of reference addresses for an Insys process

procedure create_process (user_id, acct_nr, start_exec, pros_id);
   i : = index of an unused entry in pt; {pt is the process table}
   pt[i] : = skeleton of a process control block;
   pt.pros_id[i] : = newly generated unique identification number;
   Define reference addresses for this process;
   pt.ref_add_ptr[i] : = pointer to definition of reference addresses;
   pt.IC[i] : = start_exec;
   pt.pros_state[i] : = 'ready';
   pt.user_id[i] : = user_id;
   pt.acct_nr[i] : = acct_nr;
   Put i onto ready list;
   pros_id : = pt.pros_id[i];
   return;
end.

Figure 14.3 Definition of procedure that creates a process.

is explained in Chapter 15. The information supplied by the caller is stored in the process control block. In particular, start_exec specifies the address at which execution of the process is to begin. Therefore, its value is stored in the control processor state as the value of the instruction counter (IC). When a processor is allocated to the new process, the contents of the processor state in the process control block will be loaded into the processor's registers. The next instruction to be executed after the registers are loaded will be the instruction whose address is the value of the instruction counter.

The state of the new process is set equal to 'ready', since a new process is always able to execute at least the first instruction beginning at location start_exec. However, in order for the process to actually start execution, it must be allocated a processor. To achieve this all that needs to be done is to put a pointer that points to the process control block of the new process into the ready list. From that point on the new process is treated like any other process on the ready list and will eventually be allocated a processor. The process identification of the new process is returned to the caller so that he may refer to this process in the future.

The reader should note that some process other than the process being created is doing the work of creating the new process. This is an implication of our requirement that all execution take place as part of some process. The new process cannot execute until it has been created. Thus, its creation must be done as part of another process. (Clearly there is one exception to this: the very first process. This is part of the system startup and is not the way processes normally get created.) This other process that performs the act of creation is called the *parent* process. It has life-and-death power over all of the *children* processes that it creates. Normally only the parent process, or the process itself, can destroy a process.

In general, any process may create another process. Many of the processes in Insys are created by the system answering process that was identified at the end of Section 14.1. If the system answering process determines that the user at a terminal is permitted to use the system, it creates a new process for him. This new process starts its execution in the command language interpreter by requesting that the user type a command on the terminal. This action is further explained in Section 14.6, which discusses the command language interpreter.

To destroy a process all records of the process must be removed, and all resources allocated to it must be reclaimed by the system. A process is destroyed by the call

destroy_process (pros_id);

where pros_id is the unique identification number of the process (Figure 14.4). Using this identification as a key, the process table is searched to find the process control block for the process. Once the process control block has been found, the state of the process can be determined. If the process is running its execution must be preempted (stopped) before it can be destroyed. As a result of the preemption the process's state is changed to ready. In either case, the ready list or blocked list now contains a pointer to the process control block. This pointer must be removed.

Before completely destroying the process, all resources that are currently allocated to it are released so that they may be used by other processes. The

procedure destroy_process (pros_id);
    i := index of process table entry for process whose process
      identification is pros_id;
    if pt.pros_state[i] = 'running' then Preempt the process;
    Remove pointer to pt[i] from blocked or ready list;
    Release all resources allocated to the process;
    Update the permanent accounting record by the amount of each
      resource used by the process;
    Mark pt[i] as unused;
    return;
end.

Figure 14.4 Definition of procedure that destroys a process.

method of releasing a resource is very dependent on the particular resource and can be quite difficult in some cases. Details of both allocation and release of a resource will be explored as part of the management of that resource. The amount of each resource that has been used by the process is recorded in its process control block. This usage must be recorded in more permanent form so that the user's account will be charged for the resource usage. One common technique is to maintain a permanent accounting record for each account. Each time a process is destroyed, its resource usage is recorded in this record. Then, periodically, some system process will look at all of the accounting records and issue bills to the users. The final act in destroying a process is to mark as unused the process control block in the process table. Then there is no longer a record of the process, and it ceases to exist.

Process control is also responsible for effecting the transition of a process from one state to another. Since there are three states, there are six different transitions possible. However, only four transitions will be allowed to occur. These four are shown in Figure 14.5. The transition from ready to blocked is impossible. If a process is blocked it is because the process is unable to execute until some event occurs. If a process is ready it is able to execute, but does not yet have a processor allocated to it. A ready process is not executing, and thus cannot proceed to a point where it is no longer able to execute. For a ready process to reach the blocked state it must first execute, and this means the sequence of states of the process is ready, running, and then blocked. We choose to rule out the transition from blocked to running for simplicity reasons. The fewer the allowed transitions, the simpler our system is. The remaining transitions are all required.

The transition from running to blocked is voluntary, the other transitions are not. That is, some procedure executing in a process decides that it must wait for some event to occur before execution can proceed. The system never forces a process to become blocked. On the other hand, the other transitions are forced by the system whenever it is appropriate, the transition from blocked to ready when the anticipated event occurs, the transition from running to ready (preemption) when the system decides that the processor should be allocated to some other process, and the transition from ready to running when the system allocates a processor to the process. A process can neither cause these transitions in its state to occur nor prevent them from occurring.

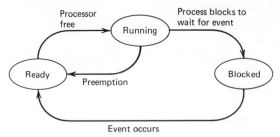

Figure 14.5 Possible state transitions for a process.

The transition from running to blocked is effected by the procedure **block**, which is defined in Figure 14.6. When a process can no longer execute until some event occurs, it executes the call

block;

The state of the process is changed by moving the pointer to it from the running list to the blocked list and updating the value of **pros_state** in its process control block to equal 'block'. The process will stay in the blocked state until the awaited event occurs. Since this process is not executing while it is blocked, some other process must discover that the event has occurred. Before calling block, this process must make arrangements with another process for it to detect the occurrence of the event. When this other process does so, it will call **wakeup**, which is the system procedure that causes the transition from blocked to ready. This action will be explained shortly.

After the state of the process has been changed, the processor that was allocated to it can be allocated to another process. Before this can be done, the total control processor usage, which is part of the resource usage totals in the process control block, must be updated by the duration of this last allocation of a control processor to the process. Also, before releasing the control processor, the contents of its registers must be saved so that the process can resume execution after the awaited event occurs. However, we want execution to resume at the return statement labeled **A** so that execution will return to the caller of **block**. The caller will then know that the event has occurred, and he can proceed with his execution. Therefore, the value of the location counter

procedure block;
    {process executing block is blocking itself}
    i := index of process table entry for this process;
    Remove i from running list;
    Put i onto blocked list;
    pt.pros_state[i] := 'blocked';
    Update total control processor usage in pt[i];
    Store current contents of control processor registers into pt[i];
    pt.IC[i] := address of A;
    start_process;
A: return;
end.

Figure 14.6 Definition of procedure to block a process.

in the control processor state is set equal to the address of A. Finally, the call to **start_process** allocates the processor being used by this process to another process.

Figure 14.7 illustrates the interaction of state transitions for a set of processes. In the example we assume that the processes exist in Insys, which has two control processors that can be used for the execution of processes. We are also assuming that the four processes are all that exist during the time period of the example. The example shows that when a process blocks itself, the control processor that it was using is immediately allocated to one of the ready processes.

The transition from running to ready is effected by the procedure **preempt,** which is defined in Figure 14.8. When a process is to be preempted, it is forced to execute a call to **preempt**. The mechanism that forces this is explained in Section 14.3. The action of **preempt** is very much the same as **block,** except for the *scheduling* of the preempted process. All decisions regarding the allocation of processors to ready processes are made by the scheduler. The scheduler will be explained in detail in Section 14.3. It decides how long a process will be allowed to execute before it is preempted in order to allocate the processor to another ready process.

The preempted process is scheduled by calling the scheduler

schedule (i);

where i is the index of the process's process control block in the process table. The scheduler puts this index into its proper place in the ready list. The reason

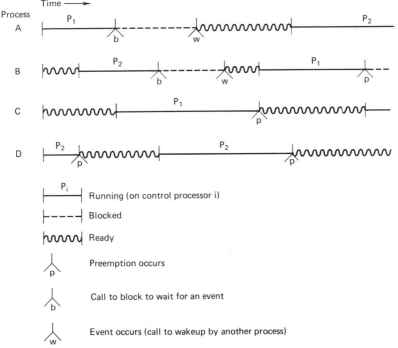

Figure 14.7 Example of state transitions for four processes using two control processors.

```
procedure preempt;
     {process executing preempt is being preempted}
     i := index of process table entry for this process;
     Remove i from running list;
     schedule (i);
     pt.pros_state[i] := 'ready';
     Update total control processor usage in pt[i];
     Store current contents of control processor registers into pt[i];
     pt.IC[i] := address of A;
     start_process;
A: return;
end.
```

Figure 14.8 Definition of procedure to preempt a process.

for this is that the pointers on the ready list are ordered by priority of the corresponding process. That is, processors will be allocated to ready processes in the order in which the processes appear on the ready list. Thus, whenever a process's pointer is to be added to the ready list, the scheduler must be called to determine where in the ready list the pointer is to be inserted. The motivation for keeping the ready list ordered by priority is that it makes processor allocation simple and efficient. This will be explored in more detail along with the scheduler in Section 14.3.

The transition from ready to running is effected by the procedure start_process, which is defined in Figure 14.9. A free processor is always allocated to the process that is at the head of the ready list. As mentioned earlier, the ready list is ordered by priority, so the first process on the list is the next process to which a processor is to be allocated. The state of this process is changed from ready to running. Execution of the process is then started by loading the registers, including the instruction counter, of the free processor with the values that are stored in the process control block for the process.

What happens if there is no process on the ready list when start_process is called? There are two ways to solve this problem. The first way is to test for this condition and, when it occurs, just let the processor sit idle until some process's state changes to ready. The idle processor is then allocated to that process. The other way of solving the problem is to arrange that the condition never occurs, that is, arrange that there is always at least one process on the ready list. This is easy to do by having special system processes, called *idle processes*, which are simply infinite loops that do nothing. Idle process will never finish executing. Thus, whenever they are not running, they will be ready.

If idle processes are assigned the lowest possible priority they will not be allocated a processor if there is any other process that is ready. Yet when there is a free processor and no ordinary process is ready, the processor will be allocated to an idle process that will execute until some ordinary process becomes ready. The idle process will then be preempted and the processor allocated to the ready, ordinary process. How many idle processes are needed? Obviously there must be as many idle processes as there are processors. Then, if all of the processors become free at the same time and there are no ordinary proc-

```
procedure start_process;
    i : = pointer at head of ready list;
    Remove i from ready list;
    Put i into unused entry in running list;
    pt.pros_state[i] : = 'running';
    Load control processor registers from pt[i];
    return;
end.
```

Figure 14.9 Definition of procedure to allocate a processor and start a process executing on it.

esses that are ready, there will be enough idle processes so that each free processor can be allocated to an idle process.

The second solution is attractive because of its simplicity. The idle processes can be treated just like any other process, except that they have a lower priority than any ordinary process. The **start_process** procedure is simpler, since no test must be made for an empty ready list. In addition, it is difficult or impossible simply to let most modern control processors sit idle. The processor cannot be stopped but must be executing some instruction, and this is equivalent to the special idle process.

The transition from blocked to ready is called *wakeup*, since a blocked process can be considered to be sleeping (not executing) until the event it is waiting for occurs. The occurrence of this event then acts like an alarm clock and wakes up the blocked process. If we adapt the idle process method discussed above, it will always be true that there will never be a free processor when a process is awakened. Therefore, the only transition from blocked is to ready. This transition is effected by the procedure **wakeup**, which is defined in Figure 14.10. When the awaited event occurs, the process that detects its occurrence executes the call

wakeup (pros_id);

where **pros_id** identifies the process that is waiting for the event that has occurred. The process table is searched to find the process control block for that process. This process is then scheduled for execution by the scheduler. The process is now unblocked and will resume execution at a later time, determined by the scheduler.

It is worthwhile to briefly consider which processor and process actually execute the procedures that effect the state transitions. The transition to blocked is executed by the process that wishes to be blocked and that called **block**. The preemption transition is executed by the process being preempted. In both cases the processor that was allocated to the process is now free and can be allocated to another process. That is, if process A executing on processor P blocks itself or is preempted, then processor P is free to be allocated to another process, say process B. Both the transition of process A from running to blocked or ready and the transition of process B from ready to running (**start_process**) will be executed by process A using processor P. The last step in **start_process**, loading the control processor registers, begins execution of process B and actually effects the allocation of processor P to process B. The sequence of instructions that accomplishes this is very dependent on the hard-

```
procedure wakeup (pros_id);
    i := index of process table entry for process whose process
        identification is pros_id;
    Remove i from blocked list;
    schedule (i);
    pt.pros_state[i] := 'ready';
    return;
end.
```

Figure 14.10 Definition of procedure to wakeup a blocked process.

ware processor. We will not attempt to discuss these instructions in this text, since detailed hardware knowledge beyond what we have assumed is required. The wakeup transition is executed by the process that "discovers" that the awaited event has occurred. This process may have no relation to the process being awakened, or it may be a cooperating concurrent process. How a process discovers that an event has occurred will be discussed in both Section 14.4 and Chapter 16.

14.3 Multiplexing of Processors

In this section we will study the allocation of processors to processes. Basically the problem is that there are not enough processors to go around. Most of the time there will be more than one process in the system. Even if we have more than one processor, it is not economically sound to have enough so that each process may have its own. Thus, the processors must be shared among the processes. The requirement for interactive response implies that preemption is required, which rules out simple first-come, first-served sharing. Preemption requires that the system be able to forcibly interrupt the execution of a process and take the processor away from that process. It is important to distinguish between the mechanics of processor multiplexing and the policy that determines which process to preempt. The mechanics consists of the state transitions that were discussed in the preceding section. The scheduler implements the policy and will be the major topic in this section.

In order to preempt a process some hardware assistance is required. Once a processor has been allocated to a process it may execute for an indefinitely long time before it calls block. The system must be able to interrupt the process's execution forcibly. Whatever method we use must work for an arbitrary number of processors, including only one processor. This rules out any method that requires that some other processor send an interrupt signal to the processor that is executing the process we wish to preempt.

The simplest method uses a hardware *interrupt timer* in each processor. The timer is set to some nonzero amount of time. Once set, the timer automatically counts down to zero. As soon as the value in the timer becomes zero, execution of the procedure currently being executed by the processor containing the timer is interrupted, and control is transferred to a fixed location in the operating system. We will call this a *timer interrupt*. In Insys the fixed location to which control is transferred is the entry point of **preempt**. (This last statement is not strictly true, as we will see when interrupts in general are discussed in

Chapter 16. However, it is accurate enough for the present context.) For example, if the timer is set to 500, every microsecond the value of the timer is decreased by one. After 500 microseconds, the value in the timer becomes zero, and control is transferred to **preempt**. Thus, once a processor is allocated to a process by **start_process**, the process executes until the value in the timer becomes zero or the process calls block.

A major duty of the scheduler is to decide the value to which the timer is to be set when a processor is allocated to a process. The scheduler may also change the value in the timer of a processor that has already been allocated. Since the value in the timer determines how long the process will execute before it is preempted, changing the value in the timer will change the execution time of the process. This capability is needed to be able to manage processes with different priorities, such as the idle processes. When an idle process is executing and an ordinary process becomes ready, the value in the timer of the processor allocated to the idle process is changed to zero by the scheduler. This immediately preempts the idle process, and its processor is free to be allocated to the ordinary process that has become ready.

As we saw in a preceding section, the order of the entries on the ready list reflects the order in which the corresponding processes will be allocated a processor. Whenever a transition to ready occurs, the scheduler is called. The scheduler then determines the priority of the process that has become ready, thus determining its position on the ready list. The scheduler must also determine the value that is to be put into the timer register when the process is started. This value will be stored as part of the control processor state in the process control block.

It should be noted that there are other ways to accomplish the scheduling. For example, the scheduler could be called only when a processor becomes free and decide at that time which process gets the processor. The trouble with this is that some of the transitions, those from blocked to ready, are missed. As a result, a high-priority process may become ready and not be allocated a processor for a longer time than is justified because the scheduler is unaware that the high-priority process is ready. Alternately, we could go to the other extreme and reschedule all processes periodically at the end of each short time interval, even if no process state transitions occur. This would allow us to have processes whose priorities change with time. The choice that we have made is a good compromise. It allows quite fine adjustment in the priorities of processes without excessive overhead.

The algorithm for ordering the ready list is an embodiment of the policy aspect of processor allocation. We can think of the ready list as a queue in which processes queue up, waiting their turn for use of a processor. The head of the ready list is then the head of the queue. The simplest allocation algorithm is *first-come, first-served*. Under this policy, a free processor is allocated to the process at the head of the queue, and this process is removed from the queue, that is, its pointer is moved from the ready list to the running list. Once it has been allocated a processor, it is allowed to execute for as long as it wishes.

When the process stops execution, either because it is destroyed or blocks itself, it relinquishes the processor, which is then allocated to the process currently at the head of the queue. A newly created process is placed at the end of the queue. A process is also placed at the end of the queue when it is unblocked. The queue is never reordered. This algorithm, except for the blocking of a process, is the scheduling algorithm used in the simple sequential batch system of Chapter 2. A variation of the first-come, first-served policy is also used in many multiprogramming batch systems.

The major deficiency of the first-come, first-served algorithm is that it does not include preemption. The simplest algorithm that does is one commonly called *round-robin*. This algorithm is the same as the first-come, first-served algorithm with one exception. When a processor is allocated to a process, the process is allowed to execute only up to a fixed maximum length of time. This maximum period of time is called a *quantum*. Thus, when a processor is allocated to the process at the head of the queue, execution continues until the process stops itself or until a quantum of time has elapsed, whichever occurs first. If the process executes for the full quantum without stopping, preemption occurs. This is achieved by setting the value of the timer equal to the length of a quantum when a processor is allocated to a process. If the process has not stopped executing when the value in the timer becomes zero, its execution is interrupted. The preempted process is placed at the end of the queue, and the processor is allocated to the process at the head of the queue.

The reader should note that neither of these two scheduling algorithms makes any assumption about the number of processors in the system. It is not only possible, but desirable for all processors in the system to use the same queue. Whenever any processor becomes free, it is allocated to the process at the head of the single queue. This is certainly simpler than having a separate queue for each processor. Having all processors service a single queue inherently results in more efficient utilization of the processors; that is, a processor is never idle as long as there is any process which is able to execute.

The round-robin algorithm insures that short execution requests will finish before long execution requests. This is a requirement for interactive systems. For example, suppose there are three processes P_1, P_2, and P_3 having execution times of 0.1 second, 0.5 second, and 0.4 second, respectively. Let the length of the quantum be 0.2 second and the queue initially have the ordering.

$$P_2(0.5) \ P_1(0.1) \ P_3(0.4)$$

where the head of the queue is at the left. The number in parentheses following the process identification is the amount of execution time remaining for the process.

Assuming only one processor, execution proceeds in the following fashion. The processor is first allocated to P_2, which executes for 0.2 second before it is preempted. Since it still requires 0.3 second of execution, it is placed at the end of the queue. The queue ordering is then

$$P_1(0.1) \ P_3(0.4) \ P_2(0.3)$$

The processor is then allocated to P_1, which completes its execution in 0.1 second.

The queue is then

$$P_3(0.4) \ P_2(0.3)$$

and the processor is allocated to P_3, which executes for one quantum before being returned to the end of the queue. The queue ordering is then

$$P_2(0.3) \ P_3(0.2)$$

and P_2 is allowed to execute for one quantum. After P_2 is returned to the end of the queue, the ordering is

$$P_3(0.2) \; P_2(0.1)$$

P_3 is then given one quantum, which allows it to complete its execution. Finally, only P_2 is on the queue

$$P_2(0.1)$$

and the processor is allocated to P_2, which completes its execution in 0.1 second. Thus, we see that the processes complete their execution in order of increasing execution time requirements, even though they were not initially ordered that way.

The above example is unrealistic in the following sense. In an actual system there will be many processes whose state is changing from blocked to ready and from running to blocked. Thus, there will be a constant flow of processes entering the queue and leaving the queue. In fact, in an interactive system, it is common to find a number of processes that enter the queue from a blocked state, execute for a quantum or two, and then leave the queue to return to the blocked state. Thus, a process in the queue with a long execution requirement will see many other processes come and go before it completes its execution.

Clearly this allocation policy can guarantee a maximum response time only if the total number of processes in the system is limited. For example, suppose the number of processes is limited to 30. The worst possible case occurs when the maximum number of processes exist, and they are all ready at the same time. Suppose that the execution time requirements of each process is 0.2 second and the length of the quantum is 0 2 second. Then the process at the head of the queue is done in 0.2 second. The second process on the queue is done in 0.4 second, and so forth, until the last process on the queue is finally done in 6 seconds (30 × 0.2 second). Thus, the response to a 0.2 second request is 6 seconds or less, giving a response ratio of 30 or better. In system operation under normal conditions the worst case seldom, if ever, occurs, and the average response ratio would be considerably better than 30.

Most existing time-sharing systems use a scheduling algorithm, which is more complicated than the round-robin algorithm. Most of these algorithms consider the priority of a process when making scheduling decisions. This priority is a function of some property, or combination of properties, of a process. In some systems the priority of a process is a function of the magnitude of the resource demands of the process. For example, in the Multics system processes that have short execution times have the highest priority. In other systems both the execution time and the memory requirements are considered in assigning a priority to a process.

Another common basis for assigning priority is job class or department number. That is, the priority of a process is determined by an administrative decision. The system administrator decides that processes owned by members of certain departments or certain types of jobs are to have higher priority. This assignment is usually not arbitrary. It may be based on the schedules and deadlines of the various departments in relation to the total computing requirements of the company, or it may simply be based on the amount of money the user is willing to pay, that is, any process can obtain a higher priority if its owner is willing to pay a higher rate for the resources that the process uses.

Ordinarily a scheduling algorithm based on priority will always select the highest priority process for execution. That is, whenever there are two processes with unequal priorities on the ready list, a free processor will be allocated to the process with the highest priority. Conceptually we can view the ready list as an ordered sequence of queues, one for each distinct priority as shown in Figure 14.11. When a processor is free, each queue is considered in turn and the processor is assigned to the process that is at the head of the first nonempty queue in the sequence. For example, the order of execution of the processes in the figure is P_4, P_2, P_8, P_3, P_7, and P_5, provided no other process becomes ready in the meantime. On the other hand, if before P_3 executes, another process P_1 becomes ready, and its priority is such that it is put on queue Q_1, then P_1 will execute before P_3, P_7, and P_5, even though these three processes have been waiting longer.

There must be one exception to always executing the highest priority process first, otherwise it is possible that a very low-priority process may never be allocated a processor. This could happen if there were many high-priority processes rapidly cycling between running, blocked, and ready so that there is always a high-priority process in the ready list. Hence, a priority-based algorithm usually has some kind of safety valve that sometimes allows a low-priority process to execute before a high-priority process.

It should be clear that a single ready list is equivalent to the multiple queues shown in Figure 14.11 as long as processes can be inserted any place in the ready list. Figure 14.12 shows the queues of Figure 14.11 organized as a single ready list. In this case it will be necessary to keep pointers to the end of each queue. Whenever a process becomes ready, it is inserted into the ready list at the place corresponding to the end of the appropriate queue for its priority.

While almost all priority-based algorithms share the idea of executing high-priority processes first, there is a wide range of variations in the basic algorithm. For example, either a first-come, first-served or round-robin queue discipline may be used to service the processes within the queue for a particular priority. The discipline used may even vary from queue to queue. If the round-robin discipline is used for more than one queue, the length of the quantum used may be the same for all of these queues, or it may be different for some queues. When a process is preempted because it has executed for a quantum, it may be placed on the end of the queue it was on when it started to execute or it may be placed on some other queue. These are only some of the ways in which priority-based scheduling algorithms may vary. Clearly not all combinations make sense. However, it is not within the scope of this text to discuss these variations any further.

Another point should be mentioned to prevent future confusion. We have been using the term preemption in a broader sense than some writers. We have defined it to mean the interruption of execution of a process by the system, for whatever reason. A common property of most priority algorithms is that

Q_0: P_4 P_2 P_8 Highest priority
Q_1:
Q_2: P_3
Q_3: P_7 P_5 Lowest priority

Figure 14.11 The ready list as a sequence of four queues.

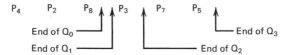

Figure 14.12 Single ready list version of queues in Figure 14.11.

they preempt an executing process whenever a process of higher priority becomes ready, even if the executing low-priority process has not yet executed for its quantum. Some writers reserve the term preemption for this type of interruption and use another term for the interruption caused by expiration of a quantum. Clearly the immediate interruption of a low-priority process by a high-priority process is required if we use idle processes as described in an earlier section. However, since the action of the system is basically the same in both cases, we feel use of the term preemption in both cases is not only reasonable but also less confusing.

14.4 Coordination of Concurrent Processes

Insys supports cooperating concurrent processes. In this section we explore the problems of coordinating the execution of several cooperating concurrent processes. As before, we make no distinction between a group of processes owned by a single user and a group containing more than one user's processes. A set of *concurrent processes* is defined as two or more processes, all of which are simultaneously in some state of execution; that is, each member process has been created and its execution has not yet been terminated. Thus, it is possible to construct some or all of these processes such that their action depends on the action of other processes. Our particular interest is the interaction between processes, that is, the action of process A depends on the action of process B, then further action of process B depends on the action of process A, and so on, back and forth.

Conceptually there is no task that a set of concurrent processes can do that a single process cannot do. However, there are three reasons why concurrent processes are desirable. It is often conceptually simpler and easier to program a complex task if it is organized as a set of concurrent processes. In addition, a task so structured may be completed more quickly. For example, if there are many processors that are not always busy, then a task organized as three concurrent processes may be able to complete in one third of the elapsed time it would require as a single process. Even if a task is not completed more rapidly, when it is organized as a set of concurrent processes, it may make more efficient use of the system's resources when executing in a multiprogramming system.

The following example of two cooperating concurrent processes will help to clarify the preceding discussion and illustrate the need for synchronization of cooperating concurrent processes. Process A computes numbers and process B prints these numbers. This is a simple version of what is known as a *producer-consumer* relationship, which is an abstraction of the essential relationship in many sets of cooperating concurrent processes. Process A is the producer, and process B is the consumer. Let us assume that process B prints six numbers per line. Thus, it cannot initiate printing of a line until process A has computed at least six numbers.

The interaction between these two processes goes as follows. Process A computes a number and gives it to process B. (In Section 14.5 we will see how this can be done.) Process A then starts computing another number, which it will try to give to process B when the number has been computed. In the meantime process B, starting as soon as it receives the first number, converts the number to the coded digit form required for printing and stores it as the first part of a print line. As soon as process B finishes this, it is ready to receive another number from process A. When process B finishes converting the sixth number, it must initiate a print operation to print the completed line before it is ready to receive the next number from process A.

During the time process A and process B exist, there are three possible states for this set of concurrent processes. In state 1 both processes are able to continue with their execution. This state includes the case where both processes are computing as well as the case where process B is ready to receive a number from process A at exactly the same time as process A is ready to give a number to process B. In state 2 process A has finished computing the next number, but process B is not yet ready to receive it. In this state process A must wait until process B is ready for the number. In state 3 process B is ready to receive the next number, but process A has not yet finished computing it. In this state process B must wait until process A has finished computing the number.

Either or both of states 2 and 3 will occur if the rates of progress of the two processes are different. Different rates of progress may be the result of the system's scheduling algorithm or it may be inherent in the two processes. For example, it may simply take longer for process A to compute some numbers than it does others. Also, even though the conversion of numbers by process B may be almost the same whatever the number, after every sixth number is converted process B will have to execute longer before it is ready for the next number, since it must initiate a print operation in addition to converting a number. In most producer-consumer problems the rates of production and consumption are unequal and fluctuate with time.

Figure 14.13 shows an example sequence of states for the set of two processes A and B. The progress of each process is shown by a horizontal line. The solid part of the line represents execution, and the dashed part represents waiting. In each case the number being computed or converted is written above the solid line segment representing the execution that computes or converts the number. The bottom line shows the sequence of states for the set of processes.

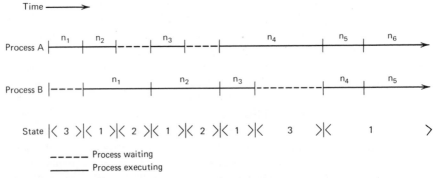

Figure 14.13 Example sequence of states for a set of two cooperating concurrent processes.

The duration of each state is indicated by a set of pointed brackets. The transition between states occurs at each vertical mark.

The reader should note that the ability to have a set of concurrent processes does not require multiprocessing, only multiprogramming. The facility for synchronization will have to include something equivalent to **block** so that one process can wait until another process has reached a certain point in its execution. The addition of multiprocessing will generally reduce the elapsed time for the set of processes by reducing the time that a blocked process will have to wait for another process to reach the required point in its execution. However, multiprocessing does not add any new capability to the system. The major contribution of multiprocessing to a system is to permit more efficient use of the system resources and increase the reliability of its operation.

Let us explore more fully the problem of coordinating concurrent processes. The basic problem is partial synchronization in an asynchronous environment. Each separate process is asynchronous with respect to all other processes; that is, each process executes at an unknown and unpredictable rate. For example, suppose A and B are cooperating processes. A executes for 1 second, then needs some data that B is computing. If it takes B 2 seconds to compute this data, A will have to wait. But the amount of time A must wait not only depends on the length of time B executes, it also depends on the number of processes on the ready list, the number of processors, and the scheduling algorithm. With a pure round-robin algorithm, if A knows that there are exactly three processes on the ready list, two processors, and the length of the quantum is 1 second, then it will be safe for A to proceed if it waits for at least 2 seconds. Process B will certainly be finished by this time. B will actually be finished in 1 second if it is the first or second process on the ready list.

In general synchronization based on elapsed time will not work because the information needed to determine how long to wait is not available when the program is written, and is not even obtainable when the program is executing. For example, suppose that process A could find out from the system the future sequence of states for every other process in the system and the length of time between each state transition. Based on this information, process A could compute how long it must wait before process B will be finished with its computation. The difficulty is that if a new process is created while process A is waiting, then the time that A must wait for B to finish may turn out to be longer than what was originally computed, since a new process is now getting its share of execution on the processors, thus delaying the execution of all of the other processes, including B. The waiting process A will have to be notified of this so that it can recompute its waiting time. Therefore, it is much simpler and more efficient to dispense with all of the complexity of computing waiting time and use a simple mechanism based entirely on the idea of notifying a waiting process when the other process finishes its computation. This should sound like **wakeup**. It is, as we will see shortly.

What we need is a simple way to coordinate (partially synchronize) a set of cooperating concurrent processes which is not sensitive to variations in timing. We also want the mechanism to be simple enough so that we are certain that it will always work. The simplest solution to the problem is to have the system provide a mechanism for one process to tell another process that it has finished the computation that the other is waiting for. The simplest conceptual mechanism of this sort uses a variable called a *semaphore*. What we use in Insys is not a semaphore, but it is equivalent to one. However, it is easier to understand the basic ideas of synchronization using semaphores, so we will initially explore the problem and its solution in terms of semaphores and later transform it into the equivalent mechanism that is used in Insys.

A semaphore is like the similar named railroad device. When a railroad semaphore is set to green, an approaching train continues on. When the semaphore is set to red, the train stops and waits until it turns green. Whenever a train passes a green semaphore, the semaphore is automatically reset from green to red. When the train has proceeded far enough down the track that it is safe for another train to follow, the semaphore is automatically reset from red to green. Our semaphore is a two-valued variable. The value 1 corresponds to green, while the value 0 corresponds to red. The best way to define the operation of our semaphores is to show how they are used.

The following discussion of the use of semaphores relates to the earlier example of two processes: A, which computes numbers, and B, which prints them. We use two semaphores, SA and SB. When SA = 1, process A has finished computing the next number and is ready to give it to process B. If SA = 0, A is still computing the next number. If SB = 1, process B is ready to accept the last number from A so that A may proceed to compute the next number. If SB = 0, B is still converting the last number and is not able to accept the next number from A.

The action of the two processes is very similar. When process A finishes computing the next number, it sets SA = 1 to indicate that it has finished. It then tests SB to see if B is able to accept the number. If SB = 0, B is still converting the last number and is not yet able to accept the next number. A must then wait until B is able to accept the number before it can start computing the next number. Thus, A continues to test SB until SB = 1, at which point B has accepted the number and A can now proceed. As soon as process A finds SB = 1, it resets SB = 0, which indicates that it recognizes that B has accepted the next number.

Process B behaves in almost identical fashion. It begins by testing SA until this semaphore has the value 1. As soon as SA = 1, process A is ready to supply the next number and B may proceed with the conversion of that number. First, it sets SA = 0, which indicates that it recognizes that A is ready to send it the next number. Then it sets SB = 1 to indicate that it has accepted the next number.

Instead of requiring a process to test the value of a semaphore continuously until it becomes equal to 1, we introduce two special operations on semaphores, called P and V.[2] When a process executes P(S), its execution is suspended until S = 1. That is, when a process executes P(S), it is allowed to proceed with its execution if S = 1. However, if S = 0, the process is not allowed to continue its execution until the value of S becomes 1. In case the process is forced to wait, it does not have to make any further tests on the value of S, nor does it use any processor time while waiting. Immediately before

---

[2] E. W. Dijkstra, "Co-operating Sequential Processes," *Programming Languages* (F. Genuys, editor), Academic Press, London, 1968, pp. 43–112.

execution of the process continues, whether the process had to wait or not, the
value of the semaphore S is reset from 1 to 0.

Figure 14.14 is a flowchart for the two concurrent processes discussed earlier
in which semaphores and the P and V operations are used for synchronization.
Prior to starting execution of the two processes, SA and SB both must be
initialized to 0, indicating that A has not yet computed the first number and
B has not yet accepted the first number. Process B begins execution with P(SA),
since it cannot convert the first number until A has computed it. Process A
begins execution with the computation of the first number. When A finishes,
it executes V(SA), which sets SA equal to 1, allowing B to proceed converting
the first number. Process A now executes P(SB) causing it to wait until B has
accepted the first number. Only when B has accepted this number may A
proceed to compute the next number. As soon as A has executed the V(SA)
operation, B may proceed, and the P(SA) operation that caused it to wait will
be completed, resetting the value of SA back to 0. B then executes V(SB) to
indicate that it has accepted the next number from A. This will allow A, which
is waiting, to proceed with computation of the next number.

The reader may already have recognized that the action of the P operation
is similar to the action of **block**. This is, in fact, how we will implement the P
operation in Insys. In like fashion the action of the V operation is similar to
the action of **wakeup**. Figure 14.15 shows the processes in Figure 14.14 converted into calls to **block** and **wakeup**. Note that no explicit semaphores are
required in this simple example when **block** and **wakeup** are used. In the
general case, discussed later, they will be required.

In order to avoid trouble we must also cover the situation where the process
to be awakened is either running or on the ready list, that is, the process has
not yet called **block**. This corresponds to the case where process A in Figure
14.14 has finished computing the next number and executes V(SA), but process
B has not finished converting the last number and therefore has not yet executed P(SA). If process control simply ignores a call to **wakeup** when the
specified process is not blocked, then later when this process does call **block**,

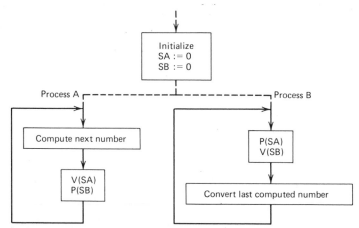

Figure 14.14 Two cooperating concurrent processes.

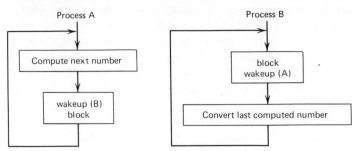

Figure 14.15 The **block** and **wakeup** analogue of the processes in Figure 14.14.

it will be put into a blocked state and remain there forever, since the wakeup that would unblock it has already occurred.

To see that this would be true, consider Figure 14.14. If the V operation is redefined so that it does not change the value of a semaphore, but merely allows a waiting process to proceed, then process B may get into an infinite wait. If B is still converting the last number when A executes V(SA), then A proceeds to execute P(SB) without changing the value of semaphore SA. Process A will now wait, since SB = 0. Process B will eventually finish converting the last number and execute P(SA). However, it cannot proceed, since SA = 0. This is a *deadlock* situation, often called a *deadly embrace*.

We did not have any trouble with the P and V operations as originally defined, since the V operation actually changed the value of a semaphore, which a P operation later tests. In other words, the semaphore is used to remember the occurrence of a V operation. We can use the same technique with **block** and **wakeup**. We call this memory the *wakeup waiting switch*. When a call to **wakeup** finds that the specified process is not blocked, its wakeup waiting switch is set. Later this process will call **block**. However, before putting the process on the blocked list, process control will test the wakeup waiting switch. If it is on, the process is not blocked, but put directly on the ready list.

As pointed out earlier in this chapter, we choose to execute all system procedures as part of the user process that requested the service. There are other alternatives. The system procedures could be executed as part of a special system process, or they could be executed in some extraordinary way that is not part of any process. We rejected the last alternative because it is conceptually unclean and complicates the management of resources. We rejected the first alternative because it is conceptually more appealing to have a process be an unbroken flow of execution (except for "invisible" delays such as blocking and preemption) from the time of its creation until it is destroyed. In the next chapter we will see that our choice leads to practical advantages with respect to referencing resources, especially data, and accounting for resource usage.

If process control executes as part of a user process a problem may arise if the process is preempted while process control is actually executing. (Recall that preemption occurs when the timer's value becomes zero.) If the preemption occurs while any of the process control data such as the process table or the ready list is being modified and the modification has not been completed, the data may be in an inconsistent state that will later cause malfunction of the system, or at least unpredictable and undesirable behavior. For example, suppose the scheduler uses a multiple-queue, priority-based, scheduling algorithm. If the multiple queues are implemented with a single structured ready list, as suggested in Figure 14.12, pointers to the end of each queue are required. When a process that has become ready is scheduled, it is added to the end of one of the queues, and the pointer to the end of that queue is updated. If preemption occurs after the pointer to the newly ready process has been added to the queue but before the pointer to the end of the queue has been updated, trouble may occur. If another process becomes ready and is added to the same queue before execution of the preempted process is resumed, the preempted process will get "lost" and will not execute again. The scheduler does not realize that the preempted process has been added to the queue because the pointer to the end of the queue did not get updated. Therefore, it will place the pointer to the newly ready process on top of the pointer to the preempted process.

We can attempt to avoid these problems by being very clever in programming process control. If we can code these procedures in such a way that the process control data is never inconsistent, then no problems can occur. This is, at best, extremely difficult, and probably impossible in some cases. It is far easier simply to prevent preemption while any of the process control data is being modified. This solution is almost universally adopted in all existing operating systems. In Insys preemption is prevented from the time a process enters process control until it exits. Most modern computers have a hardware instruction that will inhibit a timer interrupt until a companion instruction has been executed. This inhibiting of preemption will not seriously influence the response properties of the scheduling algorithm, since the execution time for any path through process control is very small.

Another problem arises in multiprocessing systems. If there is more than one processor, then more than one process will be executing at the same time. Thus, the possibility exists that more than one process may be simultaneously trying to modify the process control data. For example, two processes may be preempted at the same time and so they are both being scheduled at the same time. (We will see in a later chapter how both processes can be executing the scheduler at the same time.) As a result, both processes are trying to modify the ready list at the same time. This is a potential source of trouble. For example, suppose both preempted processes are to be put on the same queue (assuming a multiple-queue ready list). The first process is put on the end of the queue, but before the pointer to the end of the queue can be updated, the second process is also put on the end of the queue. In this case the entry for the second process will be put on top of the entry for the first process, and the first process will be "lost."

Again we can try to prevent this problem by tricky programming, but we reject this solution in favor of the simpler method of preventing simultaneous modification. How can we accomplish this? The use of a semaphore provides a straightforward solution. If the semaphore is zero, some process is currently modifying the process control data. A second process wishing to do so must now wait until the first process has finished. It seems natural to use the P and V operations to coordinate the modification of the process control data. For example, when a process wishes to modify a process table entry, it executes a

P operation on the semaphore. If the process table is currently being modified, the semaphore will be zero, and the P operation will delay the process until the other process finishes modifying the process table and executes a V operation on the semaphore.

If we attempt to use this solution by translating the P and V operations into calls to block and wakeup, as we did in preceding sections, we are led into circularity, since the need to avoid simultaneous modification of the process table actually occurs within block and wakeup. Clearly we need a more primitive mechanism. The simplest solution is to use a single memory location as a semaphore and have each process directly test and modify the value of this semaphore. When a process finds that the semaphore's value is zero, it must continuously test until the semaphore's value becomes one. The process cannot block itself while waiting, but must actually execute the test instruction repeatedly.

We have still not completely solved the problem, since we must prevent more than one process from testing the semaphore at the same time. If two processes can make the test simultaneously, the sequence shown in Figure 14.16 can occur. Both processes execute the same sequence of instructions for testing the semaphore, which is memory location S. The content of S is compared with zero. Then a conditional branch is made back to the test instruction if the content of S is zero. If S does not contain zero the process knows that the process control data is not being modified. It then stores zero into S to indicate that it is now modifying the data and continues with its execution. The reader can observe in the above example from the time at which the instructions get executed in the two processes that each process believes that it is safe for it to proceed with modification of the process control data. The failure of the testing procedure results from the fact that the first process was unable to set the semaphore to zero before the second process started testing the semaphore.

The only practical solution to this difficulty is a special hardware instruction. (Dijkstra has a procedure that does not require any special hardware instruction.[3] However, it is much too complicated and time consuming to be of any use in a part of the system like process control.) What is required is a single hardware instruction that tests the semaphore and resets its value if the test is successful. The hardware must prevent the processor from being interrupted while it is executing this instruction. In addition, other processors must be prevented from executing this instruction if it refers to the same semaphore (same memory location). Most modern computers have an instruction or some other facility which satisfies these conditions.

For example, our *test-and-set* instruction

$$\text{TS} \quad \text{S}$$

tests the contents of memory location S. If S = 0 the condition code is set to zero; otherwise, the condition code is set to one. In addition, if S $\neq$ 0 the test-and-set instruction changes the value of S to zero. Using this test-and-set instruction, we can reformulate the procedure used by a process to determine if it is safe to modify the process control data. Figure 14.17 shows the new sequence

[3] E. W. Dijkstra, "Co-operating Sequential Processes," *Programming Languages* (F. Genuys, editor), Academic Press, London, 1968, pp. 43–112

Time	First process			Second process		
t_1	SR	2,2	$G_2 := 0$			
t_2	C	2,S	S = 0?	SR	2,2	$G_2 := 0$
t_3	BE	*−4	branch if S=0	C	2,S	S = 0?
t_4	ST	2,S	S := 0	BE	*−4	branch if S=0
t_5				ST	2,S	S := 0

Figure 14.16 Unsafe sequence of instructions for testing and setting a semaphore.

Time	First process		Second process	
t_1	TS S	test & set S		
t_2	BZ *−4	branch if S=0	TS S	test & set S
t_3			BZ *−4	branch if S=0

Figure 14.17 A safe reformulation of Figure 14.16 using the test and set instruction.

of instructions. Now the first process will proceed to modify the process control data, since the testing part of the test-and-set instruction will find the semaphore S = 1. However, the second process will find S = 0 when it is tested at time t_2, since the set part of the test-and-set instruction executed by the first process has set S = 0 at time t_1, even though S = 1 when the first process began execution of its test-and-set instruction. The only way that this testing sequence could fail would be if both processes were to execute their test-and-set instructions at the same time (t_1). But this is prevented from happening by the hardware, since both of these instructions refer to the same semaphore.

14.5 Interprocess Communication

There are some situations in which **block** and **wakeup** are not sufficient to achieve the desired coordination between processes in a set of cooperating concurrent processes. When there are more than two processes in the set, a process that is awakened from the blocked state may need to know which of several processes it was that caused the wakeup. In another situation a process may need to remain blocked until a set of events has occurred. Both of these and other similar situations can be taken care of by one general mechanism, *interprocess communication*. This mechanism makes it possible for any process to send a message to any other process.

Our interprocess communication facility consists of two system procedures, **send** and **collect**, and a message storage area for each process. The call

$$\text{send (rec\_pros\_id, message);}$$

is executed by a process that wishes to send a message to another process. The arguments are the identification of the receiving process and the message that the sender wishes transmitted to the receiver. The system copies the message into the message storage area of the receiving process and prefixes the message with the process identification of the sending process. The receiving process executes the call

$$\text{collect (store);}$$

when it wishes to pick up a message that has been sent to it. A call to collect results in the oldest message in the receiver's message storage area being copied from there into the memory location specified by the argument, store.

Both of the situations described in the first paragraph of this section can be solved by using the interprocess communication facility in the following way. Immediately before a process calls wakeup, it sends a message to the process that is about to be awakened. This message identifies the event that is causing the wakeup and identifies the process that is calling wakeup. Correspondingly, whenever a process is awakened, it calls collect to pick up the message that was sent by the process that called wakeup. By examining this message, it can tell what event has occurred and which process caused the wakeup. If not all of the desired events have yet occurred or the wrong process caused the wakeup, block is called again to continue waiting.

Most interprocess communication facilities are more elaborate than the one just described. One feature that must be included in any interprocess communication facility is a way for the sending process to obtain the process identification of the receiving process. Generally all the programmer of the sending procedure knows is the name and project identification of the receiver. Therefore, at some time before the sending process calls send, it must obtain from the system the process identification of the receiver. This is a simple mapping that can be accomplished by searching the process table for an entry whose values of user_id and acct_nr match those of the desired receiver.

A desirable feature found in a few interprocess communication facilities is the ability of a process to refuse to accept messages except for those sent by one of a number of specified processes. This allows a process to refuse "junk mail," a feature that the U.S. Postal Service lacks. This feature is easily implemented by associating with each process a list of process identifications of other processes from which it will accept messages. When a process calls the send procedure, the intended receiver's list is checked to see if the identification of the sending process is included. If it is not, the message will not be sent.

So far we have avoided the question of what happens if the receiver's message storage area is full. Several solutions are possible. One solution is to throw away old messages, that is, when a new message is being sent and the receiver's area is full, the oldest message in the area is replaced by the new message. In general this is not an acceptable solution, although it may be in certain circumstances. One acceptable solution is to provide an additional argument for send, which is a status code. If the receiver's message storage area is full the message is not sent and the status code is set to indicate this. What happens then is up to the sender.

Usually, when a process sends a message to another process, the sender will not be able to do anything else until the message has been accepted. Thus, another solution is to have send block until there is empty space in the receiver's message storage area. As soon as there is a place to put the message, the sending process is awakened and the message is sent. A similar situation exists when the receiver calls collect and there are no messages in the message storage area. Again, the most desirable action may be to block until a message arrives.

There is clearly an interrelationship between the interprocess communication facility and the block and wakeup entries of process control. It is beyond the scope of this text to consider a complete integrated design of these facilities.[4] All we have tried to do is to explore the basic ideas that are involved and expose the major problems. In principle the interprocess communication procedures **send** and **collect** are sufficient for all necessary communication between cooperating processes, no matter what volume of data is involved. However, if large volumes of data must be passed between two processes, the fact that interprocess communication requires the copying of messages results in a large overhead. In Chapter 17 we will study *direct sharing* of data, which is a realistic and efficient method of transmitting large amounts of data between cooperating processes.

14.6 User Control of Processes: The Command Language

We can distinguish two major types of control for a process, *interior* and *exterior*. By interior control we mean calls on system procedures, such as **block**, by a program executing as part of a process. This type of control has been discussed earlier in this chapter. By exterior control we mean *commands* to the system that are not part of any program executing in a process. The format and meaning of these commands constitutes the *command language* of the system. Commands are the means by which the user exercises overall control of the composition and progress of a process.

In Chapter 2 we described a command language that is typical of those used in batch (noninteractive) systems, although it is much simpler than the elaborate and complex job control language used in OS/360. Commands were punched on cards and included as part of the input for a job. Cards were placed before each job step and interpreted by the system immediately prior to execution of the job step. In an interactive system the commands are usually typed on a terminal by the user as needed. That is, the commands for a job step would be typed immediately prior to execution of the job step, but not before the previous job step has completed its execution. This is a significant difference between batch and interactive systems. In a batch system all of the steps in a job must be specified before starting execution of any step in the job. In an interactive system each step is executed before the next step is specified. This makes it possible for the user to use the results of previous steps in making a decision as to what to do in the next step.

By using the word "language" in command language we imply that there exists a well-defined syntax and semantics for the command language in the sense of Chapter 7. If we recall the discussion in that chapter it is clear what functions a *command language interpreter* must perform. It must first parse a given command and then interpret the semantics of that command. As we have seen, parsing depends on syntax, and interpretation depends on semantics. Most command languages are simple enough, both syntactically and semantically, so that the commands can be interpreted one by one, that is, the meaning of a command does not directly depend on any past or future commands. If this is true, the command language interpreter can have the simple structure shown in Figure 14.18.

[4] E. I. Organick, *The Multics System: An Examination of Its Structure*, MIT Press, Cambridge, Mass., 1972.

```
procedure cl_interpret;
  repeat
    Read next command;
    Parse command;
    Interpret command;
  until forever;
end.
```

Figure 14.18 Definition of the command language interpreter.

While it is certainly possible for command languages to have any consistent syntax and semantics, most command languages are quite similar in both aspects. In most command languages a command is essentially equivalent to a procedure call. That is, syntactically the command consists of the name of a program, called the *command program*, and a set of values, which are the command program's arguments. The command language interpreter interprets a command by calling the entry point of the named command program. For example, the commands for invoking the language processors in Chapter 2 are clear examples of the correspondence between commands and procedure calls.

This correspondence of commands to command programs is often concealed by irregular syntax; that is, each command has a syntax that is, or seems to be, different from all other commands. The implementation of the command language interpreter may also conceal the correspondence between commands and programs. Often the simpler commands are interpreted directly by the command language interpreter instead of calling a separate program. An example of this is the interpretation of the job command in Chapter 2. The job command is equivalent to the *sign-on* or *log-in* command in interactive systems. Its major function is to create a process for the user. In the system of Chapter 2 the command language interpreter itself created the process control block (job control block). However, it could just as well have called a procedure such as **create_process**. The only justification for not doing so was efficiency. In a system such as Insys, where process creation is more complex than in the system of Chapter 2, the sign-on command will be interpreted by calling the sign-on command program which, in turn, will call **create_process** to create a process for the user.

In the system of Chapter 2 the notion of process was not well defined. However, in Insys we have decided to have all execution take place as part of some process. In particular, the creation of a new process takes place in some process other than the process being created. Therefore, we must decide what process interprets the sign-on command, since the user who is signing on does not yet have a process. The most reasonable solution is to have the system answering process do it.

The definition of the Insys answering procedure is shown in Figure 14.19. All unused terminals are attached to the answering process. Normally the answering process is blocked waiting for any attached terminal to ring up the system. When one does, the answering process wakes up. It reads one command from the terminal that rang up the system. The only command that will be accepted is the sign-on command. The user's name and account number are extracted from the sign-on command. If this user is authorized to use the

```
procedure answer;
  repeat
    block;
    Read command from ringing terminal;
    if command is sign on then
        if user_id and acct_nr from command are authorized then
            begin create_process(user_id, acct_nr, cl_interpret, pros_id);
                Transfer terminal to new process;
            end;
        else Print 'Unauthorized use';
    else Print 'Illegal command';
  until forever;
end.
```

Figure 14.19 Definition of system answering procedure for system answering process.

specified account, then a new process is created for him. The starting address for execution of this process is the entry point of the command language interpreter. Finally, the terminal that the user is using must be transferred to his new process. Once that is done, all commands that he types on the terminal will be interpreted as part of his process.

The command language for Insys directly exhibits the correspondence between commands and command programs. The format of a command is

$$cmd\_name \; arg1 \; arg2 \ldots argn;$$

The name of the command program, *cmd_name*, is followed by a set of values for its arguments. These argument values may be numbers, character strings, or names of objects such as files. For example, the command

delete data_1;

is a command that deletes from file storage the user's file named data_1. With this simple syntax and semantics the parsing and interpretation, which has to be done by the command language interpreter, becomes equally simple. Parsing is simply scanning the characters in the command to locate the command program name and the argument values. Interpretation then consists of constructing an argument list from the arguments in the command and calling the command program.

One of the Insys objectives was expandability, that is, the ability to add new functions to the system. For example, we may wish to add a compiler for a new language. This objective implies that the command language also needs to be expandable, that is, we may need new commands in order to invoke the new functions that have been added to the system. Some command languages have a fixed set of commands and command programs. When we add a new function to such a system, we must reprogram part of the system.

Our goal is to make it easy to add new functions to the system and to expand the command language. The correspondence between commands and command programs gives us a clue as to how to achieve this. If we allow the command program name in a command to be the name of any program stored in the file storage, whether it be a system program or a user's program, then the command language is extendable simply by putting a new command program somewhere in the file storage. Many systems do not go quite this far, but require that command programs all be stored in a special system library file. This does not limit the expandability of the command language. However, some special procedure is required in order to get a new command program into the system command library.

A common feature of command languages is a macro facility. The basic discussion of macro definition and expansion in Chapter 5 is applicable to command language macros. The definition of a macro command consists of a named collection of commands with certain arguments and command names designated as parameters of the macro. A macro command has the same format as an ordinary command. However, the command name is the name of the macro command, and the arguments are values for the macro's parameters. Expansion is done by the command language interpreter, which must recognize the name of the macro and fetch its definition. Each command in the definition is read, values substituted for its parameters, and then interpreted.

The principal difference between command macros and assembly language macros is that the definitions for command macros are usually stored in files. A command language macro definition is seldom, if ever, included as part of the set of commands for a job that uses the macro. The reason for this is that the set of commands for a job tends to be short and it is not often that the same macro will be used more than once or twice. Thus, the major benefit of macros is achieved only by storing the macro command definitions in a library from which the command language interpreter will fetch them. In OS/360 macro commands are called *catalogued procedures*. Similar advantages will accrue in assembly language if calls to system functions and other frequently used instruction sequences are stored as macro definitions in a library. In fact, most modern macro assemblers have this capability.

EXERCISES

14.1 Write an IL procedure.

wait (compound_event);

where **compound_event** is a character string that is a Boolean expression composed of event names. Once **wait** is called, it does not return until **compound_event** is true. When called, **wait** will in turn call **block**. Every time the process wakes up, **wait** should collect any messages and determine if **compound_event** is now true. If not, it calls **block** once again. For example, the call

wait('A&B||C');

will effectively block the process until either event C occurs or both events A and B occur.

14.2 Write an IL procedure for the scheduler that implements the following scheduling policy. There are four queues, Q_0 to Q_3. Q_0 is the highest priority queue. Let the quantum be q seconds. When a process is allowed to execute, it will be given $2^i * q$ seconds if it came from Q_i. No process is allowed to execute unless all of the higher priority queues are empty. Whenever a process goes from blocked to ready, it is placed at the end of Q_0. When a process that came from Q_i has executed $2^i * q$ seconds, it is preempted and placed at the end of Q_{i+1} if $i < 3$, and Q_3 if $i = 3$. When a process is placed in Q_i, if the process currently executing came from Q_j, where $j > i$, it is immediately preempted if it has already executed $2^i * q$ seconds, otherwise it is preempted as soon as it has executed $2^i * q$ seconds. When it is preempted under these conditions, it is returned to the head of Q_j. However, the next time it is allowed to execute, it will be permitted to execute only long enough to complete the original $2^j * q$ seconds that it was allotted.

14.3 Use the P and V operations to describe the synchronization required to manage a *ring buffer*. A ring buffer is an array of N words, $B_0, B_1, \ldots, B_{N-1}$ into which process P puts data, and from which process Q takes data. The rate at which P produces data may not be equal to the rate at which Q consumes data. B is called a ring buffer, since P stores words into B in the sequence, $B_0, B_1, \ldots, B_{N-1}, B_0, B_1, \ldots, B_{N-1}, B_0$, and so on. Process Q will consume data in the same sequence. Process P should be allowed to fill up the buffer with data, but not overwrite data that has not yet been consumed by Q. Process Q should be allowed to consume all of the data in B, but not try to use data that P has not yet put into B.

14.4 Write an IL procedure for your solution to Exercise 14.3, using **block** and **wakeup** instead of P and V.

15 | Memory Management

In this chapter we explore the major problems of memory management, particularly primary memory. The general use of secondary memory is treated in Chapter 16. However, a special use of secondary memory is discussed in this chapter. The problems resulting from the requirements for privacy, protection, and sharing of information are postponed until Chapter 17.

We can discover the major problems in memory management by analyzing the implications of our system's attributes. We saw in Chapter 14 that multiprogramming and multiprocessing imply that more than one process may exist at the same time. Since a process consists of at least one procedure and some data, it requires memory space in which to store the procedures and data of the process. Thus, the computer's memory must be divided up and allocated to more than one process. In particular, multiprocessing implies that primary memory must be shared by more than one process. Hence, we need a mechanism and policy for allocating memory that is more elaborate than that used in the simple batch system of Chapter 2. Another attribute of our system is permanent storage of users' information. This implies that the system be able to keep more than one user's information in memory, usually secondary memory, for a long period of time and be able to retrieve it on request. This facility is often called *the file system* and implies allocation of secondary memory.

The attributes of expandability and adaptability imply that both system and user procedures be independent of both the amount and type of memory in the hardware system. This is required if it is to be easy to expand the system by adding more primary memory, adding additional secondary memory devices of the same or new types, and removing old secondary memory devices. This implies that user references to procedures, data, and files be highly independent of their actual location in physical memory, that is, that some kind of symbolic addressing be used.

In summary, the major problems that we will consider are three. The first is the allocation of primary memory: mechanisms and policy. The second is location independent referencing of information and a generalized concept for this, which is called *virtual memory*. The third problem is that of file storage. One solution to this problem is examined here, and another solution is treated in Chapter 16.

15.1 Primary Memory Allocation

The allocation of primary memory is a problem because it is not large enough. If all of the programs and all the data that any user ever needed, both in the past and in the future, would fit into primary memory, then allocation would be trivial. Whenever a new procedure or some new data was created it would be allocated some as yet unused part of memory and would then be stored there forever. None of the memory would ever need to be reused. Allocation problems arise because primary memory is small and therefore must be reused. The reader may be tempted to assume that a superlarge memory will solve all of our problems. If it is basically unbounded it will solve the memory allocation problem; however, it introduces new problems. The worst problem is the size of the address needed to reference information stored in the memory. If the memory is unbounded, then a memory address is also unbounded. This means that address constants and all registers that are used to manipulate them also need to be unbounded in length. In addition, a large percentage of the memory will be wasted because it has not yet been allocated or the information that it holds is no longer of any use. Thus, even if it were possible to build a superlarge memory and a processor which could use such a memory, it would be excessively expensive. A far more economical solution is a memory allocation scheme that reuses memory.

Memory allocation is a problem for both the user and the system. In a multiprogramming system each process must have some primary memory when it is executing. However, when a process is not executing because it is blocked or has been preempted, it does not need any primary memory, and the memory that was allocated to it while it was executing may be reused. In fact, most of the time it will have to be reused. A user encounters memory allocation problems if the total memory he needs for the procedures and data in his process is larger than the maximum amount of primary memory that he is allowed to use. The user will also have memory allocation problems within one or more of his data areas when his program uses list processing or other techniques in which memory use varies in an unpredictable way.

At first glance it may seem that *swapping* will solve the allocation problems. In swapping, when a process is executing, it is allocated all of primary memory (other than that required by the system). When its execution is suspended, either because it called **block** or was preempted, the entire process is swapped with the process that will execute next. That is, all of the contents of primary memory that belong to the process are moved from primary memory to some secondary memory device, such as a drum or a disk. The procedures and data that make up the process that is about to execute are then moved from secondary memory into the now vacant primary memory (which had been allocated to the previously executing process).

This solution may be satisfactory in certain circumstances, principally because it is simple to implement. It was used in one of the first time-sharing systems because any other solution would have required extensive modification of a large number of programs that were already working in a sequential batch system and that were needed to make the new time-sharing system a success in a reasonable amount of time. However, there are four major disadvantages of swapping. First, the transfer time between primary memory and most secondary memory devices is rather long. In an interactive system swapping time may be as long as, or even longer than, the response time for short requests. Thus, the performance of the system may be degraded by a factor of two or more. Second, much of the information in a process is not needed every time the process executes, especially in an interactive system that uses preemption. Thus, more memory than is required by a process is allocated to it. In addition, swapping this unused information introduces unnecessary overhead. Third, only one process can execute at a time, since all of primary memory is allocated to one process. Thus, multiprocessing is impossible. Finally, swapping makes direct sharing of information extremely difficult or impossible. This difficulty will become clear with the discussion in Chapter 17.

One approach that retains much of the simplicity of swapping is to divide the primary memory up into large, fixed length blocks called *partitions*. Processes are then swapped in and out of these partitions. Several processes can be in primary memory at the same time, so multiprocessing is possible. This scheme reduces the response ratio if the scheduler shows a preference for processes that are in primary memory. Much of the swapping can then be done while some other process is executing. (Recall that the I/O processor can operate simultaneously with the control processor.) However, this modified swapping scheme does nothing about reducing the amount of unused information that occupies memory and must be unnecessarily swapped, nor does it make direct sharing possible.

A similar partial swapping scheme called *overlaying* is often used to solve the user's memory allocation problem when all of his procedures and data will not fit into the maximum primary memory that can be allocated to his process. His allocated memory is divided up into blocks. Parts of his program and data are then swapped into and out of these blocks. The procedures and data that fit into one block are called *an overlay*. The user must organize his program and overlays so that the minimum amount of swapping is required; otherwise the overhead will be crippling.

Swapping, or a variation of it, is not really a satisfactory solution to the memory allocation problem. In the system case, the wasted memory, the high overhead for unnecessary data movement, the possible degradation of response time, and the inflexibility, especially with respect to direct sharing of information, add up to a serious objection to swapping as a solution. In the user case, the effort required to plan overlays properly is often substantial. Since this is secondary to the main stream of his efforts in using the computer to solve his problem, requiring him to master the overlay mechanism is contrary to our objective of providing the user with a reasonable interface to the system.

Furthermore, swapping is no help at all in solving the memory allocation problem in the list processing type of applications. The major characteristic of the list processing type of memory use is its dynamic, unpredictable nature. The amount of memory needed, while bounded, varies almost continuously with time. The program continually makes new requests for memory to be allocated to it. It also continuously releases (frees) memory. On the average the requests and releases tend to balance each other; however, they are not matched. That is, the sequence of requests, particularly as regards the amount requested, is unpredictable. The sequence of releases is similarly unpredictable. Thus, it is impossible to decide what area of memory will be allocated in response to a request until the request is actually made. At that time an area is selected from the memory areas that are currently unallocated (free).

We reject swapping as a solution to the system's memory allocation problem and adopt the objective that a suitable solution have the property that unneeded information does not have to be kept in primary memory. In other words, a process need not be allocated any more memory than it actually needs at any given time. Thus, the amount of memory allocated will fluctuate as the process proceeds with its execution. There are several different solutions to the memory allocation problem. Unfortunately, each has advantages and disadvantages that are not completely comparable, thus making it difficult or impossible to select an algorithm that is best in all circumstances.

In the discussion that follows we are assuming a computer with memory addressing similar to that of Our 360, that is, like most contemporary computers. Our aim is to show that serious problems arise in attempting to solve memory management problems when the computer has this type of memory addressing. We contend that no really acceptable solution exists unless memory addressing in the computer hardware is modified. In the next section we describe one version of a different memory addressing mechanism that is similar to that used in a few of the more advanced computers.

There are two major classes of allocation algorithms: *contiguous* and *block*. A practical memory allocation algorithm usually clearly belongs to one of the classes, even though it may have some of the properties of the other class. In discussing different allocation algorithms in this section we will use the following terminology. A request for memory allocation will always be for a *segment*. A segment, as defined in Chapter 6, is a contiguous set of words of information. The length of a segment is arbitrary. However, once memory is allocated for it, its length remains fixed. In the context of system memory allocation a segment is a procedure or a set of data with a symbolic name, as defined in Chapter 6. Words within a segment are referenced relative to the base of the segment. A *block* is a contiguous set of memory locations. Thus, a segment is a logical unit of user information, while a block is a unit of physical memory.

In a *contiguous allocation* algorithm, a segment is stored in a block that is equal to or larger in size than the segment. Thus, the logically contiguous words of the segment are physically contiguous in memory. Hence, the memory address of a word is equal to the base address of the segment plus the relative address of the word in the segment. For example, the memory address of word 3 in segment A shown in Figure 15.1 is 486.

A *block allocation* algorithm is based on equal-sized blocks of memory. Physical memory is divided into blocks, all of which are equal in length. A segment is correspondingly divided into equal-sized logical units called *pages*. The pages are the same size as the blocks. Each page of the segment is stored in a block. Since pages and blocks are all equal in length, any page can be stored in any block. Therefore, the blocks used to store a segment need not be contiguous. That is, logically contiguous words need not all be physically contiguous. For example, in Figure 15.2 memory is divided into eight equal length blocks of 64 words. The segment shown in the figure is three pages in length. Page 0 is stored in block 3, page 1 is stored in block 6, and page 2 is stored in block 1.

Because the blocks used to store a segment need not be contiguous, it is more difficult to determine the memory address of a word in the segment than it was for contiguous memory allocation. The most common method is to use a table, called *the page table*, which contains the base addresses of the

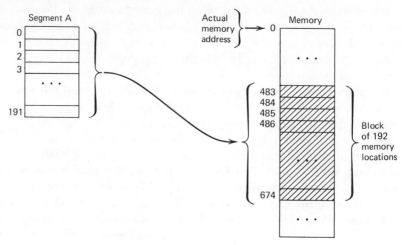

Figure 15.1 Contiguous allocation of a segment.

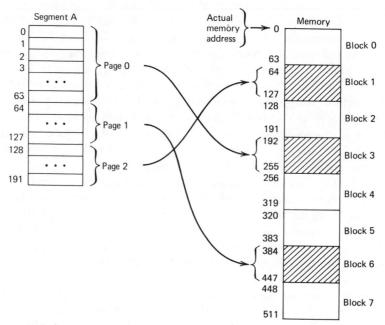

Figure 15.2 Block allocation of a segment.

blocks in which the pages are stored. For example, in Figure 15.2 the page table would contain the following three entries.

0	192
1	384
2	64

Each entry is the base address of the block in which the corresponding page is stored.

If the relative address of a word within the segment is W, then the page number, P, and word number within that page, PW, are computed by the formulas

$$P = \left[\frac{W}{K}\right] \qquad PW = W \textbf{ modulo } K$$

where K is the block and page length, the square brackets around W/K indicate the integer part of the quotient, and W **modulo** K is the remainder after division by K. The memory address, A, of word W is then computed by

$$A = PT_P + PW$$

where PT_P is the contents of entry number P in the page table. In our example, we find the memory address of word 3 in the segment by:

$$W = 3, \qquad P = 0, \qquad PW = 3, \qquad PT_0 = 192$$

and

$$A = 192 + 3 = 195$$

The memory address of word 191 of the segment is found by:

$$W = 191, \qquad P = 2, \qquad PW = 63, \qquad PT_2 = 64$$

and

$$A = 64 + 63 = 127$$

The basic problem in memory allocation is that the sequence of allocation requests usually does not match the sequence of releases. Therefore, each memory allocation algorithm must solve the problem of where to put a new segment when space for it is requested. In a block allocation algorithm, the solution is trivial. Any free block can be used for any page; therefore, if the segment is N pages in length, use the first N free pages. The solution is not so simple in contiguous allocation algorithms, since the entire segment must be stored in a single free block. The problem arises when there is no free block large enough to hold the segment, yet the sum of the lengths of all the free blocks is larger than the length of the segment. (Of course, if this sum is smaller than the length of the segment, the request for memory allocation cannot be satisfied and execution of the process that needs the memory space must be suspended. We will comment more on this later.) Various contiguous allocation algorithms attack this problem in different ways. We will briefly explore two variations. The first is a simple straightforward approach that lets the problem occur and then deals with it when it does. The other attempts to prevent the problem from arising.

The first algorithm uses a very simple allocation policy. Each request for space is satisfied by allocating a block immediately following the most recently allocated block. No attempt is made to reuse any memory that may have been released. Figure 15.3 illustrates three successive requests. Shaded blocks are blocks that have been used and then released and are currently free for reuse. Clearly, after a number of requests, the amount of free space left at the end of memory that is available for allocation will shrink to the point where it is not large enough to satisfy the next request. At this point our simple algorithm no longer works, and we must do something else. What we do is called *memory compaction*. All of the contents of all blocks still in use are moved to the beginning of memory, so there is no free space between them. Figure 15.4 illustrates this operation. Now all of free memory has been compacted into a single block at the end of memory, and our simple allocation policy will again work.

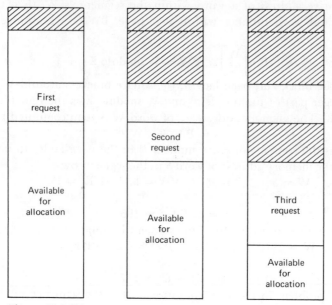

Figure 15.3 Diagram of memory for three consecutive allocation requests.

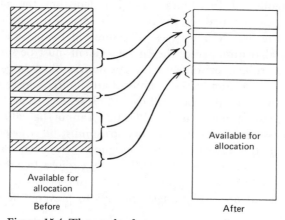

Figure 15.4 The result of memory compaction.

The other contiguous algorithm that we will explore tries to avoid having to compact memory, or at least substantially reduce the frequency of its occurrence, by using a more complicated allocation policy. The version that we describe is called *cyclic first fit*. (Knuth has analyzed and described in detail several of the most commonly used memory allocation algorithms, including this one.)[1] Each block of memory contains a header and a trailer, as shown in Figure 15.5. Both headers and trailers have the same format and contain a *use bit*, a *length field*, and a *pointer field*. The use bit indicates whether the block has been allocated (use bit = 1) or is free (use bit = 0). The length field

[1] D. E. Knuth, *The Art of Computer Programming, Volume 1: Fundamental Algorithms,* Addison-Wesley, Reading, Mass., 1968.

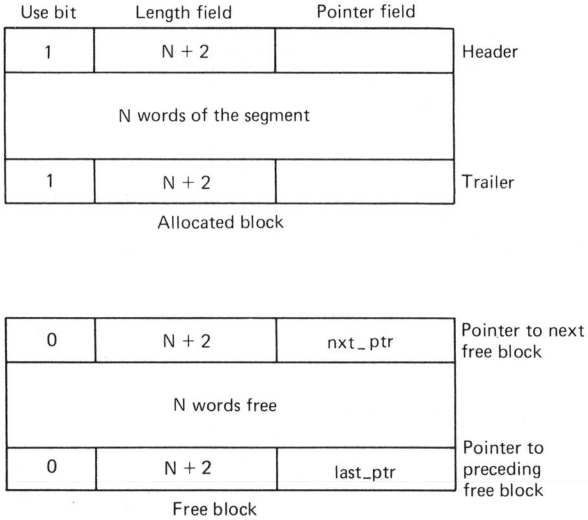

Figure 15.5 Format of blocks with headers and trailers.

contains the length of the block, including the header and trailer. The pointer field is used only if the block is free, in which case it contains the address of another free block. All free blocks are threaded together on a two-way circular list, as illustrated in Figure 15.6. The first three words of memory are reserved for special use. The first word, **start_ptr**, points to the place in the free block list to begin searching when the next allocation request is made. The next two words have the format of the header and trailer of a zero-length block. They point to the beginning and end of the free list.

Suppose a request is made for allocation of memory for a segment M words long. We must find a free block at least M words long. However, each block requires two additional words for the header and trailer, hence we must find a free block whose length field contains a value which is equal or greater than M + 2. Searching of the list of free blocks begins with the block pointed to by **start_ptr**. The **next_ptr** thread is followed until a block of the required size is found or the entire list has been searched. In the latter case, the free block list does not contain a block large enough, and memory compaction is required. Since the free block list is circular, this condition can be detected by comparing **next_ptr** from each block on the free list with the original value of **start_ptr**. When these two values are equal, the entire free block list has been searched.

If a free block is found that is large enough, it is split into two blocks, one M + 2 in length and the other consisting of the remainder of the original block. No splitting is done if the remainder block will be so small that it is useless. The minimum size of a useful block depends on the application. When used for allocation of memory to segments of a process, the minimum useful size may be as high as 512 or 1024 words. The remainder block is left on the free block list in place of the original block, and the block of M + 2 words is allocated for storage of the M word segment, as shown in Figure 15.7. The value of **start_ptr** is set to point to the block on the free list following the block that was split.

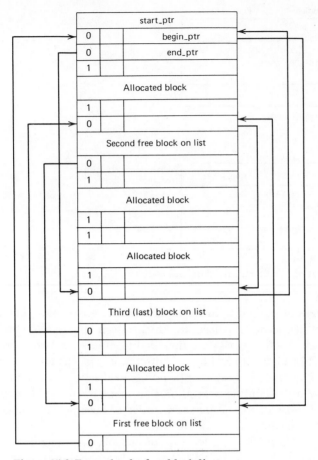

Figure 15.6 Example of a free block list.

When an allocated block is released, it is either added to the end of the free list or combined with contiguous free blocks. When it is added to the end of the free list, it is inserted in the **next_ptr** thread following the block pointed to by **end_ptr**. If either of the blocks in memory that are contiguous to the block being released is already free, the released block is combined with that block on the free list to make a larger block. If both contiguous blocks are on the free list, then one of them is removed, and it and the released block are combined with the block remaining on the free list to form a larger block. The status of a contiguous block can easily be determined by looking at either its header or trailer, one of which is always contiguous with the released block. An example of the combining of blocks is shown in Figure 15.8.

The point of all of this complexity is to avoid having to compact memory. Compaction is required when no free block is large enough to satisfy the current request. Obviously, the smaller the blocks, the more likely compaction will be required. The above algorithm has two major features that minimize the production of small blocks, *combining free blocks* and the *cyclic search*. If a released block is always combined with all contiguous free blocks, then memory is organized so that free blocks are as large as possible. The cyclic search, that

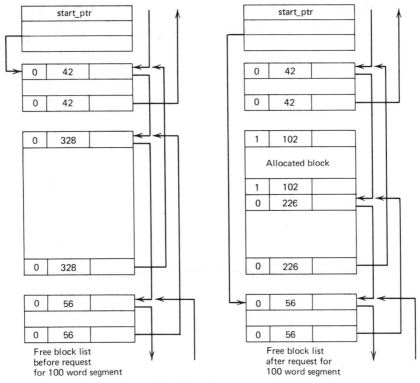

Figure 15.7 Splitting a block for allocation.

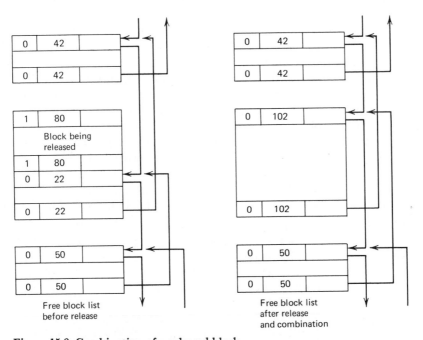

Figure 15.8 Combination of a released block.

is, always starting the search after the free block from which the previous allocation was taken, tends to distribute small blocks more uniformly through memory. This uniform distribution increases the probability that when a small block is released it will be combined with a large block. Knuth has shown from analysis and simulation studies that for some classes of segments these features are usually so effective that if the need for compaction arises, the total amount of free space is not enough to satisfy the request anyway. Therefore, it may be reasonable with this algorithm to make no provision for compaction of memory.

What do we do if no provision is made for compaction and compaction is needed or the free memory remaining after compaction is not large enough to contain the segment? It is clear that some additional memory will have to be released before the request can be granted. In our system, which has multiprogramming, the process for which the memory is needed will be blocked until some other process releases enough memory to satisfy the request. (This solution overlooks some problems, and a better solution is described later.) If the system is similar to the one in Chapter 2, which does not have multiprogramming, the job will be terminated, and the user will have to rewrite his program to use overlays.

What are the advantages and disadvantages of block and contiguous allocation schemes? Does either satisfy our objectives? The major advantage of block allocation is that the allocation policy is very simple. Since the pages of a segment need not be stored in contiguous blocks, memory compaction is never required. The major disadvantage of block allocation is the cost of computing the memory address when referencing a word in a segment. This cost is significant. Even if the block size is a power of two so that the required division can be done by shifting instead of by using the division instruction, which is very slow, several instructions are required to compute the memory address corresponding to a relative address in a segment. In some circumstances the memory address need be computed only once during execution of the program. However, in general, the use of indexing to sequence through data segments makes it necessary to compute the memory address every time a word is referenced, since the index value may be different each time. This results in an overhead that is prohibitively expensive except in very special situations.

Block allocation also costs additional memory space. Each segment requires storage for its page table in addition to the storage required for the segment itself. Additional memory may be wasted as breakage. Most segments will not have a length that is an even multiple of the block size. Therefore, the last page of a segment will usually not be a full-length page. Thus, some of the block in which that page is stored will be unused. This space is wasted and is called the *page breakage*. The average amount of space wasted in page breakage for a segment is half a block.

The size of the page table for a segment is also a function of the block size. The smaller the block (and page) size, the larger the page table for a segment needs to be. Therefore, the total cost in extra memory space is a function of both the block size and the number of segments. For any given set of segments, reducing the block size reduces the breakage. It also increases the size of the page tables. In the limiting case, a block size of one reduces the page breakage to zero and increases the page table size so that half the memory is needed for page tables. Some block allocation algorithms have a moderate block size and limit the page table to one block of memory. Nothing else is stored in the block with the page table, even if not all of the space is used. In this case the average total extra memory required is one and a half blocks per segment: one block for the page table and half a block for breakage. Another common variation is to store several page tables in one block. This can result in considerable reduction in the required extra memory, especially if the block size is rather large and the segments tend to be small.

In most contiguous allocation schemes extra space is also required. However, the breakage is either zero or very small. In the first contiguous algorithm that we discussed there was never any breakage, and no extra space was required. In our second contiguous algorithm two additional words per segment were required for the header and trailer. Some breakage results when a free block is larger than required, but the remainder block is too small to be useful.

Unless compaction can be completely avoided, which is not usually possible in the general case, contiguous algorithms also have other disadvantages. The major one is the cost of compaction. On the average, even if special hardware move instructions exist, it costs one or two memory cycles (0.5 or more microseconds, depending on the computer) to move each word when compaction is required. Thus, if only half of the memory is allocated at the time compaction is required, over 0.1 second may be required for a computer with a large memory. In general, the larger the memory, the less often compaction is required. However, when it is required, more time is required to do the compacting. Whenever compaction is done, an additional cost is incurred. Most of the segments in memory get moved when compaction occurs. All address constants that refer to any segment that was moved must be updated. This updating costs additional execution time. Extra space is also required. In order to update these address constants, a record must be kept of the location of every address constant, and the segment to which it refers must be identified.

This problem of address constant changes, which are required as a result of the movement of segments, is intensified in a multiprogramming system where the problem will exist even if compaction is not required. The system will have to move segments in and out of primary memory. When a process executes for a quantum of time, all the segments that it references during that execution must be in primary memory when they are referenced. In general, this means that some segments of processes that are not executing will have to be removed from primary memory so that enough free space is available for the segments required by the executing process. Policies for deciding which segments to remove will be discussed later in this chapter. However, it is clear that segments belonging to blocked processes should be removed before segments belonging to processes on the ready list. It should be clear that the sequence of segment movement in and out of primary memory is unpredictable. Therefore, it is unlikely that the block in which a segment was stored the last time it was in primary memory will be free the next time the segment needs to be brought into primary memory. In fact, it is likely that the block will have been split or combined in the meantime. Thus, usually every time a segment is moved into primary memory, all address constants that refer to it will have to be updated.

Special hardware for addressing memory can essentially eliminate the cost of address computation in a block allocation scheme. Similar hardware can also essentially eliminate the cost of address constant updating in a contiguous scheme. The recent tendency in operating system design is to solve the memory allocation problem with a combination of both special hardware and software. In fact, several systems combine block and contiguous ideas in an attempt to get the advantages of both without the disadvantages of either. In the next section we explore some of these solutions. In later sections and chapters we will discover some other problems whose solution is also aided by the same special hardware.

15.2 Virtual Memory

Virtual memory is a term often used in connection with a particular class of solutions to the memory allocation problem. In order to explain what virtual memory means we need to define and explore some underlying concepts. Recall that in both assembly language and compiler language the programmer used symbolic names for the memory locations that he referenced. The assembler or compiler then assigned addresses to these symbolic names. Our study of the loader showed us that the addresses that the assembler or compiler assigned were not the actual memory addresses, but were either relative to the base of the segment or still symbolic. After the loader finished relocation and linking, the address constants in the program were all actual memory addresses.

We define *name space* to be the set of symbols used in a program to reference information used in the program. We define *address space* as the set of physical memory locations, or *sites*, in which information is stored. An *address* is the symbol (number or bit pattern) used by the physical memory hardware to reference a particular site. A *name* is a symbol used by a program to reference a particular piece of information. Since a program is transformed through several forms before it actually executes, each different form of the same program may have a different name space, even though some names may be common to all of these name spaces. We are concerned here with the name space of the machine language program. In a later section we will be concerned with other name spaces.

We are especially concerned with *name-address mappings*. A name-address mapping is a mapping of a name space into an address space. It should be clear that such a mapping must be performed on any names used by an instruction before that instruction can be executed by the control processor. The function of the loader that we discussed in Chapter 6 was to perform this mapping for all of the address constants in the program. Before loading, the program consisted of object decks for all of the procedures in the program. The name space of these object procedures consisted of the symbolic names of the segments and relative addresses within the segments. Within the loader, *relocation* is that portion of the name-address mapping that maps relative addresses into absolute addresses. Relative addresses belong to the name space and absolute addresses belong to the address space. In like fashion *linking* is the portion of the name-address mapping that maps symbolic segment names into absolute addresses.

In Chapter 6 we identified relocation and linking as part of the binding of symbolic addresses. The compiler or assembler did part of the binding, and the loader did the remainder. We can now define this binding as a sequence of name space to name space mappings followed by a name-address mapping.

That is, the compiler performs a mapping from the name space of the source procedure to the name space of the object procedure. The loader then performs the name-address mapping that transforms object procedures into an executable program.

Recalling the problems of block allocation, we see that when such an algorithm is used for memory allocation, the loader often cannot perform the name-address mapping. This is because of indexing. The usual way of sequencing through the elements of an array is to use an index that is relative to the base of the array. The compiler quite naturally compiles code that loads the index value into a register that is then used as an index register relative to another register that contains the base address of the array. If the array is longer than one page its pages will most likely not be contiguous in physical memory. Therefore, the address of the base of the array and the value of the index must be combined to get the logical address of the word in the segment, then the physical address is computed according to the formulas given in Section 15.1.

This difficulty arises because the loader did not, in fact, completely perform the name-address mapping. The contents of the displacement field of an instruction and the contents of a register being used as an index register are not addresses. They are offsets relative to an address constant. This pair of values is essentially a name. When the processor computes the effective address by adding the contents of the displacement field and the index register to the contents of the base register, it is performing the name-address mapping. However, when a block allocation scheme is being used, this effective address mapping performed by the hardware is still not the name-address mapping. In computing the effective address the hardware assumes that the entire segment is stored contiguously. In fact, the hardware assumes that the entire program is stored contiguously in memory, beginning at memory address 0. When a block allocation scheme is used, this is not true. Therefore, when block allocation is used, the loader is actually performing another name space to name space mapping, since the numeric "addresses" in the loaded program have to be further transformed using the page tables in order to produce the actual memory address of the referenced information.

We have a different situation if some form of contiguous allocation is used. The loader can perform the name-address mapping on all of the address constants. The hardware computation of the effective address will work properly, since each segment is, in fact, stored contiguously, and indexing is never used to sequence from one segment to another. However, whenever the location of a segment changes because of compaction or because it was removed from primary memory and then returned to primary memory by the system, address constants referring to the moved segment are invalid and must be updated. This updating consists basically of performing the original name-address mapping again. In order to do this we will be required to retain sufficient information along with the program to identify every address constant that needs updating and to identify how to compute the name-address mapping for each address constant. This is usually quite complex, and it is much simpler to delay performing the name-address mapping until the information is actually referenced and then do it immediately preceding each reference, just as we did when using block allocation. However, if we do this, we lose the primary advantage that contiguous allocation has over block allocation, since we now have a high overhead cost in performing the name-address mapping at each reference.

There are clearly advantages in delaying the name-address mapping of a name until that name is actually used as an operand reference in an instruction being executed by the control processor. The principal advantage is the freedom that the system has to move segments around in memory. Since the name-address mapping is not performed for any name until the program actually needs to reference the named information, no unnecessary mapping is ever performed. In a highly dynamic memory-use situation such as an interactive time-sharing system, this can result in a significant saving in execution time, provided the name-address mapping can be done very rapidly by the hardware. Another advantage of delaying the mapping until the actual reference is that the system does not have to move a segment into primary memory until it is needed, since the presence or the absence of the segment can be detected as part of the name-address mapping.

A very restricted set of name-address mappings can be made very fast by the use of special hardware. Using this hardware the name-address mapping is postponed until a reference to memory occurs. Then the mapping is done by the hardware as part of the execution of the instruction making the reference. In actual computers using such hardware the execution time of each instruction is increased by less than 5 percent. There are two basic variations of this special hardware: *segmentation hardware* and *paging hardware*. Actually, both variations are quite similar, but they have very different uses.

The simplest segmentation hardware uses a *segment table*, which is similar to the page table that we used in our block allocation algorithm. The segment table has one entry for each segment in a user's process. An entry is usually called a *segment descriptor* or *descriptor*. A descriptor has three fields (we will add some more later): the *presence bit*, the *length field*, and the *address field* (Figure 15.9). The presence bit indicates whether the segment is in primary memory (presence bit = 1) or not (presence bit = 0). The length field contains the length of the segment. The address field contains the physical memory address of the base of the segment. This is an *absolute address*, that is, it is the address that the memory actually uses to reference word 0 of the segment. The descriptors are interpreted by the segmentation hardware. The length and address fields are not interpreted if the presence bit equals 0. However, these fields may be used by the operating system to locate the segment in secondary memory.

The control processor contains an additional register called the *descriptor base register*. The contents of this register are interpreted by the segment hardware as the absolute address of the base of the segment table. Figure 15.9 shows the relationship among the descriptor base register in the processor, the segment table, and the segments that are in primary memory. The segment table must be in primary memory. The segments and the segment table are shown scattered around to emphasize that they do not have to be at any particular location in primary memory. However, primary memory is linear; that is, the memory accessing hardware references the physical memory locations using integers (absolute addresses) in the range 0 to L-1, where the memory is L words in size.

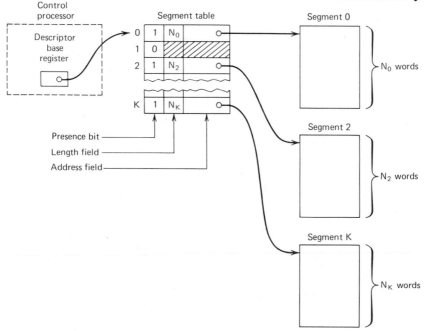

Figure 15.9 Segment table and segments in memory.

Each segment is assigned a segment number, which is simply the index of its descriptor in the segment table. No physical memory addresses ever appear in a machine language program. Any reference to a word of information by the program consists of a *segment number* and a *word number*. The segment number is the number assigned to the segment containing the information. The word number is the relative address (relative to the base of the segment) of the word of information within the segment. Ordinarily the maximum segment size is chosen to be a power of two. An address constant then has the format

Segment number	Word number

0   13   14   31

where we have assumed that the maximum segment length is 2^{18} (260,144) bytes and the maximum number of segments is 2^{14} (16,384).

The effective address is now a relative address, the *effective word number*. Its computation is performed the same as before, except that all arithmetic is done modulo 2^{18}, that is, only bits 14 through 31 of the base and index registers are added together with the displacement field of the instruction. The result is the effective word number of the referenced word in the segment. The segment number is taken from bits 0–13 of the base register. Bits 0–13 of the index register are ignored.

After the effective word number has been computed, the segment number is added to the contents of the descriptor base register. This results in the absolute address of the segment's descriptor. The segmentation hardware then fetches the segment's descriptor. The presence bit is checked to see if the referenced segment is in primary memory; if it is not, the processor transfers control to a system entry that will move the segment into primary memory (this will be discussed in more detail later). Once the segment is in primary memory, or if it is already there, the effective word number is compared to the contents of the length field of the descriptor to see if the reference is actually within the segment. If it is not, control is transferred to an error entry in the system. If the effective word number is within the segment, it is added to the contents of the address field of the descriptor yielding the absolute address of the referenced word. The segment-word number pairs form a name space for the machine language program. The segmentation hardware performs the name-address mapping from a segment-word number pair to a physical memory address by using the segment table.

What is the cost of this hardware mapping? As our description now stands, the cost is one additional cycle per memory reference. With the standard hardware the cost of referencing a word in memory is one memory cycle, the cycle required to fetch or store the word. With segmentation hardware an additional cost of one cycle is incurred in fetching the segment descriptor. This seems to increase the execution time by 100 percent. It does. However, both segmentation and paging hardware usually include a small, very fast associative memory. Before a descriptor is fetched from primary memory, the associative memory is searched to see if the descriptor is there. If it is, the memory cycle needed to fetch the descriptor from primary memory is avoided. If the descriptor is not in the associative memory it has to be fetched from primary memory. However, once it is fetched from primary memory, it is put into the associative memory for future use. When it is put into the associative memory, some other descriptor already there may have to be discarded. Actual operating experience has demonstrated that with a reasonable policy for selecting which descriptor to discard, an associative memory that can hold only 16 descriptors enables segmentation to be used with only an increase in instruction execution time of less than 5 percent.

The advantage of segmentation hardware is clear. The program uses segment numbers for all references. Each segment has the same number throughout the entire life of the process. When a segment's location changes due to movement by the system, its segment number is not changed, only the contents of the address field of its descriptor. Thus, all references are automatically updated by making a single change in the segment table and clearing the associative memory in case a copy of the segment's description is currently in the associative memory. Thus, we completely eliminate the cost and complexity of updating all of the address constants, which we pointed out as a disadvantage of contiguous allocation. Unfortunately, even with segmentation hardware, we are still faced with the problem of the potential need for compaction, which exists with any contiguous allocation scheme. Compaction can be totally avoided only with a block allocation algorithm.

Paging hardware is used to make the name-address mapping required for block allocation fast. The simplest paging hardware uses a page table almost exactly like the page table that was required for block allocation. The principal difference is that the paging hardware interprets the page table words and performs the name-address mapping in much the same way as the segmentation hardware. There is one page table for a process. It contains one entry for each page in each segment in the process. The procedures in a program must be relocated and linked exactly as was done by the loader in Chapter 6. All of the address constants in the program will contain numbers that look like absolute addresses. In fact, the name space of the machine language program is the set of integers [0, M-1], where M is the length of the program. The only real difference is that the pages of the program are not necessarily stored contiguously in physical memory. They are stored in whatever blocks are free. As with segmentation, if a presence bit is included in the page table entries, then a page of the program does not have to be moved into primary memory until it is needed by the program.

Figure 15.10 shows the relationship among the *page table base register*, the page table, and the pages in memory. We have indicated a possible segment structure where the first segment contains four pages, the second segment contains two pages, and so on. However, the paging hardware is not aware of any grouping of pages into segments. The effective address for a memory reference is computed in the normal way as the sum of the contents of the base register, the contents of the index register, and the displacement field of the instruction. If the page size is a power of two, such as 2^9 (512), then the final effective address is interpreted by the paging hardware as

Page number	Word number
0 22	23 31

where the word number selects the word of information within the page specified by the page number.

The paging hardware functions essentially the same way as the segmentation hardware. No length field is needed in the page table words, since all pages are the same length. The performance of paging hardware is identical to that of segmentation. Thus, when an associative memory is used, the increase in execution time of instructions is less than 5 percent. With paging hardware use of a block allocation scheme is simple and efficient.

We can now define *virtual memory*. A computer system is said to have a virtual memory when the executing machine language program has a name

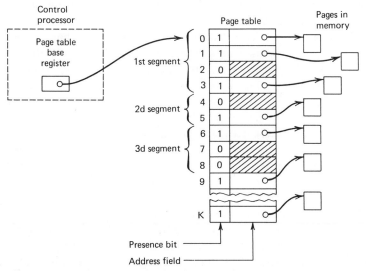

Figure 15.10 Page table and pages in memory.

space that is not identical with the address space of the program. The name-address mapping is performed during the program's execution by a combination of software and hardware, as indicated in the preceding paragraphs. Paging and segmentation hardware are sometimes referred to as *virtual memory hardware*. In addition, although our definition does not require it, virtual memory is commonly used to imply that the name space of the program may be larger than the actual address space of the computer. That is, a user's program may actually be larger than the primary memory without the user having to use any overlays. This is possible only if the program is able to execute without all of its segments or pages having to be simultaneously in primary memory. This is certainly possible when the hardware uses a presence bit, as did our segmentation and paging hardware. Lack of a presence bit does not a priori make it impossible to execute a program if part of it is missing; however, it does make it much more difficult to do so.

Our version of virtual memory will permit a user's program to be larger than the memory actually allocated to his process. This system is totally responsible for the movement of parts of the user's program in and out of primary memory. We will use the word *page* in this and the next paragraph. However, everything we say is equally applicable to segments if segmentation hardware is used. Since the user's program may be larger than primary memory, the system must be able to store some of the pages on some secondary memory device or devices. Furthermore, the system must fetch a page from secondary memory when it is needed. If the primary memory is full, one of the pages already in primary memory will have to be removed and stored in secondary memory. All of this is done automatically; the user's program does not need to do anything explicit in order to make it happen. In fact, it is unaware of when it happens. The portion of the system responsible for this movement of pages is usually called the *pager*.

The term *multilevel store* is sometimes used to refer to a virtual memory. This is because the pages of a user's program may be stored in primary memory and one or more secondary memory devices while it is executing. For example, virtual memory systems often use either a drum and a disk or bulk core and a disk for the storage of pages in addition to primary memory while a program is executing. The disk has a larger capacity than either drum or bulk core. However, it takes longer to fetch a page from the disk. Thus, the pager selects storage locations for the pages, depending on their use, in an attempt to achieve the most efficient use of storage and at the same time minimize the execution time of the program. This is a complex problem and will be explored in greater depth in Section 15.5.

The reader may be left with the impression that paging hardware has such a significant advantage over segmentation hardware that there is no reason to consider segmentation. This is not the case. In the first place, paging lacks the flexibility of segmentation. With pure paging, segments must be rigidly mapped into a linear name space (i.e., relocation by the loader). Once mapped into this name space, they cannot grow in size without being moved, in which case all address constants that reference the moved segment must be updated. When segmentation hardware is used, no relocation is required. In addition, a segment may grow up to the maximum segment size without affecting any of the address constants that refer to it. Segmentation hardware makes direct sharing of information possible in a straightforward, simple way. Paging hardware alone does not. Segmentation hardware also provides reasonable access control for the purposes of privacy and protection. These two latter topics will be discussed in detail in Chapter 17.

For the above reasons, some systems combine segmentation and paging hardware in order to gain the advantages of both. The most successful system to do this is the Honeywell 645/6180, which is used by Multics.[2] The Burroughs B6700/B6700[3] systems combine segmentation with a limited form of paging. Our Insys will use a combination of segmentation and paging hardware. Each segment will have its own page table. The address field of a segment descriptor contains the absolute address of the base of the page table for the segment. Figure 15.11 shows this relationship. The presence bit in the segment descriptor is interpreted as indicating the presence of the page table for the segment. If the presence bit equals one, the page table for the segment is in primary memory, although none of its pages need be. If the presence bit equals zero, then the segment's page table is not in primary memory and, of course, none of its pages are, either.

The next three sections discuss the implementation of virtual memory and the use of segmentation and paging hardware in Insys. We hope to show the power and flexibility of a segmentation and paging-based virtual memory. In addition, we will expose the most significant problems in memory management, which segmentation and paging hardware help to solve.

15.3 File Storage and Segment Management

The term *file storage* usually implies more or less permanent storage of large amounts of information. In modern computer systems file storage always involves the use of disks or tapes, or both. The basic unit of information is a *file*, which is a collection of words of information. Typically, a file may contain one or more procedure segments or data segments. Depending on the particular file system, a file may have additional internal structure. For example, a file may consist of a set of *records*, each of which is a collection of words.

A user file that is given to the file system for storage is written onto some secondary memory device by the file system and kept there until the user needs to use the file again. This implies that the file system must keep a record of the files with which it has been entrusted. This record we will call a *file directory*. In some systems a file directory is called a *catalog*. Every file system must maintain a file directory. At the very least each entry in the file directory contains the name of the file and its location. The name of a file is some symbol by which the user is able to later identify the particular file that he wishes to use.

Ordinarily the directory entry for a file will contain additional information about the file, sometimes referred to as the *attributes of the file*. For example, a directory entry may contain the following information: the file's name, its

---

[2] E. I. Organick, *The Multics System: An Examination of Its Structure*, MIT Press, Cambridge, Mass., 1972.

[3] E. I. Organick, *Computer System Organization: The B5700/B6700 Series*, Academic Press, N.Y., 1973.

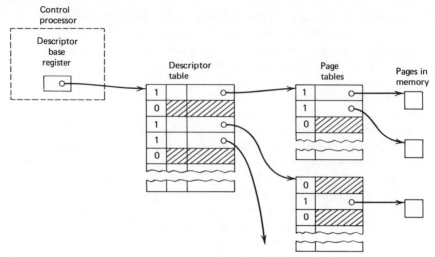

Figure 15.11 Combined segmentation and paging hardware.

length, its location in secondary memory, its type (procedure or data), the date when it was created, the date when it was last used, and a date when it may be destroyed by the system.

Most file systems maintain many file directories, not just one. There are three major reasons for this. If there are many directories then none of them needs to be large, which would be the case if all files had to be recorded in a single directory. The search time for small directories is much shorter than for large directories. Use of small directories is therefore more efficient than a single large directory.

The second advantage of many file directories is that the avoidance of name conflicts is simple. In a multiuser system, if each user has his own directory, then he does not have to be aware of the names used by other users for their files. If all files are recorded in the same directory, a user must not name any of his files with a name that is the same as any of the names used by any other user. The file system can take care of this for the user and allow him to use names that other users have used, but any solution is basically equivalent to multiple directories without any of the other benefits that accrue from the use of multiple directories.

The third advantage of many directories, especially if each user may have more than one directory, is that a user can easily group his files into meaningful subcollections. For example, files containing mathematical procedures might be recorded in one directory, old programs in a second directory, and new programs currently under development in a third directory.

Many file systems that use multiple directories organize all of the directories into a *hierarchical structure*. Sometimes the directories for each user are organized into separate hierarchical structures. A hierarchical structure can be graphically depicted as a tree (which always seems to be drawn upside down) similar to that shown in Figure 15.12. When organized this way, a directory may contain entries for other directories as well as for files. The values of the type attribute in a directory entry would then be extended to include 'directory' as a type indicating that the entry corresponded to another directory instead of to a file containing procedures or data.

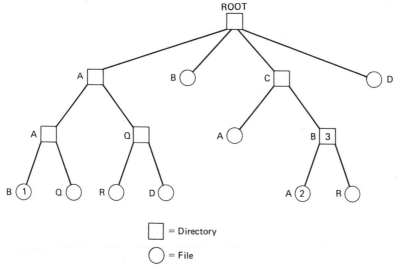

Figure 15.12 Hierarchical structure of file directories.

Only the names within a single directory need be unique. As illustrated in the figure, the same name may appear in many different directories (the name of each file and directory is written beside it in the diagram). If the names do not need to be unique other than in a single directory, how do we identify a particular file? By a compound name called the *tree name*. A tree name consists of the concatenation, separated by some marker, of the names of all of the ancestors of the file being referenced. For example, the tree name of the file numbered 1 is ROOT.A.A.B (using the PL/1 notation for compound names), the tree name of the file numbered 2 is ROOT.C.B.A, and the tree name of the directory numbered 3 is ROOT.C.B. The tree name of every file and directory is unique as long as the names in a single directory are unique.

We define a *file system* as a set of directories and files, and a set of operations that can be applied to the directories and files. What kind of operations do we need? We must be able to create and destroy files. We also must be able to read the contents of a file and write into a file. Most file systems also include an open and close operation for files and do not permit a file to be read or written unless it is open. These last two operations are not a logical necessity, but they are a significant help in providing information about the use of files and often allow the file system to do a more efficient job of file management.

We need similar operations for directories. We must be able to create and destroy directories. We need to both read and modify the contents of a directory entry. In addition, we need to add new entries to, and delete existing entries from a directory. These last two operations for directories are implied by creation and destruction of files. When a user creates a new file a new entry must be added to a directory to account for the file. When a user destroys a file the directory entry for the file must be deleted.

In this text we will explore two quite different types of file systems. In one file reading and writing is explicitly specified in the user's program. In the other file reading and writing is implicit. The file system with explicit reading and writing is the *IBM OS/360 Data Management*. These facilities actually constitute a very elaborate file system. In OS/360 the user's program must explicitly specify when a file is to be read or written and how many words to read or write, very much like input and output were specified in Chapter 2. In fact, input and output in OS/360 are part of data management. Therefore, we will discuss OS/360 data management in Chapter 16 as part of our discussion of the problems of input and output.

In Multics file reading and writing is implicit. To read from a file in Multics a program executes a fetch-type instruction, such as add register or load register. The program executes a store-type instruction when it wishes to write into a file. This implicit file reading and writing is implemented by coupling the file system to a virtual memory. The name space of an executing program potentially includes all of the files in the file system, even though most programs are permitted access to only a small subset of these files. In fact, in Multics a segment and a file are identical. The name space is segment based. An object program name consists of a symbolic segment name and a word number of a word within that segment. Thus, to reference any word of information in a file (segment), a Multics program simply uses its name. The Multics software maps a symbolic segment name and word number pair into a name space consisting of segment-word number pairs. The segmentation and paging hardware then maps a segment-word number pair into the absolute memory address of the referenced word. If the page containing a referenced word is not in primary memory, the pager moves it from the file storage into primary memory.

In the remainder of this section we present a description of segment management in Insys. This is similar to a greatly simplified version of the Multics segment manager. We define a *known segment*, with respect to a given process, to be a segment that has been referenced at least once in the past by that process. A segment may be known to more than one process, but this will not be discussed until Chapter 17. The *known segment table* has an entry containing the name of each segment known to the process. This table grows as execution of the process proceeds. The *segment table* contains a descriptor for each known segment. An *active segment* is a segment that has a page table in primary memory. The *active segment table* contains an entry for each active segment. Not all known segments must be active. However, all active segments must be known. If we do not make this distinction between known and active segments, then all known segments would have to have a page table in primary memory, and it is more likely that the primary memory would become full of page tables leaving no space for those pages of segments that are needed by the process.

The easiest way to understand how the Insys file system works is to follow what happens when a process makes a reference to a word in a segment. Any reference to information, data or instructions has the form (segment name, word number). The first time such a reference is used it is replaced by the pair (segment number, word number). The reference then remains in that form as long as the process exists. This mapping is done by a part of the system that we call the *linker*, since it is basically *linking* as defined in Chapter 6. Before this mapping can be done, the referenced segment must be known; that is, it must have been assigned a segment number and have a descriptor in the segment table. The details of the linker and dynamic linking will be discussed in Section 15.4. Making a segment known will be discussed shortly.

Once a segment is known, it remains known the rest of the process's life. However, the segment need not be active during this entire time. When it is actually referenced, it must be active, since there must be a page table in order for the hardware to reference pages in the segment. When a segment is not being referenced, it may become *inactive*. The presence bit in a segment's descriptor indicates whether the segment is active. If a reference is made to a segment and the presence bit in its descriptor is zero, the segment is not active. Before it can be referenced it must be made active. A reference to a segment whose descriptor has its presence bit equal to zero will cause a *missing segment interrupt*. Control is then transferred to the *segment manager* in the system, which makes the segment active. Making a segment active and inactive will be discussed shortly.

When a segment is active, it has a page table in primary memory. Thus, its pages can also be in primary memory. However, not all of its pages have to be there. The presence bit in a page table word is used to indicate whether the corresponding page is in primary memory. A reference to a page whose page table word has its presence bit equal to zero will cause a *missing page interrupt*. Control is then transferred to the *pager*, which brings the referenced page into primary memory from secondary memory. The pager is discussed in detail in Section 15.5.

The linker calls **make_known**, which is a component of the segment manager, with the tree name of a segment as an argument. The result of this call will be the segment number of the named segment or an error code, indicating that there is no such segment in the directory hierarchy. The procedure **make_known** is defined in Figure 15.13. If the segment is already known there will be an entry for it in the known segment table (**KST**). This entry contains the segment's tree name. The segment number of a segment is the index of its entry in the known segment table. Thus, a known segment's number is found by searching the known segment table. If the tree name of the segment

```
procedure make_known (tree_name, seg_nr, status);
    if tree_name is in KST then
        begin seg_nr : = index of KST entry for tree_name;
            return;
        end;
    if tree_name is not in directory hierarchy then
        begin status : = 1;
            return;
        end;
    k : = index of an unused entry in KST;
    KST[k] : = tree_name;
    ST.presence[k] : = 0;
    ST.length[k] : = length of segment from directory entry for tree_name;
    seg_nr : = k;
    return;
end.
```

Figure 15.13 Definition of procedure to make a segment known to a process.

cannot be found in the table, the segment is not known, and so no segment number has been assigned to it yet.

The operation of making a segment known consists of assigning a segment number to the segment. Before this is done the directory hierarchy is searched to see if the referenced segment really exists. If no entry for it is found, an error code is returned. Thus, the user can be informed, allowing him to take corrective action. If the segment does exist, its tree name is entered into the known segment table, and the index of that entry is assigned as its segment number. In addition, a descriptor for the segment must be added to the segment table (ST). Since there is no page table yet for the segment, the presence bit of the descriptor is set equal to zero.

The operation of making a segment active consists mostly of creating a page table for the segment. However, because of the limited size of primary memory, it may be necessary to make some active segment inactive. A segment is made active as a result of the segmentation hardware trying to interpret a descriptor that has its presence bit equal to zero. When this situation is encountered, the hardware forces control to the **make_active** entry in the segment manager, which makes a segment active. This forced change of control is an interrupt and is done in such a way that when the segment has been made active, the control processor can resume execution of the instruction it was executing when the segmentation hardware detected the presence bit equal to zero. The interrupt also occurs in such a way that when the segment manager is given control it can tell what segment was being referenced and caused the interrupt.

Figure 15.14 defines the **make_active** component of the segment manager. When a segment is active, it has an entry in the active segment table (AST) and a page table in primary memory. The active segment table entry for a segment contains its segment number and the location of each of its pages. A page of an active segment may be either in primary or secondary memory. Wherever it is located, its location is recorded in the active segment table entry for the segment. If a segment is not active, all of its pages must be in secondary memory. In that case, the location of each page in the segment is stored in the file directory entry for the segment.

procedure make_active;
    k := number of segment which caused the interrupt;
    if AST full **then** Select an active segment and deactivate it
                      (see Figure 15.15);
    j := index of an unused entry in **AST**;
    tree_name := **KST[k];**
    Find directory entry for **tree_name;**
    AST.pages[j] := page location information from directory entry;
    AST.seg[j] := k;
    Call pager to allocate one block (512 words) of primary memory
      for the page table;
    ST.address[k] := location of page table;
    Set presence bit equal to zero in all page table words in new page table;
    ST.presence[k] := 1;
    return;
  end.

Figure 15.14 Definition of procedure to make a segment active.

If the active segment table is full, some currently active segment must be made inactive before an entry can be made for the new active segment. Once an empty entry in the active segment table is obtained, the entry for the new active segment can be created. In making a segment active, the segment manager uses the segment number of the segment that caused the interrupt as an index in the known segment table in order to find the tree name of the segment that is to be made active. The directory entry for this segment is then located by searching the directory hierarchy. This directory entry contains the page location information needed to make the active segment table entry for the segment and to build its page table. This information is copied into the active segment table entry. The segment number, which is already known, is also copied into the active segment table entry. (This will not quite work if sharing of segments is allowed. However, the full solution is explored in Chapter 17 along with the sharing of segments.)

Before the segment's page table can be built, primary memory space must be allocated for it. The maximum segment size is 2^{18}, and the page and block size is 2^9. Thus, the maximum number of pages in a segment is 2^9, and a page table will always fit into one block. One free block is allocated for the page table by calling the pager (see Section 15.5). (In the interest of simplicity we put only a single page table in a block, no matter how small the page table is.) Once the block is obtained, its address is put into the address field of the segment's descriptor.

The final task is to build the page table. The presence bit in each page table word is set equal to zero, indicating that the corresponding page is not in primary memory. When a reference is made to information in a particular page, a missing page interrupt will occur due to the presence bit in its page table word being equal to zero. The pager, described in Section 15.5, will then get control and fetch the page from secondary memory.

Deactivation of a segment is shown in Figure 15.15. Before destroying the segment's page table and deleting its entry in the active segment table, any pages of the segment that still remain in primary memory must be put back

```
procedure deactivate (j);
    n : = number of pages in segment corresponding to AST[j] − 1;
    k : = AST.seg[j];
    Let PT be address from ST.address[k];
    repeat
       if PT.presence[n] = 1
       then Call pager to possibly write out into secondary memory
               page n of segment and release block containing it;
       n : = n−1;
    until n < 0;
    Call pager to release block containing page table;
    ST.presence[k] : = 0;
    Mark AST[k] as unused;
    return;
end.
```
Figure 15.15 Definition of procedure to deactivate a segment.

into their proper place in secondary memory. Each word in the page table is examined. If the presence bit equals zero, the corresponding page is not in primary memory. If, however, the presence bit equals one, the corresponding page is in primary memory. In this case the pager is called to force the page to be written into secondary memory if an exact copy no longer exists there. Once all of the page table words have been examined and the corresponding pages paged out if necessary, the page table can be destroyed. This is done by releasing the space used by the page table so that it is available for reallocation. The presence bit in the segment's descriptor is then set to zero, indicating that the segment is not active and thus no longer has a page table.

15.4 Dynamic Loading and Linking

In this section we are concerned with *linking*, that is, binding of an inter-segment reference. In a segment-based system like Insys (or Multics), all external references by a segment have the form (segment name, W), where W is the address of a word relative to the base of the segment. This corresponds to the external symbol described in Chapter 6. The loader described in Chapter 6 linked all segments in a process before execution began, that is, all external symbols were replaced by their corresponding absolute addresses before starting to execute the program. Linking in Insys consists of replacing the segment's name by its segment number; that is, (segment name, W) is replaced by (S,W), where S is the segment number assigned to the named segment.

Linking can be done by a loader similar to the one described in Chapter 6. The principal difference is that the loader would not assign storage. Instead, it would make each segment known by calling **make_known** in the segment manager. This entry returns the segment number of the segment and the loader will store it in its symbol table, just as it did the absolute address in Chapter 6. Relocation remains a subcase of linking as in Chapter 6. Our new loader will not allocate storage to any segment, since this is done by the pager without any help from the loader. The pager will also load the segment, one page at a time as needed. Thus, the loader is reduced to performing two functions, *linking* and *library search*.

One of the goals of Insys is to avoid any unnecessary work by postponing many functions, such as loading the text of a segment, as long as possible. Using the paging hardware, no page of a segment is loaded into primary memory until it is actually referenced. The same goal is applied to linking and library search. In Insys no external reference by a procedure is linked until the first time the procedure actually uses the external reference. This is called *dynamic linking* (or *dynamic loading* by some). Dynamic linking and loading is not limited to virtual memory systems such as Insys and Multics. It can be implemented in a system such as OS/360 (extremely primitive facilities of this type actually do exist in OS/360).

If we interpret the word loading as reading a segment's text into primary memory when speaking of a nonvirtual memory system and as making a segment known when speaking of a virtual memory system such as Insys, then we make some general statements about dynamic loading and linking. There are two common alternates. Suppose segment A refers to segments B, C, and D. The first time A is called by any procedure in the process, all external references of A are linked and all referenced segments are loaded if they are not already. Since A refers to B, C, and D, the first call to A causes segments B, C, and D to be loaded and A to be linked to each of these three segments. Alternately, linking and loading can be delayed further and not done until the actual reference (this is what Insys does). For example, even though A is called, A is not linked to B until B is actually referenced. B is loaded at that time if it has not yet been loaded. In like fashion A is not linked to C or D until each segment is actually referenced.

In order to delay the linking until during execution, the information that defines the external references must be accessible during execution. We will call the address constant corresponding to an external reference a *link* and the information that defines it a *link definition*. All of a procedure's links and link definitions are collected together into the procedure's *linkage section*. For reasons that will be explained in Chapter 17, the linkage section for a procedure segment cannot be part of the segment. In Insys all linkage sections for a process are collected into a single segment called the *linkage segment*.

In order to accomplish dynamic linking there must be a way for a procedure to tell when an external reference has not yet been linked. That is, a link is not a machine address constant until it has been linked. After linking, the link will be a machine address, that is, it will be a pair of integers (S,W). The obvious way to accomplish this is to define a link so that it contains an interrupt bit, as shown in Figure 15.16a. Whenever an attempt is made to reference a segment using an unlinked link, a *linkage interrupt* occurs. This interrupt transfers control to the linker. The linker then makes the segment known if it is not already known and replaces the unlinked link by its corresponding machine address, as shown in Figure 15.16b. Note that we must decrease the maximum number of segments from 2^{14} to 2^{13} in order to accommodate the interrupt bit.

The link definition contains a segment name and the number of a word in that segment. The diagram of a linkage section in Figure 15.17 shows the relationship between a link and its definition. The pointer in an unlinked link is the address of the first word of the link definition. The link definition contains the tree name of the segment (with ROOT omitted, since every tree name begins with ROOT) and the relative address of a word in that segment. In Figure 15.17 we see an unlinked link for an external reference to word 3 of the segment whose tree name is ROOT.X.AB. Since segment tree names can

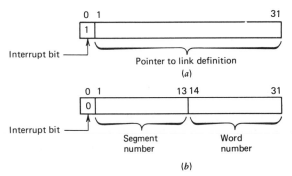

Figure 15.16 Formats of unlinked and linked links. (*a*) Unlinked link. (*b*) Linked link.

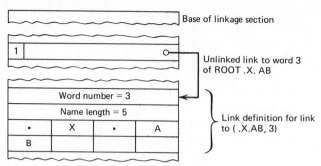

Figure 15.17 Example linkage section.

be arbitrarily long, the link definition contains the length of the name (number of characters).

The action of the linker as a result of a linkage interrupt is shown in Figure 15.18. When a linkage interrupt occurs, the hardware makes available the address of the unlinked link that caused the interrupt. The information required to restart execution of the instruction that was making the reference which caused the interrupt is also stored when the interrupt occurs. After the link is linked, execution is continued at the point where the interrupt occurred.

Using the address of the unlinked link, which was saved when the interrupt occurred, the linker obtains the link definition pointer from the unlinked link that caused the interrupt. The link definition pointer is used to locate the beginning of the link definition information. The tree name of the segment is extracted and used as the argument in a call to the segment manager to make the segment known. This call returns the segment number of the segment after making the segment known if it was not already known. This segment number is combined with the word number from the link definition to form a machine address that replaces the unlinked link. After the unlinked link has been replaced by a linked link, control is returned to the instruction that caused the linkage interrupt.

There are two major objections to the requirement that the tree name of a segment be required in a link definition. The first objection is that much unnecessary information must be written in the source procedure and stored in the linkage section of the object procedure. In many cases the referenced

```
procedure linker;
    Get location of unlinked link;
    Get link definition pointer from unlinked link;
    Get segment name from link definition;
    Call make_known to get segment number of referenced segment;
    Get word number from link definition;
    Combine segment number and word number to form a link;
    Replace unlinked link by the newly formed link;
    return;
end.
```

Figure 15.18 Definition of the linker.

segments are either in a single user's directory or in the system library. With appropriate conventions, most of the time the name of the directory could be omitted from the name in the link definition, just as we have already omitted ROOT. The second objection is that there are distinct advantages in not specifying the directory. With appropriate conventions, a procedure name would be able to be used in several different contexts without having to change the procedure name each time.

One convention and a slight modification of the linker will answer both of these objections. We adopt the convention that if the segment name in a link definition does not begin with a period, then it is interpreted as a *relative name*. Whenever the linker encounters a relative name, it follows a set of standard search rules in an attempt to find the corresponding segment. The obvious set of search rules is first to try the user's directory and then the system library directory. A simple extension is to allow the user to specify a number of additional directories to try after trying the user's directory, but before trying the system library directory. Thus, the search for a segment may try several private user libraries before trying the system library.

The straightforward implementation is to use a *search list*, which is a list of tree names of directories to search. The linker tries the directories on this list, one after the other. This modification to the linker is shown in Figure 15.19. If the segment name in the link definition begins with a period, it is a tree name, and the linker does exactly as it did in Figure 15.18. However, if the segment name does not begin with a period, it is a relative name (and is not a complete tree name). The linker must try each of the directories whose names are on the search list. These directories are tried in the order in which they appear on the search list. In each case the linker creates a complete tree name by prefixing a directory name to the segment name from the link definition. This tree name is then used as an argument in a call to the segment manager. If a segment exists with that tree name, its segment number is returned, otherwise a status code is returned indicating that no such segment exists.

if segment name begins with '.' **then**
    begin Call **make_known** to get segment number of referenced segment;
      if not found **then** Error;
    end;
    else
      begin $i := 1$;
        while $i \leqslant$ length of **search_list do**
          tree_name $:=$ directory name from **search_list[i]** prefixed
            to segment name from link definition;
          Call **make_known** to get segment number of **tree_name;**
          if segment found **then unloop;**
          $i := i+1$;
        end;
        if $i >$ length of **search_list then** Error;
      end;

Figure 15.19 Replacement for "Call **make_known** to get segment number of referenced segment;" in Figure 15.18 to implement relative names.

Unless the prefixed directory name is the last name on the search list, this is not an error. The next directory on the list is tried to see if the segment is in that directory.

Suppose the hierarchy of segments is as shown in Figure 15.20, and a procedure makes external references to the relative names A and Q. If U1, U2, and U3 are user directories, Figure 15.21 shows the actual segments referenced by the relative names A and Q for several different search lists. In the first three cases only two directories are in the search list, the user's directory and the system library directory. In the last case, user U3's private library directory is also in the search list. Observe that normally each user gets the same segment Q, the one in the system library. However, if a user wishes to use his own version of Q, he either puts it in a private library and puts the name of this library directory on the search list, or he puts his own version of Q in his user directory. On the other hand, each user gets his own version of segment A, which is normally the case if A is a data segment.

15.5 Paging

In this section we discuss the *pager*. The pager is a collection of procedures that carry out two major functions: management of primary memory and movement of a segment's pages between primary memory and secondary mem-

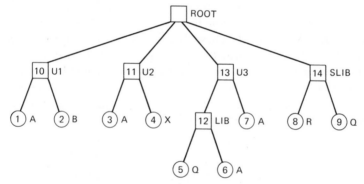

Figure 15.20 Example directory hierarchy.

	Search list	Directory number	Segment number of A	Segment number of Q
i)	.U1	10	1	9
	.SLIB	14		
ii)	.U2	11	3	9
	.SLIB	14		
iii)	.U3	13	7	9
	.SLIB	14		
iv)	.U3	13	7	5
	.U3.LIB	12		
	.SLIB	14		

Figure 15.21 Examples of effect of search list on relative names.

ory. The segment manager calls the pager to force the pages of a segment that is being deactivated to be moved out of primary memory. Pages are moved into primary memory as a result of missing page faults. The procedure page_in moves pages into primary memory from secondary memory, while page_out moves pages from primary memory to secondary memory. The pager also manages primary memory space. The segment manager calls the pager to get space for the page table when making a segment active. It also calls the pager to release the space occupied by the page table when deactivating a segment. The procedure assign_block reserves a block of primary memory, while free_block releases a block.

The segment manager does not actually move any pages in or out of primary memory. When a segment is made active its page table is built and the presence bit set to zero in all of the page table words, but none of its pages are read in. Later, when the process attempts to reference some information in a page of the segment, a missing page interrupt occurs, since the presence bit is equal to zero. This interrupt causes control to be transferred to page_in. This procedure then reads the referenced page into primary memory from secondary memory and sets its page table word so that the process can use the page. When a segment is being deactivated, the segment manager calls page_out once for each page of the segment still in primary memory. The procedure page_out writes out each of these pages from primary memory to secondary memory unless an up-to-date copy already exists in secondary memory. This procedure is also called by assign_block whenever there are no longer any free blocks in primary memory.

The definition of page_in is shown in Figure 15.22. It is called as a result of a missing page interrupt. When this interrupt occurs, the segment number and page number of the missing page are saved. The information needed to restart the instruction that caused the interrupt is also saved. After the missing page has been read into primary memory and the presence bit in its page table word has been set to one, the instruction that caused the interrupt will be restarted. This time the instruction will be executed without causing an interrupt.

```
procedure page_in;
    j : = segment number of segment which contains referenced
        missing page;
    k : = page number of missing page;
    assign_block (loca);       {get free block for missing page}
    Let PT be the page table whose address is in ST.address[j];
    PT.address[k] : = loca;
    Get secondary memory address of page k from AST.pages[j];
    Read missing page from secondary memory into primary memory
        at loca;
    PT.presence[k] : = 1;
    return;
end.
```

Figure 15.22 Definition of procedure to read pages into primary memory from secondary memory.

After identifying the missing page, page_in needs to find a free block of primary memory in which to put the missing page. The procedure assign_block is called to obtain the required space. The address of the free block that is assigned is put into the address field of the missing page's page table word so that the process will eventually be able to reference the contents of the page.

Now that a block has been allocated, the missing page must be read from secondary memory into primary memory. The location of the missing page in secondary memory is obtained from the page location information, which was stored in the active segment table when the segment was made active. The details of I/O programs and I/O processor operation are discussed in Chapter 16. The process that caused the missing page interrupt can do nothing until the missing page has been read into primary memory; therefore the process waits until the read operation has completed. Actually, some other process can now execute until the page has been read in; however, some care must be used in selecting the other process to execute. This problem is discussed later in this section. Once the read operation has completed, the missing page is in primary memory and can be referenced by the process. Hence, the presence bit in its page table word is set to one, the pager returns control to the instruction that caused the missing page interrupt, and normal execution is resumed.

The procedure page_out is defined in Figure 15.23. It is the inverse of page_in. When this procedure is called, the segment number and page number of the page to be written out are supplied as arguments. The presence bit in this page's page table word is set equal to zero so that it can no longer be referenced. The primary memory address of the block in which the page is stored is found in the address field of the page's page table word. The secondary memory address from whence the page originally came is found in the active segment table entry for the segment. With these two addresses the required write operation can be accomplished.

The procedure assign_block is defined in Figure 15.24. It is responsible for allocating the blocks of primary memory. It maintains a list of available blocks, called the *free block list* (free_list). This list is formed by chaining together all of the blocks that are currently available. A call to assign_block allocates one block whose absolute address is returned as the procedure's single argument. This block, which is first on the free block list, is allocated and removed from the list.

The free block list may be empty. In this case some page in primary memory must be moved to secondary memory. To accomplish this, page_out is called

```
procedure page_out (j,k);
    Let PT be page table whose address is in ST.address[j];
    PT.presence[k] := 0;
    loca := PT.address[k];
    Get secondary memory address of page k from AST.pages[j];
    Write page from primary memory at loca into secondary memory;
    free_block (loca);
    return;
end.
```

Figure 15.23 Definition of procedure to page out a page from primary memory to secondary memory.

```
procedure assign_block (loca);
    if free_list empty then
        begin Select page to be paged out;
            j : = number of segment containing selected page;
            k : = number of selected page;
            page_out (j,k);
        end;
    loca : = address of first block on free_list;
    Remove first block from free_list;
    return;
end.
```

Figure 15.24 Definition of procedure to assign a free block of primary memory.

with the segment number and page number of the selected page as arguments. Once the page has been written out, page_out releases the block in which the page was stored. Thus, when control is returned to assign_block, there will be a block on the free block list. The procedure free_block is defined in Figure 15.25. It releases a block by adding it to the end of the free block list. The address of the block to be released is supplied as the argument.

In assign_block a page was selected to be removed from primary memory. How is this page selected? This is one of the major problems in paging. So far we have looked at the mechanics of paging. Selection of a page to remove is governed by a policy called the *page replacement policy*. There are many conceivable different page replacement policies. In general they all attempt to select a page that no process will reference in the very near future. Removing any page of any process not currently executing would satisfy this goal. However, if the process to which the selected page belongs resumes execution shortly after its page has been paged out, it is likely that this page will be needed almost immediately. Thus, a page fault will occur and the page will be read back into primary memory right after it was written out. Therefore, the real goal of a page replacement policy is to select that page in primary memory that will not be referenced by any process for the longest time; that is, every other page in primary memory will be referenced by some process before the selected one is referenced by any process.

Unfortunately, there is no practical way to tell when in the future the pages of a process will be referenced. To predict correctly the future page reference pattern would require more intimate knowledge of a process's behavior than is available to the system. It has been suggested that the user supply this information. In many cases it is extremely difficult or impossible for the user to derive this information from his program. This is especially true when the user's program is written in a high-level source language and makes extensive use of procedures written by others (for example, the procedures in the system library).

Denning has abstracted the properties of *locality* that are characteristic of most programs and formulated the *principle of locality*.[4] In part, the principle

[4] P. J. Denning, "Virtual Memory," *Computing Surveys 2*, (3) (September 1970), pp. 153–189, and E. G. Coffman, Jr. and P. J. Denning, *Operating Systems Theory*, Prentice-Hall, Englewood Cliffs, N.J., 1973.

procedure free_block (loca);
  Add block whose address is **loca** to end of **free_list**;
  return;
end.

Figure 15.25 Definition of procedure to free a block of primary memory.

of locality states that during a small time period the references that a process makes tend to cluster into a few small intervals in the process's address space; that is, all references are confined to a few pages. Furthermore, these pages will be the most recently referenced. This suggests that a *least recently used* page replacement policy achieves our goal more consistently than any other policy. This policy selects for replacement that page that has been least recently used. This policy will be explored in more detail later.

The principle of locality seems reasonable if one considers the nature of programs. The normal sequence of instruction execution is sequential; after executing an instruction, the instruction immediately following it in memory is executed next. The only exception is when a branch instruction is executed. However, the most common use of branch instructions is for loops and procedure calls. In all of these cases the sequence of instructions executed in a small time interval are all in a small number of pages. Straight line code and loops will be confined to a few pages. Procedure calls result in a few pages for each procedure called. However, in a very short time interval, typically only one or two procedures will be called. The same confinement to a few pages is normally true for data references. Procedures that reference many data items are actually using indexing to step through the items in a small number of arrays.

It must be emphasized that these observations apply to most procedures that comprise the large majority of the programs that are executed. There are a few special procedures that do not have these characteristics. However, they are a small enough percentage of the total number of procedures that we can reasonably assume all procedures are "normal." Denning further defines the *working set at time T* of a process to be those pages that were referenced during a small time interval, $\triangle T$, which began at time $(T - \triangle T)$.

We can now restate our goal in terms of working sets. A desirable page replacement policy is one that selects pages that are not in the working set of any process on the ready list. The actual composition of the working set depends on the size of the time interval $\triangle T$; that is, there are many working sets for a process, each of which is determined using a different value for $\triangle T$. The least recently used (LRU) policy does not use a fixed value for $\triangle T$. Instead, it directly determines some working set for a process and does not actually ever have a value for $\triangle T$.

A working set is determined by ordering all of the pages in primary memory according to the time for their last reference, beginning with the page that was just referenced. Then any sequence of pages beginning with the first on the list is a working set for some $\triangle T$. For example, suppose there are six pages in the system and the sequence of references is as shown in Figure 15.26a. If we order these pages according to the time of last reference, we get the sequence shown in Figure 15.26b. For this sequence of page references there are six work-

(*a*) Sequence of page references

time (T)	0	1	2	3	4	5	6	7	8	9	10	11	12	13	14
page referenced	p_2	p_1	p_6	p_1	p_3	p_2	p_6	p_4	p_5	p_4	p_5		p_2	p_4	p_2	p_5

(*b*) Pages ordered according to recency of reference.

p_5 p_2 p_4 p_6 p_3 p_1
↑___ Most recently   ↑___ Least recently
      referenced (used)   referenced

(*c*) List of the six working sets at time T = 15.

Working set	Corresponding $\triangle$T
p_5	1
p_5 p_2	2
p_5 p_2 p_4	3–8
p_5 p_2 p_4 p_6	9–10
p_5 p_2 p_4 p_6 p_3	11
p_5 p_2 p_4 p_6 p_3 p_1	12–15

Figure 15.26 Example of determination of working sets.

ing sets at time T=15, as shown in Figure 15.26*c*. Note that the same working set may result from different values of $\triangle$T.

This observation that working sets can remain constant over a period of time is the key to a least recently used page replacement policy. The working set that we need is the one based on the time period from T to (T+$\triangle$T) when we need to replace a page at time T. However, the only information we have concerns past references. Thus, we can only determine a working set based on the time period (T−$\triangle$T) to T. In order to determine the desired working set we would need information on future references. Fortunately, for a normal process the current working set remains constant for a relatively long time. Furthermore, when its current working set changes the change is gradual, that is, if an N page working set changes to another N page working set containing N completely different pages, this change usually takes place over a much longer time period than the time for N memory references.

Given this property of normal processes—that their working sets are relatively constant and change slowly when they change—we can approximate the working set (T, T+$\triangle$T) by the working set (T−$\triangle$T, T). The more constant and slow changing a process's working set, the better this approximation. Using this policy, the page selected to be replaced is the least recently used page, that is, the page at the end of the list of pages ordered according to time of last reference, which is p_1 in the previous example. As can be seen in that example, the remaining pages form a working set. This set of pages has a higher probability of being the working set in the next $\triangle$T time interval than any other set of pages.

Implementation of the least recently used policy requires that pages be ordered by time of last reference. We will store this ordering in a *page use list*. Given such a list, implementation of the selection in assign_block is straightforward. The identification of the page that we select to page out is contained in the last entry of the page use list. The most difficult part of the implementation is ordering the page use list. We must be able to tell when a page is referenced. A little reflection will show that unless we have some fairly elaborate additional hardware, we cannot reorder this list after each primary memory reference.

Detection of page usage can be accomplished only by some kind of addition to the hardware. A simple solution is to include a *usage bit* in the page table word

Page table word

This is the solution adopted in Multics, TSS, Insys, and several other systems. The usage bit can be examined and changed by the pager. Periodically, the usage bit is examined and set to zero. If this examination finds the value of the usage bit is equal to one, then the page has been referenced since the last examination. A value of zero indicates no reference to the page since the last examination.

We could examine the use bit in the page table words after execution of each instruction. In this case the one or two pages referenced by the last instruction, the only ones whose usage bits are equal to one, are moved to the head of the page use list, the others remain in their existing order on the list. Thus, the pages are always properly ordered according to last reference time. However, this would be much too inefficient. Instead, we must be satisfied with examinations at much longer time intervals. At the end of an interval $\triangle T$ all pages whose usage bit equals one are moved to the top of the page use list, the others remain in their existing order. Among those pages referenced during $\triangle T$, there is no way to tell the order of reference. This is not a serious problem unless $\triangle T$ is so large that all pages were referenced during that time interval.

The example in Figure 15.27 shows how, even with $\triangle T$ greater than one, the page use list becomes a reasonable reflection of the time sequence of last references. Let $\triangle T$ equal five and assume six pages in primary memory. The page use list is updated every five time units. Notice that even though there is no ordering within the set of pages that was referenced during the T interval

T = page references during $\triangle T$					Page use list					
0					P_1 P_2 P_3 P_4 P_5 P_6					
P_1 P_4 P_1 P_4 P_2										
1					P_1 P_2 P_4 $\mid$ P_3 P_5 P_6					
P_6 P_4 P_2 P_4 P_6										
2					P_2 P_4 P_6 $\mid$ P_1 $\mid$ P_3 P_5					
P_4 P_6 P_5 P_5 P_6										
3					P_4 P_6 P_5 $\mid$ P_2 $\mid$ P_1 $\mid$ P_3					
P_5 P_6 P_6 P_5 P_5										
4					P_6 P_5 $\mid$ P_4 $\mid$ P_2 $\mid$ P_1 $\mid$ P_3					

Figure 15.27 Example of updating the page use list.

(pages to the left of the leftmost vertical line), there is an ordering among the unreferenced pages. This order is shown by the vertical lines. Thus, p_3 is definitely the least recently used, p_1 the next least recently used, and so forth.

It is not possible to discuss other page replacement policies in this text. We merely comment that most have serious defects. For example, one policy is *first-in, first-out*, which is similar to the *first-come, first-serve* scheduling policy that we saw in Chapter 14. Using this policy, the page selected for replacement is the page that was referenced first of all the pages in primary memory. However, there is no a priori reason to believe that pages first referenced will be any less apt to be referenced in the future than any other page. In fact, some common conventions used in systems are such that some, but not all, of the first referenced pages are continuously referenced. For example, in a system that uses a stack for local storage, the current stack frame is the first page referenced after entry to a procedure in order to initialize it, and it is continually referenced thereafter, since it is used to store intermediate results during computation of all arithmetic expressions. Experience with the Multics system, which uses stacks in such a way, confirms this defect in the first-in, first-out policy. Experiments with the Multics system have shown that use of a least recently used policy results in at least 50 percent fewer missing page faults compared with use of a first-in, first-out policy.

Most virtual memory systems use more than one type of secondary memory device and are sometimes referred to as *multilevel memory systems*. For example, Multics uses drums, disks, and magnetic tape as secondary storage for segments. With primary memory this gives four levels of memory, each level having a longer access time than the preceding level. This situation calls for some form of page replacement policy between each pair of levels, for example, between primary memory and drum, drum and disk, and disk and tape. In general some variation of a least recently used policy is used. As time passes, an unused segment is progressively moved to slower and slower devices. Thus, when a page is first removed from primary memory, it is moved to drum. If it continues unused for a much longer time, then it is moved from drum to disk. Finally, if it continues unused for an even longer time, it is moved to tape. In principle this movement could be on a single page basis; however, in most systems, movement from drum to disk and disk to tape is always the entire segment.

In a previous section we saw that making a segment active may require deactivating an active segment. At that time we did not discuss how a segment was selected to be deactivated. Obviously, any segment that has no pages in primary memory is chosen first. If there are none, then choose the one segment, S, such that every other active segment has at least one page that is higher on the page use list than all pages of S on the list.

We have tacitly assumed that no page is read into primary memory until it is actually needed. This is called *demand paging*. A rule for deciding when to read in a page is called *a page fetch policy*. There are other alternatives to a pure demand paging policy. For example, we could try to anticipate which pages will be used in the near future and read in each page before it is needed. However, the same lack of information about a process's behavior that caused us trouble in deriving a page replacement policy also prevents us from formulating any practical anticipatory page fetch policy.

A limited amount of anticipation is possible and even practical in a few cases. For example, when execution of a ready process begins, the page containing the first instruction to be executed will certainly be needed. Depending on a system's conventions for composition of a process, it may be obvious that certain other pages will always be needed. For example, in the Multics system, whenever a process is executing it always needs, in addition to the page containing the current instruction, the pages containing the executing procedure's linkage section, its current stack frame, and some other pages containing information private to the process. In addition, when execution of a process that was preempted at the end of its allocated time is resumed, the probability is very high that the process will reference those pages that it referenced just prior to the preemption. Multics is able to determine this set of pages by examining the contents of the address mapping hardware's associative memory at the time the preemption occurs. It must be emphasized that these are special cases and that general anticipatory page fetch policies are not practical.

One serious pitfall must be mentioned. Denning has explored *thrashing*. This is a phenomenon that can easily occur unless some safeguard is built into the paging and scheduling policies. Thrashing occurs in the following way. The ready list will often contain a sufficiently large number of processes that the total number of pages in all of their working sets exceeds the size of primary memory. One process starts to execute. Almost immediately it causes a missing page fault. This process is put in blocked state while its missing page is read in. In the meantime the next process on the ready list starts to execute. It, too, almost immediately causes a missing page fault and ends up on the blocked list. This same sequence happens for all processes on the ready list. By the time the first process's page has been read in and it gets to execute, some other page in its working set has been paged out because of all of the intervening missing page faults, each of which required allocation of a memory block for its missing page.

The only way to be sure that thrashing does not occur is to limit the number of processes that are considered for execution. In Multics these are called *eligible processes*. The number of eligible processes is then limited to a small enough number that it is certain that primary memory is larger than the sum total of all their working sets. Multics implements this by defining two additional states for a process: *eligible* and *wait*. The wait state is similar to block, except that a process in wait state is waiting for a missing page to be read in.

The state diagram in Figure 15.28 shows the permitted state transitions among the five states of a process. As before, a running process goes to the blocked state when it calls the block entry of process control. When the event it is waiting for occurs (wakeup), the process goes to the ready state. However, now a process does not go directly from ready to running. Before a process can run, it must become eligible. The number of eligible processes is fixed. A ready process becomes eligible only when some eligible process goes blocked or is preempted because its time allocation has been exceeded or a higher priority process has become ready. Whenever a control processor becomes free, it is allocated to the process on the top of the eligible list, unless this list is empty. If the eligible list is empty, the control processor is not allocated to any process, even if there are processes on the ready list. When a running process causes

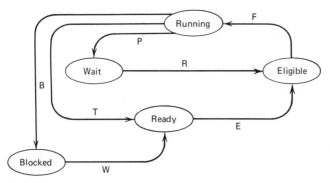

W – Wakeup
B – Call to block
T – Preemption due to time exceeded or higher priority process
    becoming ready
P – Missing page interrupt
R – Reading in of missing page completed
E – Process becomes eligible
F – Control processor free

Figure 15.28 Process state transitions to avoid thrashing.

a missing page fault, it goes to wait state until the page is read in, then it goes back to the eligible state so that it can execute again.

If the primary memory is large enough and the eligible list short enough the working sets of all eligible processes will fit together in primary memory. Consequently, when a missing page fault occurs and space is needed for the missing page, no page of any eligible process need be removed. The net result is that once an eligible process's working set is in primary memory, no missing page faults will occur because some of its pages have been removed. Missing page faults will occur only when its working set changes. The least recently used page replacement policy should be modified slightly to consider first only pages of blocked processes, then consider only pages belonging to ready processes. Only as a last resort will the pager consider pages belonging to eligible processes. However, this will be necessary at times, since an executing process may change its working set. The system must adjust to this change in working set by replacement of pages belonging to eligible processes. Under the above assumptions, these pages are no longer in its working set.

EXERCISES

15.1 Write two IL procedures, **request** and **release**. These procedures obtain and release space for a segment, using the cyclic, first-fit allocation policy. Blocks should have the format shown in Figure 15.5. The free list should be maintained as shown in Figure 15.6. The call

request(N,A);

should allocate a block for a segment of N words, as in Figure 15.7. No block smaller than K words should ever be generated. The value of A should be set to the address of the first data word in the allocated block. The call

release(A);

will release the block whose first data word's address is the value of **A**. The released block should be combined with any adjacent free blocks, as in Figure 15.8.

15.2 Write a set of IL procedures that manage directories in the hierarchy. The call

create(tree_name);

should create a new directory. Its name and its location in the hierarchy (i.e., the directory in which an entry for the new directory will be made) are specified by **tree_name**, which is the tree name of the new directory. The call

delete(tree_name);

should delete the directory whose tree name is the value of **tree_name**. All segments and directories with entries in the directory being deleted should also be deleted. The call

find(tree_name,A);

should set the value of **A** equal to the address of the segment or directory whose tree name is the value of **tree_name**. For the purpose of this exercise, assume that each directory is stored in a separate segment.

15.3 Write an IL procedure to reorder the page use list according to the least recently used policy. The usage bit of all page table words for pages that are present in memory should be examined. Any usage bit that is one should be reset to zero after the corresponding page is moved to the head of the page use list.

16 | Input and Output

In this chapter we discuss the management of the third major system resource: I/O devices. In particular we consider the operation and control of the I/O processor and its attached devices, allocation of devices to processes, and certain common methods for storing and accessing information on these devices. Most modern hardware is built so that the control processor and the devices are able to operate concurrently. The Insys attributes of conversational response ratio and simultaneous use of the system by many users imply that the I/O processor and devices be managed so as to achieve a high degree of this concurrency. The attribute of simultaneous use also implies the need for a mechanism and policy for device allocation. This allocation should be dynamic in order to achieve reasonably efficient use of the devices. Finally, the existence of an extremely wide variety of devices and the attributes of expandability and adaptability all demand a high degree of device independence at the user level.

16.1 Operation and Control of I/O Processors and Devices

Early computers were relatively simple and included only a few simple I/O devices, for example, a card reader, a card punch, and a printer. These devices were attached directly to the control processor and controlled by it. As computers evolved, the control processor became much faster than the devices and the number of different devices increased by an order of magnitude. In order to accommodate this great variety of devices and their wide range of characteristics and to make efficient use of the control processor, most computer systems now include one or more special purpose processors whose function is to control the I/O devices.

All I/O devices are similar in one respect. Each has some basic unit of information that is the smallest amount of information that can be transferred to or from the device in a single operation. This unit is usually called a *record* (or less often, a *block*). For some devices a record has a fixed length. This may reflect some physical property of the device. For example, a record on a typical card reader is always 80 bytes because there are 80 positions on a card in which characters can be punched. For other devices records may vary in length. The major differences between devices are in the details of their operation, particularly the way in which they store records and the way in which these records are read from or written onto the device. One significant difference is the *access time* of a device. The access time of a device is the time it takes to read a record from that device. The variation in access time is extreme, the difference between the slowest and fastest being as much as six orders of magnitude.

Basically devices can be classified as *sequential access* or *random access.* Records are stored sequentially on a sequential access device. In order to reach a specific record the device must read (i.e., skip over) all records preceding it in sequence. Magnetic tape units, card readers, and line printers are typical sequential access devices. A sequential access device can be viewed as a continuous strip of storage medium that moves past a fixed read-write head, as shown in Figure 16.1. For magnetic tape units the medium is a strip of magnetic tape rolled onto a reel. For a card reader the medium is a deck of cards that can be considered as being pasted end to end. For a line printer the medium is the strip of paper that moves through the printing mechanism.

Random access devices are similar to primary memory in that each record has an address associated with it, just as does each word in primary memory. Any record can be read directly, without having to read explicitly any previous records, by presenting the device with the address of the desired record. Drums and disk units are typical random access devices. We could view a random access device as a nonmoving medium with a moving read-write head, where the head moves to the given address in order to read a record. However, since the medium in most random access devices does, in fact, move, our representation of a random access device must be slightly more complicated to capture certain important properties of this type of device. We will picture a random access device as one or more relatively short continuous loops of medium that are constantly in motion. There are also one or more read-write heads that may or may not be able to move.

A drum is an example of a *fixed head, random access device.* The strips of medium are pasted on the outside of a cylinder, side by side. There is one read-write head for each strip of medium, as in Figure 16.2. The strips of medium are called *tracks.* A record address specifies one of the tracks and also the location of a record within that track. A record is read by selecting the read-write head corresponding to the specified track and then waiting until the desired record passes under the head. Selection of the head is essentially instantaneous; however, waiting for the record to pass under the head results in an average delay that is equal to half the time for one rotation of the drum. This delay is called the *latency.* Since a typical drum makes one complete rotation in 17 milliseconds, the latency is around 8 milliseconds on the average, with a maximum of 17 milliseconds.

A typical disk unit is an example of a *moving head, random access device.*

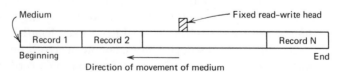

Figure 16.1 Representation of a sequential access device.

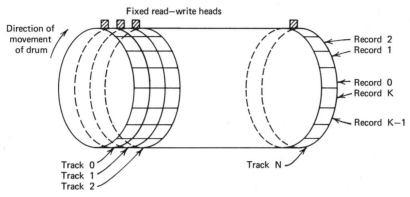

Figure 16.2 Diagram of a typical drum.

The strips of media are pasted onto both sides of a *platter* or *disk,* which closely resembles a large phonograph record, side by side in the manner of concentric circles, as shown in Figure 16.3. There is one read-write head for each side of each platter. All of these heads are able to move along a rod that extends inward toward the center of the platter like a radius of a circle. Normally, all heads move at once. As with a drum, the record address specifies a track and also the location of a record within that track. To read a record, the read-write heads must first be moved so that one of them is opposite the specified track. That head is then selected, and the disk unit waits until the desired record passes under the selected head. As with the drum, there is an average delay equal to half of the rotation time of the disk. Typical average delays are on the order of 12 to 16 milliseconds. Unlike the drum, there is an additional delay caused by having to move the read-write head. This delay is called the *seek time* and ranges from 25 to 180 milliseconds if any movement is required, otherwise it is zero.

There is also a delay when accessing a record on a sequential device. This delay depends on the number of records that must be skipped over in order to get to the desired record. For example, consider magnetic tape. A reel of mag-

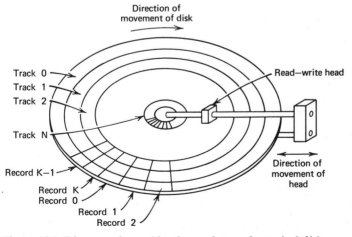

Figure 16.3 Diagram of one side of one platter of a typical disk.

netic tape is normally 2400 feet in length. If the average record size is 3000 bytes, then the tape reel can hold 12,000 records. The delay is equal to the record passing time multiplied by the number of records passed over. The record passing time for a 3000 byte record is 20 to 80 milliseconds, depending on the speed of the tape unit. Thus, there is a delay of more than 2 to 8 seconds if there are 100 records preceding the desired record.

Because of the differences in the way the various devices operate, each device or group of similar devices is interfaced to the I/O processor by a *control unit*. Logically this control is independent from either the device or the I/O processor. However, in some computers it may be physically part of the device or the I/O processor. The functional relationship between devices, control units, I/O processor, control processor, and the primary memory is shown in Figure 16.4. The I/O processor is the master controller of the control units. In addition it assists in the transfer of information between the devices and the primary memory. In both cases the I/O processor acts as an intermediary between the primary memory or control processor and the control units. Information being read is transferred from a device to the primary memory. Information being written is transferred from primary memory to a device. The I/O processor is usually capable of maintaining several independent information transfer paths, called *channels*. (In the IBM 360 the entire I/O processor is confusingly called a channel, even though it may contain several independent information transfer paths.) For each channel in use, the I/O processor keeps track of what device is using the channel and what location in primary memory that channel is currently reading information into or writing information from.

The control processor, the I/O processor, the control units, and the devices are all capable of operating concurrently. Thus, it is possible for a magnetic tape unit to be rewinding a tape while its control unit is engaged in reading information from another tape unit attached to the same control unit. At the same time that the I/O processor is assisting in transmitting the information being read from the tape, it may also be engaged in transmitting information to or from other devices or issuing orders to other control units. All of the time

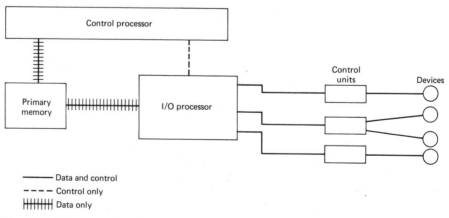

Figure 16.4 Computer hardware system structure.

this activity is going on in the I/O processor, control units, and devices, the control processor may be executing a program that is doing some kind of computation. All of this concurrent operation requires a certain amount of synchronization. In general the required synchronization of I/O processor, control units, and the devices is wired into the hardware. On the other hand, the synchronization required between the control processor and the I/O processor is not fixed in the hardware and must be completely programmed as part of the operating system.

All of the earliest computers and some that are currently being built (in general, only minicomputers) have direct I/O, that is, there is no I/O processor and the control processor communicates directly with the device control units. In this case operation of the device control units and most of the devices is not concurrent with operation of the control processor. The control processor starts the operation of a device and then waits until the device finishes its operation. The principal reason for development of the I/O processor was to achieve concurrent operation of the control processor and the I/O devices.

We cannot hope to describe all variations of the division of labor between the control processor, the I/O processor, the control units, and the devices. What we will do is to describe the operation of a specific structure that contains the main features found in most of the large, modern computer systems. The specific details of this structure are not any real computer's I/O hardware, but a consistent, abstracted, realistic model. Our purpose is to give the reader a flavor of the kind of hardware he will find in real life, but in a simplified setting where the problems can be clearly exposed. The I/O processor that we will use in the remainder of this chapter is part of Our 360. A complete description of it is contained in Appendix F.

The I/O processor is subordinate to the control processor in the following sense. The I/O processor is capable of performing one or more concurrent *I/O tasks*. Execution of each separate I/O task by the I/O processor must be initiated by the control processor. The I/O processor cannot initiate execution of any I/O tasks on its own. The control processor starts the I/O processor executing an I/O task and is then free to do something else until that task is either finished or the I/O processor encounters trouble that it is unable to handle. An I/O task is defined by an *I/O program*. Since the I/O processor has no internal memory, except for a few registers, I/O programs are stored in primary memory. The I/O processor fetches instructions from primary memory as it executes them, just as the control processor does.

An I/O program specifies the locations in primary memory between which information is to be transferred, the direction of transfer, the amount of information to be transferred, and any special information that the control unit needs to control the device. A control unit cannot directly access primary memory, so the I/O processor must transfer any device control information from the I/O program in primary memory to the control unit. For a read operation, the device passes the data to the control unit, which passes it on to the I/O processor, which passes it on to primary memory. For a write operation the sequence is reversed.

Each I/O task has associated with it a *device control block* that identifies the device to be used and gives the location of the I/O program. Under certain conditions, information concerning the status of the I/O task will be stored in the device control block by the I/O processor. The device control block consists of four full words with the following format:

0	Device identification
4	I/O program location
8	Status
12	

The device identification specifies both a channel and a device that is attached to that channel. The I/O program location is the memory address of the first *command* in the I/O program that is to be executed by the I/O processor. An I/O program is a sequence of commands. Each command is a pair of full words with the format:

0	Operation	Flags	Count
4	Address		

One of the flags is the *program end flag*. The last command in an I/O program has this flag set equal to one. In all other commands of the I/O program this flag must have a value of zero.

There are three different commands: READ, WRITE, and CONTROL. The READ and WRITE commands transfer information between devices and primary memory. The count field of a READ or WRITE command specifies the number of bytes of information to be transferred, and the address field contains the memory address of the first byte to be read into or written from. If an I/O program contains any READ or WRITE commands, then one or more records are read or written. The number of records read or written depends on the *record end flag*. All commands in sequence read or write the same record until a command is executed that has the record end flag equal to one. For example, the following I/O program consisting of six commands writes three records.

Operation	Program end flag	Record end flag	Count	Address
WRITE	0	0	20	813
WRITE	0	0	40	1034
WRITE	0	1	10	5830
WRITE	0	1	300	2000
WRITE	0	0	250	1850
WRITE	1	1	250	720

The first record written contains 70 bytes; 20 from primary memory locations 813 to 832, 40 from locations 1034 to 1073, and 10 from locations 5830 to 5839. The second record contains 300 bytes, and the third contains 500 bytes.

The third command, the CONTROL command, is used for passing control information to a control unit, for example, backspace or rewind a tape. The count and address fields combined are transmitted to the control unit by the I/O processor. The meaning of this information depends on the device that is addressed. For example, if the device that is specified in the device control block is a magnetic tape, the following I/O program will backspace the tape one record, read a record of 1000 bytes, and rewind the tape.

Operation	Program end flag	Record end flag	Count	Address
CONTROL	0	0	0	4
READ	0	1	1000	5050
CONTROL	1	0	0	3

In this example we are assuming that the tape control unit interprets the control code of 4 as backspace one record and the control code of 3 as rewind.

There are two control processor instructions for controlling the I/O processor: STARTIO and TESTIO. The STARTIO instruction starts the I/O processor executing an I/O task. The TESTIO instruction checks the status of an I/O task. In both instructions the effective address is the location of the first word of the device control block for the I/O task that is being started or tested. When the control processor executes one of these two instructions, it sends the address of the device control block along with the start or test operation code to the I/O processor. The I/O processor then returns information that the control processor uses to set the value of the condition code (CC). This information indicates the result of the operation requested of the I/O processor. Thus, after the control processor has completed execution of either a STARTIO or TESTIO instruction, the value of the condition code indicates the result. The control processor then immediately executes the next instruction in sequence.

For the STARTIO instruction the possible results are:

Value of CC	Result
0	Execution of I/O task successfully started.
1	Channel or device addressed is busy.
2	Status stored; indicates unusual condition.
3	I/O processor is inoperative.

A condition code of 1 indicates that either the channel or device addressed in the device control block is busy and the I/O task specified in the STARTIO instruction cannot be executed yet. In order to start this I/O task the control processor must execute another STARTIO instruction sometime in the future. A condition code of 3 indicates that the I/O processor is inoperative; that is, the control processor gets no response from the I/O processor within a reasonable amount of time. No further information is available. A condition code of 2 indicates that some condition exists that prevents the I/O task from being started. For example, the addressed device or channel is inoperative. Status information that identifies the exact problem is stored in the device control block by the I/O processor. A condition code of 0 indicates that the I/O task has been successfully started by the I/O processor. The control processor is now free to execute further instructions while the I/O processor continues executing commands in the I/O program.

The principal function of the TESTIO instruction is to cause the I/O processor to store status information into the device control block. This information indicates the current status of the channel and device that are being used by the I/O task defined by the device control block addressed by the test instruction. After the control processor executes a TESTIO instruction, the condition code indicates the result:

Value of CC	Result
0	Status information successfully stored.
1	Channel busy; status stored.
2	Device busy; channel free; status stored.
3	I/O processor inoperative; status not stored.

Except when the I/O processor is inoperative, status information is always stored by the I/O processor when the control processor executes the TESTIO instruction. This information is stored in the last two full words of the device control block that is addressed by the TESTIO instruction. Even though the status information indicates if the device or channel is busy, the condition code also indicates this so that it can be rapidly checked. If the condition code is 0 neither the channel nor the device is busy.

In addition to indicating whether the channel and device is busy, the status information indicates the state of the I/O task. In particular the status information includes the memory address of the next command to be executed. If the I/O processor has completed execution of the I/O program this address will be 8 greater than the address of the last command in the I/O program. If the last command to be executed was terminated before it was finished and was a READ or WRITE command, the status information contains the *residual count*, that is, the number of bytes remaining to be read or written out of the total number specified by the count field of the last command. The status information also contains flags that indicate if certain errors occurred during execution of the I/O task, such as garbling of the data during reading or writing. Other status information indicates errors or unusual conditions that are specific to particular devices, such as card hopper empty, card jam in the card reader or card punch, or write protect ring on a magnetic tape addressed by a write command.

Figure 16.5 shows a complete program for reading five cards from the card reader. Figure 16.5a shows the control processor program, Figure 16.5b shows the device control block, and Figure 16.5c shows the I/O program. All three of these must be in primary memory. The I/O task is defined by the device control block at RDCARD and the associated I/O program beginning at READCDS. The first full word in the device control block specifies the card

(*a*) Control processor program to control I/O processor in reading cards
      from card reader.

BUSY	STARTIO	RDCARD	ask I/O processor to start I/O task
	BC	4,BUSY	try again if channel or device busy (CC=1)
	BC	3,TROUBLE	branch if some problem (CC=2 or CC=3)

```
/*
/* I/O task successfully started.
/* Test for completion before information can be used.
/*
```

NOTYET	TESTIO	RDCARD	cause I/O processor to store status
	BC	6,NOTYET	try again if channel or device busy (task not completed) (CC=1 or CC=2)
	BC	1,TROUBLE	branch if I/O processor inoperative (CC=3)

```
/*
/* I/O task completed; five cards have now been read.
```

(*b*) Device control block for I/O task that reads cards from the card reader.

RDCARD	DC	F'42'	device identification of the card reader
	DC	A'READCDS'	address of I/O program
	DS	2F	reserve two full words for status

(*c*) I/O program that reads five cards.

READCDS	READ	0,1,240,X	read three cards
	READ	0,0,40,A	read half a card
	READ	0,1,40,B	read the other half card
	READ	1,1,80,W	read another card

Figure 16.5 Example control processor and I/O programs to read cards.

reader as the device to be used in this I/O task. The second full word contains
the address of the associated I/O program, which begins at READCDS. This
I/O program consists of four commands that read five cards. Figure 16.6 shows
the relation between the action of the control processor and the I/O processor.

    The instructions executed by the control processor to initiate this I/O task
begin at BUSY. The first instruction is a STARTIO instruction whose effective
address is the location of the device control block for the I/O task. This instruction signals the I/O processor to begin execution of the I/O task. Before the
control processor completes execution of the STARTIO instruction, the condition code is set to indicate the result of the instruction. If the control processor does not receive any reply from the I/O processor within a reasonable
time (equal to the execution time of a few instructions), the I/O processor is
inoperative and the control processor sets the condition code equal to 3.
Otherwise the I/O processor replies to the original signal by returning to the
control processor the value for the condition code.

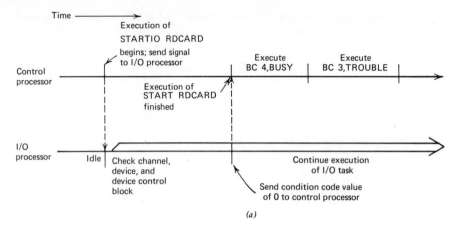

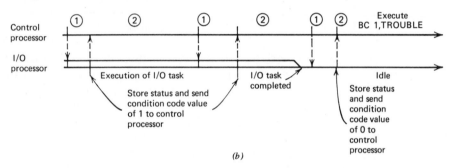

Figure 16.6 Time sequence relationship of control processor and I/O processor for program in Figure 16.5. (a) Execution of STARTIO instruction that successfully initiates the I/O task. (b) Execution of TESTIO instruction to wait for completion of I/O task.

In order to determine this value the I/O processor fetches the first two full words of the device control block. Both the channel and device are checked. If either of them is busy a condition code value of 1 is immediately returned. If the specified device is inoperative or some other problem such as an invalid program address is detected, the status is stored into the last two full words of the device control block, and a condition code value of 2 is returned. Otherwise a condition code value of 0 is returned and the I/O processor proceeds with execution of the I/O program.

Once the condition code has been set, the control processor proceeds to execute the instruction following the STARTIO instruction. In our example the next instruction is a branch on condition instruction that tests the value of the condition code to see if either the channel or device is busy, in which case the STARTIO instruction is executed repeatedly until both the channel and device are free. When neither the channel nor the device is busy, we know for certain that the I/O processor has at least tried to start execution of the I/O task that was specified by the start instruction at BUSY. However, the I/O processor may not have been successful because it encountered some difficulty that prevents it from proceeding any further. The second branch on condition instruction tests for further problems. When we find the condition code equal to 0, that is, both branch on condition instructions fail to branch, we know that the I/O processor has successfully started execution of the I/O program beginning at READCDS.

Once the I/O task has been successfully initiated, the control processor can do something else. However, before the information being read can be used, we must be sure that it is in the primary memory. The only way to be sure is to wait for completion of the I/O task. (In the next section we will see how we can avoid wasting time while waiting.) The TESTIO instruction at NOTYET is used to test for completion of the I/O task. When this instruction is executed the status is stored and the condition code is set. The branch on condition instruction following the TESTIO instruction tests the condition code to see if either the channel or device is busy; if either one is, then the I/O task has not yet been completed. When a condition code value of 0 is detected, the task has been completed and status information for the task has been stored in the device control block. By examining this status information, the program being executed by the control processor can determine if the I/O task was successfully completed or if some problem has been encountered. If a problem has been detected, then some corrective action must be initiated, or the executing process is terminated and the system executes some other process.

16.2 Interrupts and the I/O Process

It was pointed out earlier that both the control processor and the I/O processor can be executing programs concurrently. However, in the example in Figure 16.5, we made no attempt to take advantage of this capability. Instead, we simply had the control processor loop on the TESTIO instruction until the I/O processor finished executing the I/O task. The easiest way of taking advantage of the capability for concurrent execution is to use the *I/O interrupt* feature of the I/O processor; in fact, this is the normal mode of operation. Use of the I/O interrupt feature allows the I/O processor to tell the control processor when it has completed an I/O task or encountered a problem that prevents it from completing an I/O task.

An I/O interrupt, initiated by the I/O processor, functions just like the timer interrupt feature described in Chapter 14, which is used by process control to preempt a process that has used up its time allotment without completing execution or going blocked. When using the I/O interrupt feature, the control processor starts an I/O task and then starts doing something else; that is, it does not loop on a test instruction. When the I/O task has been completed, or a problem encountered, the I/O processor sends an *interrupt signal* to the control processor. Before sending the interrupt signal, the I/O processor stores some *interrupt information* in primary memory that identifies the I/O task that caused the I/O interrupt and indicates the reason for the interrupt. When this signal is received, execution of the current process by the control processor is interrupted, and control is transferred to a fixed location in pri mary memory. The instructions beginning at that location will then determine if the I/O task has been successfully completed or if a problem has been encountered and take the appropriate action. When this action has been finished, the control processor will resume execution of the process at the point where it was interrupted.

Since an I/O interrupt is similar to preemption, it is not surprising that similar problems exist. In particular, when the system is responding to one interrupt, it cannot be interrupted again if we are to prevent inconsistent data and the resultant unreliable behavior of the system. Of course, if we allow only one I/O task to be executed at a time, this will not happen. However, normally the I/O processor has several channels and, especially in a multiprogramming time-sharing system, several I/O tasks will be executing concurrently. Thus, frequently some of the interrupt signals that result from these I/O tasks will occur before the system has completed its response to a previous interrupt.

To solve this problem there is one bit in the program status word of the control processor that can be set to prevent I/O interrupt signals from affecting the execution of the control processor. When this bit is equal to one, I/O interrupts are *disabled*. When I/O interrupts are disabled, the control processor ignores any I/O interrupt signals that are received from the I/O processor except to set a hardware switch that records the fact that one or more such signals have occurred. This switch is like the wakeup waiting switch used by process control. In addition, the interrupt information for each I/O interrupt signal that occurs while I/O interrupts are disabled is automatically stored in a queue in primary memory so that none of it will be lost. I/O interrupts are *reenabled* when the I/O interrupt bit in the program status word is set to zero. If any I/O interrupt signals were received from the I/O processor while I/O interrupts were disabled, a single I/O interrupt will occur immediately after I/O interrupts are reenabled.

It is normally the case that after a process initiates some I/O task, it cannot proceed in its execution until that task has been completed. This being the case, in a multiprogramming system the process will normally go blocked after starting an I/O task. Thus, when the I/O interrupt occurs that signals the end of an I/O task, the process that started the task will not be executing, and so it cannot be interrupted. Whatever process is currently executing on the control processor will be interrupted when the I/O interrupt signal occurs. This process will not know anything at all about the I/O task that caused the I/O interrupt. Thus, the interrupted process will seldom be in a position to respond in any sensible way. What process does respond? Although the obvious candidate is the process that initiated the I/O task, there are three major reasons why this is not a good choice. First, identification of the process that initiated the I/O task that generated the I/O interrupt is not trivial and therefore is an excessive burden to place on the process that actually gets interrupted. Second, response to the I/O interrupt requires intimate knowledge of the I/O processor and devices. Finally, concurrent execution of several I/O tasks must be coordinated.

The best solution seems to be a special system process that is solely responsible for the management of the I/O processor and all of the devices. Hence, the response to all I/O interrupts is carried out by this special *I/O process*. This process mirrors the action of the hardware I/O processor in that it executes concurrently with other processes in the system just as the hardware I/O processor operates concurrently with the control processors. All I/O needed by any process is routed through the I/O process, which acts as the master controller for all of the I/O. The I/O process is the only process that actually initiates operation of the I/O processor. It is also the only process that needs to know any details about the operation of the I/O processor or the devices.

Normally the I/O process is blocked waiting for a wakeup that is either a request by another process for some I/O or an I/O interrupt from the I/O processor. An I/O interrupt will usually occur when some process other than the I/O process is executing. Because the I/O process will be executing only when it has something to do, we cannot assume that it will be executing when an I/O interrupt occurs. Thus, what a process does in response to an I/O interrupt is to wake up the I/O process. However, before calling **wakeup**, it must send a message to the I/O process to indicate that the wakeup is because of an I/O interrupt. If the interrupted process happens to be the I/O process the result is simply that the wakeup waiting switch gets set. This action is shown in more detail in Figure 16.7.

Several questions arise at this point. How does the I/O process get into execution? The simplest method is to treat it like any other process and use the normal scheduling procedure. When it is awakened, it is put on the ready list. Eventually it will be allocated a processor and be able to execute. The only difficulty with this is that there is apt to be too long a delay before the I/O process is able to execute, which results in poor response or the inability to respond properly to high-speed devices. Therefore, the I/O process should probably have the highest priority and always be eligible (Figure 15.28). In this case the I/O process always cycles between the wait, eligible, and running states. It will be in the wait state when it has nothing to do. A wakeup causes a transition from wait state to eligible state. Since it has the highest priority, it will usually preempt a running process so that it can begin execution immediately.

When a user's process wants some I/O, it must send a message to the I/O process and then wake up the I/O process. The message describes the I/O being requested. The call to **wakeup** will cause the I/O process to call the system to collect all of its messages, examine them, and carry out the I/O requested by each message. Figure 16.8 shows the action of a user process in

```
procedure io_interrupt;
    Disable I/O interrupts;
    send ('io_process', message_containing_interrupt_information);
    wakeup ('io_process');
    Re-enable I/O interrupts;
    return;
end.
```

Figure 16.7 Conversion of an I/O interrupt (hardware) to a call to **wakeup** (software).

```
send ('io_process', message_describing_I/O_request);
wakeup ('io_process');
block;
```

Figure 16.8 Request by user process for I/O.

requesting I/O. After calling **wakeup**, the user process goes blocked to wait for completion of the I/O. When the user process is awakened from the blocked state, the I/O will have been completed and the user process can now continue with its execution.

The actions of the I/O process are shown in Figure 16.9. The I/O process starts by initializing itself and then going into an infinite loop that responds to wakeups resulting from either I/O requests by other processes or I/O interrupts. The normal state of the I/O process is blocked. Whenever there is nothing for it to do it calls **block**. When it wakes up from the blocked state, it first must determine why it is being awakened, for an I/O interrupt or for an I/O request. In the first case the interrupt information, which was stored by the I/O processor before it sent the interrupt signal to the control processor, must be examined to determine which I/O task caused the interrupt. This interrupt information consists of two full words. The first full word contains the address of the device control block for the I/O task that caused the interrupt. The second full word contains a code that identifies the cause of the interrupt.

After identifying the I/O task by getting the address of its device control block, the cause of the interrupt needs to be determined. The interrupt code gives us this information. If the interrupt resulted from normal completion of the I/O task, then all that needs to be done is to determine which process

procedure io_process;
   Initialize I/O process;
   while forever do
     collect (message, status);
     if status indicates no messages **then**
       begin Reenable I/O interrupts;
         block;
         Disable I/O interrupts;
         collect (message, status);
       end;
     if message indicates interrupt **then**
       begin Identify I/O task that caused interrupt;
         if interrupt resulted from normal completion **then**
           begin pros_id : = identification of process that requested this
                I/O task;
             wakeup (pros_id);
          end;
         else Take appropriate error action;
       end;
     else begin Transform I/O request into an I/O program;
           Build device control block;
           Execute STARTIO instruction;
           if STARTIO instruction not successful **then** Take action;
       end;
   end;
end.

Figure 16.9 Definition of the I/O process.

requested the corresponding I/O and wake up that process. On the other hand, if the interrupt occurred because of an error or unusual condition, appropriate action must be taken. The exact problem may be known from the interrupt code or the I/O process may have to execute a TESTIO instruction and analyze the status information in order to identify the problem completely. The action taken depends on the problem, and detailed discussion is beyond the scope of this text.

If the wakeup is due to an I/O request by another process, the I/O process must initiate an I/O task to carry out the request. Using the information supplied in the message that the requesting process sent to the I/O process, the required I/O program and the corresponding device control block are constructed. A STARTIO instruction is then executed that will cause the I/O processor to begin execution of the I/O task. As we know, the start instruction may not be successfully executed because of some problem or simply because the channel or device is busy. The only general solution to the problem of a busy device or channel is to put the I/O task onto a queue for the channel or device and then continue normally. Then, whenever an I/O interrupt is processed that signals completion of an I/O task, the queues for the device and channel that are now free are examined to see if there are any I/O tasks waiting. If there are, the task at the head of the queue is started.

All of the various possible cases of action terminate at the end of the infinite loop. This loop is then repeated. At the beginning of the repetition, a call to collect is executed to see if there are any more messages. If there are, then some additional I/O interrupt or I/O requests still must be taken care of before the I/O process blocks itself.

Since additional I/O interrupt signals may occur while I/O interrupts are disabled, the interrupt information must be saved on a queue. The interrupt information is queued automatically by the hardware whether or not I/O interrupts are disabled. The *I/O interrupt exchange area* for I/O interrupts consists of six full words having the following format.

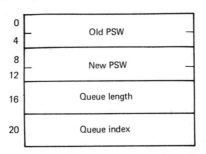

When an interrupt occurs, the current contents of the program status word is stored in the first and second full words (old PSW) and the program status word is loaded from the third and fourth full words (new PSW). This transfers control to the beginning of the instructions shown in Figure 16.7. This action happens only if I/O interrupts are not disabled.

The I/O interrupt exchange area is located at a fixed place in primary memory, and its base address is known to the I/O processor. When an I/O interrupt occurs, the I/O processor stores the associated interrupt information on the I/O interrupt queue at the location whose address is stored in the sixth full word of the interrupt exchange area (queue index). The I/O processor then increments the queue index by eight in anticipation of the next interrupt. The fifth word (queue length) contains the address of the last location reserved for the I/O interrupt queue. If the queue index becomes larger than this address a trouble interrupt occurs, and the system must take corrective action immediately in order to continue to guarantee reliable operation. This is only a safety valve; ordinarily the I/O process will be able to process the interrupts fast enough so the queue will not fill up. Normally, when the I/O process examines the queue, it will remove all of the current entries and reset the queue index back to the beginning of the queue.

Aside from the details of this particular implementation, the preceding discussion was concerned with one key idea for the management of the I/O processor, especially with respect to the interrupts that it generates during its operation. This was to transform as rapidly as possible a hardware I/O interrupt into a call to **wakeup** and then use the normal process control primitives. The process that is actually interrupted is involved only long enough to achieve this transformation of an interrupt into a call to **wakeup**. This takes a minimal amount of time and does not cause a noticeable perturbation of the interrupted process's execution. The transformation of I/O interrupts into calls to **wakeup** makes the I/O processor look very much like a process. Thus, the I/O processor, the I/O process, and a user process requesting I/O form a set of three cooperatinig concurrent processes.

16.3 Device Independence

In order to achieve the high degree of adaptability and expandability that we desire, user programs must be written to be independent of particular devices. This requires *symbolic device names* and a *standard device interface*. We use the concept of a *virtual device* to achieve this. In order to actually execute, a process must use real devices, just as it needs real memory. However, the process references virtual devices, just as it references virtual memory instead of real memory. Each virtual device has a symbolic name that the procedures in the process use to reference it. Some system procedure is responsible for mapping references to a virtual device into references to the corresponding real device.

A virtual device functions as a standard device interface in the sense that the characteristics of a virtual device are an abstraction of only those characteristics of a class of real devices that are relevant to user procedures. This makes it possible for a process to execute using any one of a class of real devices instead of being restricted to just a single device. In addition, the user's procedures do not need to have detailed knowledge of the operational characteristics of the various real devices.

Use of virtual devices makes *dynamic device allocation* fairly straightforward. Dynamic device allocation is desirable both for improving the efficiency of resource usage and for expandability and adaptability. For example, if the program requires the use of a magnetic tape unit and the hardware address is used in the program, then execution of the process is held up when use of this tape unit is requested if that one particular tape unit is in use by another process. This is true even if all of the remaining tape units are free. If, however, the program uses a symbolic name for the tape unit, any free tape unit can be allocated to the process. Thus, execution can proceed unless all tape units are in use. With respect to expandability, a program that uses hardware device addresses cannot take advantage of devices that are added to the system unless the program is modified. In fact, considering adaptability, not only is a program that does not use symbolic device addresses unable to take advantage of new devices without being modified, it may not even be able to execute if some of the old devices have been replaced.

Implementation of symbolic device names is relatively easy and straightforward. All that is needed is a mapping between a virtual device name and the identification of the corresponding real device. This can be easily implemented by use of a symbol table. Let us call this symbol table the *process attachment table*. It contains one entry for each virtual device used by the process. Each such entry consists of the name of a virtual device and the hardware address of the currently corresponding real device, if there is any. If there is a corresponding real device we say that that real device is *attached* to the corresponding virtual device and vice versa. Real devices that are attached to virtual devices are *allocated* to the process. A real device may be allocated to the process but not attached to any virtual device. Such a real device is, in effect, reserved for use by the process at some other time.

Implementation of a standard device interface can be only partially achieved because programs often wish to make use of the unique properties of different classes of devices, for example, sequential access devices versus random access devices. Use of these different properties should not be prohibited. However, the majority of I/O is either simple sequential access or simple random access. We have discussed random access in the last chapter, since random access devices are largely used as storage for file systems. We will also discuss random access later in this chapter when we explore another file system. The remainder of this section will be concerned only with sequential access I/O.

What constitutes a standard device interface? A *format for information* and a *set of operations*. The format defines the size of records, the location of data within the record, and any control information in the record. In addition, if the data consists of characters, the format specifies the codes for the characters. In the remainder of this section we will only be concerned with character data, which makes up a large majority of the sequentially accessed data. The set of operations is used by a procedure when it refers to a virtual device. These operations provide a procedure with the means to read and write data as well as other common housekeeping functions. A standard device interface permits a program to use a number of different real devices without modification of the program. This is accomplished by letting the program talk in generalities such as read and write, in much the same way that the I/O commands READ and WRITE can be used to read or write any device.

In order to clarify these ideas we will describe a specific standard device interface and examine its implementation. A virtual device is used for either input or output at any given time, but not both at the same time. A virtual input device serves as an infinite source of characters and will be called an *input stream*. There is no further structure to an input stream. The length of a record is variable and essentially unbounded. When a program invokes a read operation, it receives one or more characters from the input stream. A virtual output device is essentially an infinite sink and is called an *output stream*. An output record may also be of any length. An input stream is basically an abstraction of the card reader (ignoring card boundaries), while an output stream is an abstraction of the line printer (ignoring line boundaries).

The read and write operations are invoked by calling the appropriate procedure

read (in_name,N,A);
write (out_name,N,A);

The read procedure reads N characters into primary memory, beginning at memory address A, from the input stream whose name is in_name. The write procedure writes N characters from primary memory, beginning at memory address A, into the output stream named out_name. The characters that are read or written are represented using USASCII character codes and are stored in memory as one character per byte.

The streams described above are basically sequential access, nonreversible devices. As such they can be attached to any of the following real devices: card reader, card punch, line printer, typewriterlike terminal, paper tape reader, paper tape punch, magnetic tape unit, and even a disk or drum. The function of the I/O control part of the system is to transform a real device into a virtual device and vice versa. For our example of character streams this means, first, a transformation between USASCII character code and the character code used by a real device and, second, a transformation between the essentially record free format of the character stream and the natural record structure of a real device. In general this latter transformation will require some form of *buffering*. For example, a user procedure may read characters from an input stream in arbitrarily sized groups. However, I/O control must read records from a real device, and the size of these records may not be the same size as the number of characters requested by the user's program. Records from the real device will be read into a buffer and then transferred to the user in whatever size group is requested.

Since the above described transformations are different for each different kind of real device, the simplest implementation is to have a separate procedure for each kind of real device. We will call these procedures *device interface modules*. A device interface module is responsible for performing the transformations between character streams and some particular real device or class of similar real devices. It is responsible for conversion between USASCII character code and the device's character code. The device interface module is also responsible for the buffering and any other work required in the transformation between character streams and the real device's native record format. Finally, it is responsible for construction of the I/O programs that are required to read or write the records containing the characters as well as any other control functions that are required to make the real device look like the virtual device.

16.4 Device Allocation, Attachment, and Use

Before a process can use (read or write) a virtual device it must be attached to a real device. *Attachment* is the coupling of a virtual device with a real device. This association is recorded in the *process attachment table*. In our example this attachment must be explicitly requested by the process. This is accomplished by the call

attach (virtual_name, real_name);

where virtual_name is the name of the virtual device to be attached and real_name is the name of the real device to which it is to be attached. By convention, if the virtual device is already attached to another real device, it is detached before being attached to the new real device; that is, a virtual device is never attached to more than one real device at a time. Note that this is not a logical necessity. It would be meaningful to permit a virtual device to be simultaneously attached to more than one real device when doing output. This would permit data to be both printed and punched at the same time, for example. However, providing this capability would make I/O control more complicated.

Before a real device can be attached, it must be allocated to the process. Allocation of a real device to a process means that the device is reserved for exclusive use of the process, and no other process may use it until it is released. This is another instance of resource allocation; hence, we should expect to be concerned with both an allocation policy and the mechanics of allocation. Device allocation is centered around the *system device table*. This table accounts for every device in the hardware system. Its major function is to record the status of these devices, that is, if the device is operational or not and, if it is, to which process it is allocated, if it is allocated. In addition, it records limits on the current allocation and any other constraints on its use.

Allocation of a real device is explicitly requested by

request (device_type, real_name);

This call requests I/O control to allocate a real device of type device_type to the calling process. Ordinarily a request is not made for a specific real device, only for some one device of a particular type. This allows the system much more flexibility in its allocation policy and greatly increases the probability of a process obtaining the requested device without delay. The system will allocate a specific device to the process and return the name of that device as the value of the argument real_name.

Just as was the case with other resources, there are many possible allocation policies, and they cannot all be considered in this text. Many of the considerations discussed with regard to allocation policies for other resources are also pertinent to allocation policies for device allocation. In fact, many of these considerations are pertinent to the allocation of any resource without regard to its type. One of the simplest allocation policies is demand allocation, that is, a device is not allocated until a process requests it. If no device of the requested type is currently available the requesting process is blocked until one is released by another process. This allocation policy is such that it is possible for deadlock to occur, just as is possible with shared data. For example, processes A and B both need the use of a card reader and a line printer in order to complete their execution. Process A requests and is allocated the card reader. Process B then requests and is allocated the line printer. Process A then requests the line printer. Since the printer is not free, the process blocks to wait until it is free. Process B then requests the card reader that is not free, so it also blocks. The two processes are now deadlocked.

The most common method used by existing systems to avoid this kind of deadlock is to require that before a process begins execution it be allocated all of the devices that it will need during its life. Thus, a process waits to begin execution until all of the devices that it will ever require during its execution are free. Then these devices are allocated to the process, and it is allowed to begin its execution. The disadvantage of this allocation policy is that it defeats part of the advantages of dynamic device allocation. All needed devices will be allocated to a process throughout its entire life, even though a particular device may not be needed except for a short time. The majority of the time this device could be used by another process. Thus, the efficient use of devices is often impaired.

Habermann[1] has developed an alternate allocation policy that largely corrects the difficulty in the previous policy. Using Habermann's policy, a process must declare its needs when it begins its execution; however, the required devices are not allocated to the process until it requests them, which is usually not until each device is needed. Except for certain restrictions, all of the unallocated devices that a process may need can be allocated to other processes for their use. The constraints, while not excessively complex, must be formulated very precisely and somewhat abstractly and will not be stated here. The important thing about these constraints is that Habermann has proved that if they are observed, deadlock cannot possibly occur. Even though the constraints are severe enough that they do not always allow the most efficient use possible of devices, they do permit a significant improvement over the first allocation policy.

16.5 A Complete I/O Control System

In this section we give a fairly detailed description, similar to the one used by the sequential batch system in Chapter 2, of a complete I/O control system. However, this I/O control system is designed to operate within Insys, our model time-sharing system. This description will draw together all of the ideas discussed in the last few sections. By presenting a complete system, we can see how all of these ideas relate to each other and how they function together to achieve our objectives of concurrent use, dynamic device allocation, and device independence.

Two major data bases play a central role in I/O control: the *system device table* (SDT) and the *process attachment table* (PAT). The system device table is a systemwide data base that records the status of every device physically connected to the I/O processor. It contains one entry for each (real) device currently connected. An entry consists of four items:

> type = the device type, such as magnetic tape, printer,
> or card reader
> process_id = process identification of process to which
> device is allocated
> = 0 if device not allocated
> = −1 if device not operational

[1] A. N. Habermann, "Prevention of System Deadlock," *Communications of the ACM, 12* (7), July 1969, pp. 373–377.

dim_id = identification of the device interface module
for this device

device_id = device identification of this device for use
in a device control block

The process attachment table is a per process data base, that is, each process has its own private process attachment table. This table contains one entry for each (real) device currently allocated to the process. An entry consists of three items:

sdt_pointer = index of entry in the system device
table for corresponding (real) device

name = name of the virtual device to which this device
is currently attached

= blanks if this device is not currently attached

direction = 0 if attached for input

= 1 if attached for output

The first two entries in the process attachment table are permanent entries for two reserved virtual devices named IN and OUT. The virtual device IN is permanently attached to the user's terminal for input, and the virtual device OUT is permanently attached to the user's terminal for output. If we were actually using this I/O control in the batch system of Chapter 2 these would be permanently attached to the card reader and printer, respectively.

Each virtual device is a one-way stream of characters represented in the ASCII code; that is, at any particular time a given virtual device may be used for input or output of characters, but not both, depending on how it is attached. In addition, a real device may be attached to at most one virtual device at any given time, with the exception of the user's terminal. As pointed out earlier, this restriction is not a logical necessity, it merely simplifies the design. Inversely, a virtual device may be attached to at most one real device at any given time.

There are six entry points to I/O control: **request, release, attach, detach, read**, and **write**. The entry **request** is used to ask that some real device of a particular type be allocated to the calling process. If the call is successfully completed a real device has been allocated to the process and remains allocated until explicitly released. The entry **release** is called to deallocate an allocated real device. If that real device is attached to a virtual device, it is also detached as well as released. The entry **attach** is called to attach an allocated real device to a virtual device. The entry **detach** is used to detach a virtual device from a real device. The entry **read** is called to read characters from an attached virtual device, while the entry **write** is called to write characters onto an attached virtual device.

In the definition of **request** shown in Figure 16.10, we assume that the Habermann allocation policy is being used. Thus, immediately after the process was created, it must have given I/O control a list of all the real devices it will need during its lifetime. Immediately after entry to **request**, it is determined if any real device of the requested type is connected and operational. If not, we return to the caller with that information in **status**. If there are operational real devices of the requested type that are connected, but they are all allocated, then the process must wait, by going blocked, until a real device of that type becomes free. Our allocation policy guarantees that no deadlock will occur, so

```
procedure request (requested_type, allocated_device, status);
  while forever do
  Search SDT for device of requested type;
    if no device of requested type is connected and operational
      then begin status : = 'none';
             return;
           end;
    if unallocated device of requested type then
      begin Compute Habermann constraints;
        if safe to allocate this device then
          begin i : = index of entry for this device in SDT;
            SDT.process_id[i] : = identification of process
               requesting device;
            allocated_device : = i;
            j : = index of unused PAT entry;
            PAT.sdt_pointer[j] : = i:
            PAT.name[j] : = ' ';
            PAT.direction[j] : = 0;
            return;
          end;
      end;
    Put identification of requesting process on queue for requested
      device type;
    block;
  end;
end.
```

Figure 16.10 Definition of procedure that allocates devices.

we know that eventually a real device of the requested type will be released by some other process. In order to be awakened when this happens, we put the identification of this process onto a queue for the requested real device type. There will be a queue for each type of real device, since it is fruitless to awaken the process when other types of real devices become free.

Even if a real device of the requested type is free, the Habermann constraints may not be satisfied. In this case the process also blocks to wait until an additional real device is released in the hope that the constraints will then be satisfied. Our allocation policy guarantees that eventually the contraints will be satisfied and some real device allocated. When this occurs, the selected free real device is allocated to this process. This is accomplished by storing the process's identification in process_id of the system device table entry for the real device being allocated. The index of this entry is returned to the caller so that he may attach the real device to one of his virtual devices. The final step in allocation is to add an entry for the allocated real device to the process attachment table of this process, since all allocated real devices must be accounted for in that table. However, the real device is still not attached to any specific virtual device. This the process must do explicitly by calling attach.

Once a real device has been allocated to a process, attach can be called to attach it to a virtual device. Figure 16.11 shows the definition of attach. The

```
procedure attach (virtual_device, real_device, in_out, status);
    if virtual_device = 'IN' or virtual_device = 'OUT' then
        begin status : = 'reserved virtual device';
            return;
        end;
    if entry for virtual_device in PAT then
        begin i : = index of that entry;
            PAT.name[i] : = ' ';
        end;
    if entry for real_device in PAT then
        begin i : = index of that entry;
            if PAT.name[i] = ' ' then
                begin PAT.name[i] : = virtual_device;
                    PAT.direction[i] : = in_out;
                    status : = 'ok';
                end;
            else status : = 'device currently attached';
        end;
    else status : = 'device not allocated';
    return;
end.
```

Figure 16.11 Definition of procedure to attach a virtual device to a real device.

reserved virtual devices IN and OUT are permanently attached to the user's terminal and cannot be reattached to some other device. If the virtual device specified in the call to **attach** is not reserved, the process attachment table is searched to see if the virtual device is currently attached to some real device. If it is, then we detach it from that real device by blanking out **name** in that real device's entry in the process attachment table.

The virtual device can now be attached. However, two situations may still prevent the attachment. If no entry for the real device can be found in the process attachment table, then we cannot make the attachment because the specified real device is not allocated to this process. Even if the real device is allocated to this process, it may already be attached to another virtual device as indicated by **name** being nonblank in the real device's entry in the process attachment table. Since we are not permitting more than one virtual device to be attached to a real device, with the exception of the reserved virtual devices IN and OUT, we cannot make the requested attachment in this case either. If all of the restrictions are satisfied we make the attachment by filling in **name** and **direction** in the process attachment table entry for the real device to which the virtual device is being attached. The procedure to detach a virtual device, **detach**, is defined in Figure 16.12. There will be at most one entry for the virtual device in the process attachment table. If there is no entry, **virtual_device** is already detached, so nothing needs to be done. An attached virtual device is detached by simply setting **name** to blanks in the process attachment table entry for the real device to which it is attached.

A real device is deallocated by calling **release**, which is defined in Figure 16.13. A real device cannot be released if it is not connected. A user cannot release his terminal. Since the first two entries in the process attachment table

```
procedure detach (virtual_device);
   if entry for virtual_device in PAT then
      begin i := index of that entry;
         PAT.name[i] := ' ';
      end;
   return;
end.
```

Figure 16.12 Definition of procedure to detach a virtual device from a real device.

```
procedure release (real_device);
   if entry for real_device in SDT then
      begin i := real_device;
         if PAT.sdt_pointer[1] = i or PAT.sdt_pointer[2] = i
            then return;
         if SDT.process_id[i] ≠ identification of calling process
            then return;
         SDT.process_id[i] := 0;
         if PAT contains entry with sdt_pointer = i
            then Remove this entry from PAT;
         if queue for devices of type SDT.type[i] is not empty then
            begin pros_id := identification of process at head of queue;
               wakeup (pros_id);
               Remove process from head of queue;
            end;
      end;
   return;
end.
```

Figure 16.13 Definition of procedure to release an allocated device.

are for IN and OUT, which are attached to his terminal, we can prevent its release by checking that the entry for **real_device** is not one of these first two entries. If a real device is not allocated to this process it cannot be released. In all three cases we simply ignore the release request. Otherwise, the real device is released by setting **process_id** in the system device table entry for the real device equal to zero and removing its associated entry from the process attachment table. The final task is to check the queue for real devices of this type to see if any other process is waiting for such a real device. If there are any processes waiting, we wake up the process at the head of the queue.

Both **read** and **write** are similar except for verification of the direction for which the virtual device is attached and possibly the values of the status codes used when errors or other problems are encountered. Therefore, we discuss only **read**, which is defined in Figure 16.14. If the specified virtual device is not attached to any real device or not attached for input, then we cannot do the requested read. Otherwise, we need to get the index of the corresponding real device's entry in the system device table, which will be passed along to the I/O process. This index is found in the process attachment table for the virtual device.

```
procedure read (virtual_device, N, A, status);
    if entry for virtual_device in PAT then
        begin i : = index of that entry;
            if PAT.direction[i] ≠ 0 then
                begin status : = 'not attached for input';
                    return;
                end;
            j : = PAT.sdt_pointer[i];
            message : = (j, N, A, 'in');
            send ('io_process', message);
            wakeup ('io_process');
            block;
            collect (message, status);
            if reading was successful then status : = 'ok';
                else status : = code indicating problem encountered;
        end;
        else status : = 'virtual device not attached';
    return;
end.
```

Figure 16.14 Definition of procedure to read N characters from a virtual device.

Before invoking the I/O process, a message containing all of the relevant information must be constructed and sent to the I/O process. After sending this message, the process blocks until the read has been completed or some unresolvable problem is encountered. When the process wakes up from being blocked, it will find a message from the I/O process that explains the outcome of the read request. This outcome is reported back to the caller via status.

The process attachment table must be searched for each call to read or write. This may amount to an unacceptably large cost if the number of calls to read and write is large. We briefly sketch how this search can be eliminated without giving complete details. A *virtual device name* will be an external symbol (just like the name of the entry point of a procedure). The *value* of this symbol is the address of a memory location that normally contains a pointer to the process attachment table entry for the real device to which the virtual device is currently attached. The argument of attach, detach, read, and write, which specifies the virtual device, will be the value of the corresponding external symbol, that is, it will be the address of the memory location in which is stored a pointer to a process attachment table entry. The procedure attach sets the contents of the memory location whose address is the value of the external symbol after the process attachment table table entry for the real device to which the virtual device is being attached has been found. Now the detach, read, and write procedures can locate this entry immediately, since one of their arguments is a pointer to this entry.

This mechanism still retains the symbolic nature of virtual device names, since source programs are written using external symbols and the linker fills in the address of the memory location that will contain the pointer to the corresponding process attachment table entry. This decision of whether to introduce additional complexity for the sake of improved performance is typical of the engineering-type decisions that frequently must be made by the system designers. In this case the additional complexity is not large. However, many of these "small" increases in complexity often have a cumulative effect that is exponential.

Since the device interface modules are part of the I/O process, the design

```
procedure io_process;
    Initialize I/O process;
    while forever do
        collect (message, status);
        if status indicates no messages then
            begin Reenable I/O interrupts;
                block;
                Disable I/O interrupts;
                collect (message, status);
            end;
        if message indicates interrupt
            then Process I/O interrupt;
            else Process I/O request;
    end;
end.
```

Figure 16.15 Definition of control procedure in new version of the I/O process.

```
i := j from message;
dim_id := SDT.dim_id[i];
pros_id := process identification of process that requested I/O;
case dim_id of
    'card_read': card_read (message, dcb, status);
    'card_punch': card_punch (message, dcb, status);
    'printer': printer (message, dcb, status);
    'tape': tape (message, dcb, status);
    'tty': tty (message, dcb, status);
end;
if status ≠ 0 then
    begin message := error information from status;
        send (pros_id, message);
        wakeup (pros_id);
    end; else
    begin chan_id := channel used for I/O with specified device,
                which is found in dcb;
        Put (dcb, pros_id) on queue for chan_id;
        if length of queue for chan_id = 1 then
            begin Execute STARTIO instruction for I/O task on queue;
                if not successfully started then Take action;
            end;
    end;
```

Figure 16.16 Expansion of "Process I/O request;" from Figure 16.15.

dcb : = location of device control block for I/O task causing interrupt;
dim_id : = identification of device control module responsible
   for interrupting I/O task;
chan_id : = channel used by interrupting I/O task;
pros_id : = process identification from entry at head of queue
   for channel chan_id;
case dim_id of
   'card_read': card_read_end(dcb, interrupt_info, status, ok);
   'card_punch': card_punch_end(dcb, interrupt_info, status, ok);
   'printer': printer_end (dcb, interrupt_info, status, ok);
   'tape': tape_end (dcb, interrupt_info, status, ok);
   'tty': tty_end (dcb, interrupt_info, status, ok);
end;
if ok then
   begin message : = information in status;
      send (pros_id, message);
      wakeup (pros_id);
      Remove entry at head of queue for channel chan_id;
   end;
if queue for channel chan_id not empty then
   begin Execute STARTIO instruction for I/O task at head of queue
            for channel chan_id;
      if start instruction not successful then Take action;
   end;

Figure 16.17 Expansion of "Process I/O interrupt;" from Figure 16.15.

shown in Figure 16.9 needs to be revised. Basically, the main procedure in the
new I/O process acts as a dispatcher to the device interface modules for all of
the device specific activities (see Figures 16.15 to 16.17). In addition, it manages
the I/O task queues for the channels and initiates all I/O operations. The
control procedure defined in Figure 16.15 is essentially as before. When the
I/O process wakes up from its call to block, it must determine whether an I/O
interrupt or an I/O request from another process has occurred. This difference
is determined by examination of the message that has been received. After
processing the interrupt or I/O request, the loop is repeated, either processing
an additional message, if there are any, or blocking.

   Processing of an I/O request is shown in Figure 16.16. It is the responsibility
of each device interface module to build the I/O program that is required to
operate its corresponding device. The device interface module, dim_id, that
is required for this request is obtained from the system device table entry
whose index is part of the I/O request message that the requesting process sent
to the I/O process. The system added the process identification, pros_id, of
the sending process to the message when it was transmitted to the I/O process.
The device interface module identification is extracted from the message and
used to select the proper device interface module to call. The I/O request
message is passed on to the device interface module, which builds the appropriate I/O program and device control block. The location of this device control block, dcb, is returned by the device control module. The value of status

indicates whether the device module was able to construct a valid I/O program. If it was not, an error message is sent to the process that requested the I/O and that process is awakened.

If a valid I/O program has been built, its device control block is added to the queue of I/O tasks for the channel needed for the referenced device. There is one queue for each channel. The channel that must be used for the I/O program can be determined from the device identification in the device control block. If the newly added device control block is the only one on the queue for its channel, then that I/O task is started by executing a start instruction. If the start instruction fails, then the problem must be diagnosed and some action taken. We assume that the device interface module detected all errors in the I/O request. Thus, failure of the start instruction indicates a system (hardware or software) failure and only the system can do something about it. If there were already other device control blocks on the channel queue, the channel is already busy, so no attempt is made to start the new I/O task.

Interrupt processing is shown in Figure 16.17. The interrupt information that is available as a result of an interrupt contains an interrupt code that identifies the cause of the interrupt and the location of the device control block, dcb, for the I/O task that is responsible for the interrupt. This device control block contains the identification of the responsible device interface module, dim_id and the channel used, chan_id. The entry at the head of the queue for this channel is for the I/O task that caused the interrupt. The identification of the process that requested that I/O, pros_id, can be obtained from the entry at the head of that queue.

If the interrupt was caused by some condition or problem that is not device specific, then the appropriate corrective action can be taken at this point. Otherwise, the device interface module for the device causing the interrupt must be called so that it can take corrective action, or decide that the I/O request has been successfully completed. The corrective action that the device interface module determines is necessary may require execution of an additional I/O program, for example, rereading the record just read. If this is the case, the device interface module replaces the device control block for the old I/O program, which is at the head of the channel queue, by the device control block for the new I/O program and sets the value of the argument ok to false.

Upon return from the device interface module, if no additional I/O program needs to be executed, then the value of status is used to construct a message, which is sent to the process that requested this I/O. This message will indicate either successful completion or an unrepairable error, such as a bad tape. After this message has been sent to the process and wakeup called, the device control block for the I/O request is removed from the head of the queue. Whatever the value of ok, the last action in processing the interrupt is to start the I/O task defined by the entry at the head of the channel queue, if that queue is not empty.

In the description of the main procedure in the I/O process we have shown calls to only five device interface modules. If the system has additional devices of different types, then additional device interface modules will be required. As an example, the definition of the device interface module for the card reader is shown in Figure 16.18. All device interface modules are similar in structure. Each has two entries, one for initially generating an I/O program

```
procedure read_card (message, dcb, status);
    direction := direction of I/O specified in message;
    if direction = 'output' then
        begin status := 'wrong direction';
            return;
        end;
    Construct I/O program for card reader from N and A in message;
    Construct device control block for card reader and I/O program
        just constructed;
    dcb := location of device control block just constructed;
    status := 0;
    return;
end.
procedure read_card_end (dcb, interrupt_info, status, ok);
    if interrupt_info indicates successful read then
        begin status := 'success';
            ok := true;
            return;
        end;
    if problem implies execution of additional I/O program then
        begin Construct new I/O program;
            Construct new device control block at location specified by dcb;
            ok := false;
        end;
    else begin status := identification of problem;
            ok := true;
        end;
    return;
end.
```

Figure 16.18 Definition of device interface module for card reader.

for an I/O request and the other for action when an interrupt occurs. The request message included as an argument contains all of the information required to construct the initial I/O program. If there is any restriction on the direction, as is the case with the card reader, this must be checked first. The I/O program required to read the specified number of characters from the card reader is constructed. Finally, the device control block for the I/O task is constructed, and the device interface module returns to the main procedure of the I/O process.

The second entry to the device interface module is called in response to an interrupt. If the interrupt has resulted from successful completion of the I/O program, then nothing needs to be done except to set status to indicate this and return to the main procedure of the I/O process. If the read operation specified in the I/O program was not successful, then there are two cases. In the first case the device interface module decides that perhaps it can recover from the problem if another I/O program is executed. So a new I/O program and its device control block are constructed. The old device control block is replaced by the new one and control is returned to the main procedure of the

I/O process so that it may start execution of the new I/O program. In the second case the device interface module decides that nothing more can be done to recover from the problem, so the value of **status** is set to identify the problem, and control is returned to the main procedure of the I/O process.

In describing the device interface module for the card reader we have not included any buffering. In general, each device interface module is required to provide any buffering required for its device. Buffering is usually required, since there is no guarantee that every I/O request will be for a number of characters that is an integral multiple of the device's record size. Most devices must read an entire record at one time. They cannot read a lesser amount. The card reader must always read a complete card whenever it performs a read operation. This is a record containing 80 characters. However, an I/O request may be for more or less than 80 characters. Suppose it is for less than 80 characters. A card is read into a buffer, and the requested characters are copied into the area specified by the I/O request. The remaining characters from the card are left in the buffer to be copied, or partially copied, when the next I/O request for the card reader is processed.

16.6 OS/360 Data Management

In this section we give a very much simplified description of the data management facilities available in the IBM OS/360. What we wish to do is simply to highlight the essential characteristics of this approach to file storage and I/O. Both permanent file storage and I/O are encompassed within a single, "unified" approach. Therefore, OS data management has some of the characteristics of both the segment type file system we studied in Chapter 15 and I/O control described in the preceding section. Unfortunately, as is often the case when two quite different functions are combined, some of the significant advantages of both are missing.

The basic unit corresponding to both a virtual device and a segment (file) is called a *data set*, which may have a symbolic name. All transfer of information between primary memory and secondary memory must be explicitly specified by the user. That is, when he wishes to reference some data in a data set, he must first call a procedure that will read it into some work area in primary memory. No dynamic device allocation is permitted. All needed devices are allocated at the beginning of a job step by the system. The device requirements are specified by control cards that define the job step. Device attachment is accomplished during program execution by calling **open**. A device may be detached by calling **close**. Direct sharing of data is possible. However, it is very awkward and possible only in very limited situations.

A data set can, but is not required to be, *catalogued* in a hierarchical set of *catalogs*. A catalog is similar to a file directory. However, not all catalogs are always stored in the system. Data sets are stored on *volumes*. A volume is a physical unit such as a reel of tape, a drum, or a disk pack. Some of these volumes can be removed and carried away from the system. Each volume contains a *table of contents*, which is basically a catalog of all the data sets stored on that volume.

Data sets have structure. Each data set is composed of one or more *logical records* that may be organized in different ways. For example, the records in a data set may be ordered according to the value of some key. The user can choose one of several organizations, depending on how he expects to use the data. The data in a data set is physically stored in *blocks*. The size of a block depends more on the device than on the size of the records in the data set, although block sizes are normally an integral multiple of the record size. Reading and writing of a data set may be automatically buffered or not, whichever is most convenient or efficient for the user.

Three types of data organization are available: *sequential, indexed sequential,* and *direct.* Sequential organization is tapelike. The records are arranged in any sequence of the user's choosing. When a particular record is desired, all records that precede it in the sequence must be read (skipped over) before the desired record can be read. The only thing a user can do is to read (or skip) the next record in the sequence. Direct organization permits the user to use any organization of his choosing. When reading, the user supplies the address of the desired record. Direct organization can be used only with random access volumes—disk or drum, but not tape. Indexed sequential has characteristics of both sequential and direct organizations. The records in a data set are ordered in sequence according to the value of some key field in the record. The user may specify which field is the key. The user can treat the data set as sequential, always reading the next record in sequence, or he can treat it as direct, except that he supplies the value of the key for the desired record instead of the address.

Data sets may be accessed in two different ways: *queued* and *basic.* When using queued access, automatic buffering is provided by the system. This type of access applies only to sequential or indexed sequential organizations. The user calls **get** or **put** to read or write a logical record of data. If basic access is used, no automatic buffering is provided. The user must program any buffering that is required. He calls **read** or **write** to read or write a physical block of data.

Combining an access type and a data organization gives an *access method,* for example, *queued sequential access method* (QSAM) and *basic sequential access method* (BSAM). With queued sequential **get** is used to read a record. Automatic buffering with read-ahead is provided. Since the organization is sequential, it is valid to assume that the next record in sequence will soon be read. Therefore, the system always reads as many records as the input buffers will hold, before the records are needed. When **get** is called, the next record is usually available; if it is not, **get** waits until it is. Therefore, when control is returned from **get**, the user knows that the next record is actually available. With basic sequential **read** is used to read the next block in sequence. This may contain one or more records. No buffering or read-ahead is provided. The user's program may continue its execution after control is returned from **read**; however, the actual reading of the requested block from the device is taking place concurrently. Therefore, when the user needs any of the data in this block, he must check with the system to see if the system has finished reading the block. If not, he must wait until the reading is finished.

When using queued type access, the user may select one of two *buffering techniques* and one of three *transmittal modes.* The buffering techniques are called *simple* and *exchange*. In simple buffering one or more buffers are used exclusively by each data set. In exchange buffering, buffers are exchanged between two different data sets and a work area. Each *buffer segment,* holding a single logical record, is first an input buffer, then a work area, and finally an output buffer. After the record has been written, its buffer segment is available for reading a new record, and the cycle of use repeats itself. The contents of an input record are thus modified in place, that is, there is no movement of data from one memory area to another.

The three transmittal modes are called *move, locate,* and *substitute*. In move mode **get** and **put** transfer data from the input buffer to the work area and then from the work area to the output buffer. In locate mode **get** and **put** do not move data, but provide a pointer to the location of the record in the buffer. In substitute mode, which is similar to the locate mode, the user must provide a work area in storage equal in size to the current buffer. Then, after a call to **put** or **get**, the buffer and user-supplied work area have changed roles. The old work area is now the buffer and the old buffer is now the work area. Clearly, the normal IBM 360 addressing relative to a base register makes this technique easy and relatively straightforward.

The user must specify both a buffering technique and a transmittal mode when using a queued type access method. The combination of simple buffering and move mode is illustrated in Figure 16.19. The buffers and the work area are always used for the same purpose. The data in the input record is copied twice before it is written on the output device. This is the simplest form of buffering, but often not the most efficient. When using this combination, there is no restriction on the sizes of the buffers or work area, that is, all three may be a different size. The combination of exchange buffering and substitute mode, shown in Figure 16.20, is the most efficient form of buffering. However, it requires that the input buffer, the work area, and the output buffer all be the same size. The figure shows the different use of three memory areas at two different times, illustrating their changing role. In order for the program to refer to the proper area three pointers must be maintained that locate the current input buffer, the current work area, and the current output buffer.

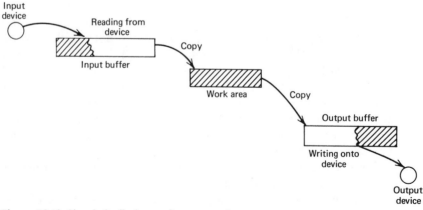

Figure 16.19 Simple buffering and move mode.

Pointer to current
input buffer

Pointer to current
work area

Pointer to current
output buffer

Writing onto
device

Reading from
device

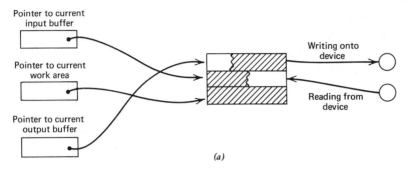

(a)

Pointer to current
input buffer

Pointer to current
work area

Pointer to current
output buffer

Reading from
device

Writing onto
device

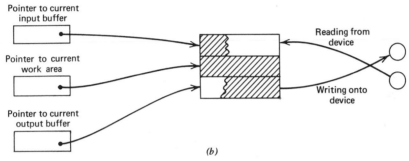

(b)

Figure 16.20 Exchange buffering and substitute mode. (a) Use of buffer areas at time t_1. (b) Use of buffer areas at time $t_2 > t_1$.

EXERCISES

16.1 The device interface module for the card reader in Figure 16.18 does not do any buffering. Modify it so that it does. A card reader record is 80 bytes in length and contains exactly one card. Use one or more buffers of sufficient size so that at least five cards can be in the buffer(s) at once. The device interface module should try to keep the buffer(s) full, immediately starting to read another record as soon as there is room for it.

16.2 We have assumed in the text that the address used in an I/O processor command for reading or writing is an absolute address. However, the user's program cannot generate any absolute addresses. The only kind of address that a user program can generate has the form

segment number	word number
0 13	14 31

as explained in Chapter 15. Write an IL procedure that will compute an absolute address, given a segment-type address as shown above. Assume that this procedure will be executed by the system, since the user's program will not have access to the information needed to do the computation. This procedure will have to be called by **read** and **write**, since segment numbers used in a process do not necessarily correspond to anything meaningful in any other process. Therefore, a segment-type address must be converted to an absolute address before being passed to the I/O process.

16.3 Assume that the I/O processor uses segment and paging hardware to reference primary memory, just as the control processor does. Now the address in an I/O processor command to read or write will have to be a segment-type address. However, the I/O process will not normally have descriptors in its segment table for all of the segments that all of the user processes wish to use for input or output. Even if it did, they would not have the same segment numbers. How can a segment-type address be passed from a user process to the I/O process? Modify read, write, and the I/O process so that this communication is accomplished.

17 | Sharing, Privacy, and Protection

In this chapter we study two problems that are inherently related, even though they may not seem to be when first considered. These two problems are *the sharing of information* and *protection*, especially the protection required to insure *privacy* of a user's information. These two problems are inherently related because the two objectives of sharing and privacy basically contradict each other. In our model system we wish to satisfy both objectives. Sharing of information is required if the system is to be useful for the cooperative efforts of more than one user. However, the requirement that the privacy of a user's information be guaranteed demands that the sharing be controlled, that is, a user's information is shared only if he permits it to be and only with those other users whom he specifies.

In order to enforce privacy and controlled sharing certain forms of protection are required. A user's information must be protected from modification and examination by other users. In addition, the system procedures that enforce protection and control sharing must themselves be protected from unauthorized modification or use. The requirement for reliable system operation also implies protection. To insure reliable operation, all of the system procedures and system data must be protected from misuse and accidental as well as deliberate damage. This requirement also implies that the system may at times have to repair or replace damaged information.

17.1 Sharing of Information

Both procedures and data may be shared. There are two forms of sharing: *direct* and *indirect*. By direct sharing we mean that all sharers of a segment of information actually reference the same copy of the information; that is, there is essentially only one copy of the shared information. This single copy belongs to one user who is the owner of the information. In indirect sharing each sharer gets his own copy of the information. Thus, there is an original copy belonging to the owner, and each user who is sharing the information has an additional copy belonging to him. This difference between the two forms of sharing results in a significant difference in the difficulty of implementation of the two forms, direct sharing being more difficult.

Indirect sharing is straightforward when the sharer only wishes to use, but not modify, the information. When he requests the shared information, the system must find the original and make a copy of it for him. Thus, one problem is to find the master copy. Since finding the segment to be shared is also a problem in direct sharing, we will defer its discussion and consider it in that context. Once the master is located, the system creates a new empty segment for the sharer. A copy of the master is then placed into this new segment.

There is one additional problem. What happens if the owner of a segment is in the process of modifying it when another user wants a copy of it for the purpose of sharing? We can simply ignore this question and let the sharer make his copy whenever he wishes. However, this is not satisfactory, since the result may be inconsistent information in the copy. For example, suppose the shared data is a company's telephone directory with each entry containing an employee's name, his room number, and his telephone number. Suppose the owner is in the process of modifying this telephone directory and has changed the room number for an employee who has moved. If a copy of the directory is now made by a sharer before the owner changes the telephone number for the employee who moved, the contents of the copy will be inconsistent. The result of this inconsistency may not be disastrous, but it certainly can cause undesirable confusion. The only thing that can be done to prevent this inconsistency from occurring is to prevent a copy from being made while the original is being modified. When the owner wants to modify the original, he *locks* it, modifies it, and then *unlocks* it. If a sharer requests a copy while it is locked he must wait until it is unlocked. In order to wait, the sharer's process blocks. When the original is unlocked, the sharer's process is awakened.

What if a sharer modifies his copy? Since it is a copy, when he modifies it, the original is unaffected. This may be the intended effect. However, in a large majority of cases, data is shared just so that each of the sharers may modify it and these modifications are immediately reflected in the original. That is, all parties concerned want the original to be modified whenever one of the sharers modifies a copy of it. For example, if the shared data is the record of booked seats in an airline reservation system or the process table and other information defining processes states (ready list) in an operating system, then it is essential that any modifications of the data by any of the users sharing it are immediately reflected in the original, since these modifications will affect the behavior of other users.

To achieve the latter effect the original will have to be replaced by the modified copy. However, unless severe restrictions are placed on the user of the common data, simply replacing the original with the modified copy will not work. For example, suppose both user A and user B are sharing the data segment X. Consider the following sequence.

1. A makes a copy, XA, of X.
2. B makes a copy, XB, of X.
3. A modifies XA.
4. B modifies XB.
5. X is replaced by XA.
6. X is replaced by XB.

Both users think they have modified X. However, in step 6, when the current contents of X are replaced by XB, A's modifications are lost. This is basically identical to one of the problems in testing and modifying semaphores, which was discussed in Chapter 14, except on a larger scale.

The only way out of this problem is to prohibit more than one user from modifying the original or any copy at any given time. That is, when a user makes a copy for the purpose of modifying the data, all other users, including the owner, are prohibited from modifying the original or copying the original for the purpose of modifying it until the first user has finished his modifications and replaced the copy with the modified original. This clearly introduces delays that can be excessive if the data is frequently modified. In addition, there is a great deal of copying of the data that is costly in both processor execution time and memory space. We observe that if the data is frequently modified, then almost all of the time only one user is using any copy of the data. Therefore, why have more than one copy? Direct sharing evolved for just this purpose, to avoid having more than one copy. The only reasons for not having direct sharing are that it is somewhat more complicated to implement and too inefficient unless the computer has appropriate segmentation or paging hardware.

The central problem of direct sharing is that there must be a mechanism that makes it possible for all users who are sharing a segment to reference the same copy of the segment. That is, the single original copy of the segment must be in the address space of all processes that are sharing the information. Thus, when a user first makes reference to a shared segment, the system must find the segment and then add to the address space of the user's process the information necessary for the process to be able to reference the segment.

In Insys, our Multics-like model, finding the segment is straightforward. The first reference to any segment results in a linkage interrupt. As a result, the linker calls the entry of the segment manager, which makes a segment known (see Figure 15.13). The argument for this entry is always a complete tree name. Since all segments belonging to any user of the system are cataloged in the hierarchy, any shared segment can be named by a complete tree name. If a segment exists with this name it will be made known to the sharing user's process. That is, its complete tree name will be entered in the process's known segment table, and a descriptor for the segment will be put into the segment table of the process. The descriptor generated by this procedure will have its presence bit set so that an interrupt will occur when the process uses the segment number in a reference. When the interrupt occurs, the segment will be made active. It is at this point that modifications are required to accomplish direct sharing.

The most efficient way to achieve direct sharing is for each sharer to have his own descriptor that contains the address of a single page table for the shared segment, as diagramed in Figure 17.1. Note that it is possible but less efficient to have a page table for each user who is sharing the segment. This is inefficient because whenever a page is paged in or out, many page table entries may need to be changed instead of just one, which is the case if there is only a single page table for the segment. Also, if each process has its own descriptor in its own segment table, the descriptors need not be in the same position in each table, that is, each process may reference the shared segment using different segment numbers. For example, in Figure 17.1, process P_1 uses segment number i

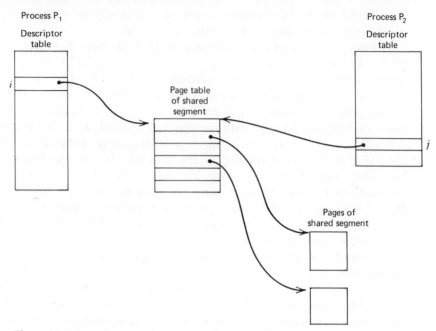

Figure 17.1 Descriptors and page table for a shared segment.

and process P_2 uses segment number j when referencing the shared segment.

Returning now to the problem. When a process first uses the segment number of a shared segment, a missing segment interrupt will occur, and the segment manager will be called to make the segment active. Referring to Figure 15.14, we see that part of making a segment active is to construct a page table for the segment. What if some other user who is sharing the segment is currently referencing the segment? If this is so, then the segment will be active for that user, and a page table for the segment will already exist. Since we want at most only one page table per segment, whether it is shared or not, we cannot let this user's process build an additional page table. This is easily accomplished by having only one active segment table. That is, we will have a single active segment table that all processes in the system will use whenever a missing segment interrupt occurs. Thus, if the segment is already active because some other user's process is currently using it, the interrupted process does not build a new page table. Instead, the descriptor that caused the interrupt is set to point to the existing page table for the segment.

Use of a single active segment table requires minor modifications in both the algorithms for making a segment active and for deactivating a segment. When a segment is deactivated, the presence bit in its descriptor is set equal to zero. However, if the segment is a shared segment, there may be other descriptors for it whose presence bit is not equal to zero. All of these descriptors must have their presence bits set to zero. One field in the active segment table entry for a segment contains that segment's number. However, the segment number in each of the sharing processes may be different. So this field must be expanded so that it may contain a list of segment numbers along with the process identification of the corresponding processes. Now, when a segment is deactivated, this list must be scanned. For each entry (s,p) on the list, the presence bit in descriptor number s of process p is set to zero. The corresponding modification in making a segment active is to add to this list a pair consisting of the process identification of the process that caused the missing segment interrupt along with the number, in this process, of the descriptor that caused the interrupt.

Direct sharing is very complicated and effectively impossible in a computer that does not have some kind of address mapping hardware. By effectively impossible we mean that the cost in execution time overhead and complexity of the system and user programs is excessive. The heart of the difficulty is that, in general, it is impossible to have a given shared segment of information at exactly the same memory address in each process which shares it. However, just this is required if there is no address mapping hardware. In the absence of such hardware and because a segment cannot be at the same memory address in all processes, we are forced to keep moving the segment around in memory. When the system switches from one process to another, all of the segments that the first process shares with the second will have to be moved in memory so that they will have the required addresses in the second process. The situation is even worse if the system has more than one control processor. In this case it is possible that two processes that share a segment may actually execute simultaneously, thus requiring the segment to be at two different memory locations at the same time. The bookkeeping required to avoid this is even more complex than the bookkeeping required to determine which segments need to be moved when execution is switched from one process to another. The only other alternative is to simulate address mapping hardware, which is also quite inefficient.

The segmentation hardware described in Chapter 15 is sufficient to solve the problems that arise in the mechanics of sharing data segments (policy will be discussed shortly). Note that no attempt is made by the system to prevent the use of a shared segment while it is being modified by one or more of the sharers. Since one of the major reasons for direct sharing is to avoid unnecessarily long lockups of a data segment while it is being modified, all locking and unlocking of shared data that is required to preserve its consistency is left up to the users of the data. This is reasonable, since only the users can know the conditions under which such locking is required. This enables them to plan their modifications so that the data is locked the least possible amount of time. In addition, users may subdivide a segment into several pieces, each with its own lock. These pieces then can be modified independently of each other. Any simple automatic locking imposed by the system would have to be limited to entire segments and would inevitably be for time periods much longer than are required. In general, mutual exclusion will be required whenever a subsegment of data is being modified if inconsistent data is to be avoided. But it is seldom required for the entire segment for the entire time a user is referencing it. As with the sharing of any resources, deadlock can occur (see Chapter 14). Thus, the users must follow some algorithm with respect to locking and unlocking that will guarantee that deadlock will not occur.

Additional problems arise when we attempt to directly share procedure segments. When two users are sharing a procedure segment, they both want the same thing done, but to their own data, not to someone else's. For example, if two users are sharing the procedure SIN(X), both compute the sine function, but the value of X is different for each user. This means that the location of X must be different for each of the two users. If not, since the two processes may execute simultaneously on two different control processors, we are faced with the impossible task of having two different values in the same location at the same time. Thus, we see that we need a mechanism by which a single copy of a procedure can use two different addresses for an argument's location at the same time, depending on which process is making the reference.

The base registers, which are a part of each control processor, along with some conventions and a little software will take care of this problem. If only one argument was involved, then we could use a base register to hold its address, and all references to it by the shared procedure would be made relative to the base register. This form of addressing relative to a base register is the normal form of addressing anyway, so that very little else would be needed to make it work for the argument of a shared procedure. However, a procedure may have an arbitrary number of arguments, and most have more than one. Hence, the only solution is to put the addresses of all the arguments in a block of memory and use one base register to point to this block.

The same considerations that lead us to the need for private addresses for the arguments of a shared procedure apply to all of the addresses used by the procedure: addresses of other data segments that are referenced using external symbols instead of having their addresses passed as arguments, addresses of other procedures that it calls, and even addresses which refer to itself. The reason for this is that each address contains a segment number, and there is no guarantee that any segment, even shared ones, will have the same segment number in all processes. Thus, the addresses used by a shared procedure will be different for every process sharing the procedure.

The addresses that a procedure uses are of two types: those that remain fixed throughout the life of the process and those that may change between one call to the procedure and the next. The latter set of addresses includes the arguments of the procedure and the address of the return point in the calling procedure. If our system is going to be capable of supporting recursive procedure calls, these addresses must be stored in a pushdown stack. The first set of addresses includes the addresses that correspond to external symbols, that is, the links. These addresses are stored in the procedure's linkage section (see Chapter 15). Obviously, each process must have its own private copy of the linkage section for each procedure in the process.

Figure 17.2 diagrams the relationship of a shared procedure to its linkage section and current stack frame. The convention used is that G_{13} contains the address of the base of the current stack frame, while G_{12} contains the address of the base of the linkage section for the currently executing procedure. We are also assuming that when a procedure is called, the calling procedure puts the address of its return point into the first word of the called procedure's stack frame and the addresses of the arguments in the following words of the stack frame. In the example, the first instruction in the shared procedure,

$$L \quad 3,8(,13)$$

loads the address of the second argument into G_3 in preparation for referencing its value. The second instruction,

$$L \quad 4,4(,12)$$

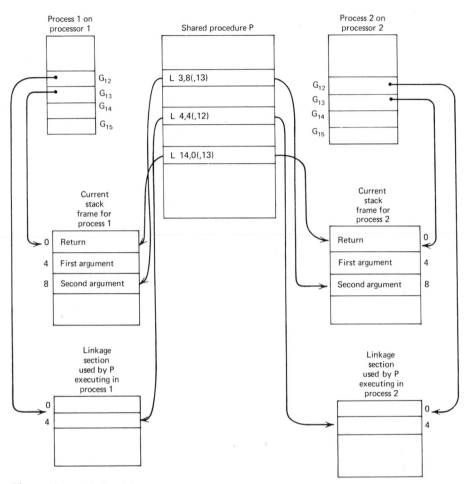

Figure 17.2 Relationship of a shared procedure to its stack frame and linkage section.

loads the value of the link that is the second word in the linkage section. This value is the address corresponding to some external symbol in the source program. It may be either the address of some datum or the entry point of some procedure. The third instruction,

$$L \quad 14,0(,13)$$

loads the address of the return point in the calling procedure in preparation for returning to the caller.

The conventions required to make shared procedures work properly are:

1. Every procedure must use the current stack frame or the linkage section for the storage of all address constants that it needs, depending on the type of the address constant.

2. Two registers must be reserved for the sole purpose of holding the base address of the linkage section and the current stack frame.

3. Every procedure must use a standard calling sequence that puts the address of its return point and the addresses of the arguments for the called procedure in a known place in the stack frame. Furthermore, the standard calling

sequence must update the contents of the two reserved registers so that they contain the addresses of the stack frame and the linkage section for the called procedure.

4. Every procedure must use a standard return sequence that restores the contents of the reserved registers so that they contain the addresses of the stack frame and the linkage section of the calling procedure.

5. None of the information in a procedure segment may be modified. This means that none of the instructions or constants may be modified. It also means that nothing except the instructions and constants that were originally part of a procedure may be stored in a procedure segment. Such a procedure is sometimes called "pure."

In addition to requiring that all procedures observe the above conventions, we require that the system provide a private stack for each process and a private copy of the linkage section for each procedure in the process. If all of these conditions are met, then every procedure is shareable. Furthermore, the decision as to whether a procedure is actually shared may be made after the procedure has been compiled or assembled. Since the conventions that are required are mostly needed for other reasons (e.g., the stack is needed for recursive calls), and they are also good programming practice, we see that no "special" coding is required to make a procedure shareable.

17.2 System Requirements for Protection

Certain requirements must be satisfied by the hardware and the software before protection can be achieved. An essential goal of protection is to prevent unauthorized access to a user's information, wherever it is. When in primary memory, the area that the information occupies must be made unaccessible to all processes other than the user's process. When in secondary memory, all other user's processes must be prevented from reading or writing that area of secondary memory. This implies two hardware requirements: subdivision of primary memory with respect to access and restriction on the use of certain of the computer's instructions.

The memory subdivision capability provided by the hardware must be such that the subdivisions can be changed. Not all user's programs are the same size; therefore, a fixed subdivision of primary memory would be awkward and generally inefficient with respect to space utilization. It should be clear that the segmentation hardware described in Chapter 15 provides the required memory subdivision capability. The block that contains a segment is a subdivision of memory. There is no way for a process to reference a location in primary memory unless it is part of a segment that is pointed to by a descriptor in the process's segment table. The set of memory subdivisions (segments) that is accessible by a process is identified by the entries in the process's segment table. We will call this set of segments the process's *domain of reference*. Since the descriptor base register in a control processor points to the segment table for the process that is executing on that processor, the domain of reference for an executing process is defined by its processor. Thus, each control processor in the system may be executing different processes with different domains of reference, as illustrated in Figure 17.3.

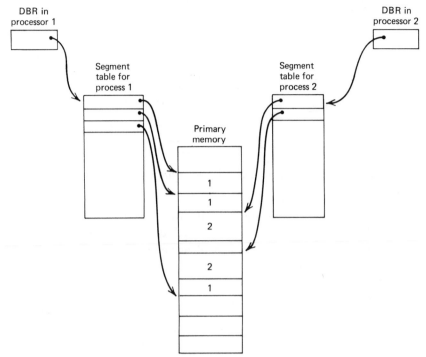

Figure 17.3 Memory partitioning for two processes.

Most modern computers, even those without any form of segmentation hardware, have some primitive mechanism for subdivision of primary memory. Memory can be subdivided easily into two parts by having a pair of registers in the control processor. One register in the pair will contain the address of the beginning of a user's area, and the second register will contain either the address of the end of the user's area or its length. The remainder of the memory is unaccessible to the process executing on the processor. This essentially provides each process with a domain of reference consisting of a single segment. Even though, from the viewpoint of a process, memory is divided into only two subdivisions, actually more than two subdivisions may be in effect. Since the pair of registers that define the domain of reference of an executing process are in the processor, each processor can have a different domain, as illustrated in Figure 17.4. It should be clear that with this type of memory subdivision hardware, the direct sharing of some data may be possible. However, in general, direct sharing of data is impossible. Any two processes's memory areas can overlap, thereby sharing the overlapping area. However, the memory areas for three processes cannot all overlap without at least one process being completely contained in one or both of the other processes.

The idea of a pair of registers is often extended to several pairs of registers or something equivalent. Having several pairs of registers is some help in getting away from the requirement that all of the user's information must be packed into one block of contiguous memory locations. However, the number of register pairs is usually small, and so the problem remains of packing all of the user's information into a small number of blocks of contiguous memory

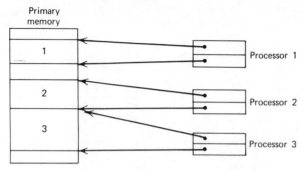

Figure 17.4 Single segment memory subdivision with three processors.

locations, which is basically not an easier problem. Limited direct sharing is possible when the domains of reference consist of more than one segment but, as long as the maximum number of segments in a domain is small, the potential benefits of direct sharing are severely crippled.

In the IBM 360 primary memory is subdivided into a number of fixed length blocks. Associated with each such block is a protection key whose value may be set by the system. Each executing process has an associated protection key value. The domain of access for the process consists of all those blocks of memory whose protection key value is equal to the protection key value associated with the process. This is equivalent to a set of registers that define a multisegment domain, except for the following two disadvantages. First, the blocks in the domain are all of fixed length, whereas the segments defined by a pair of registers can be of any length. Second, any reasonable form of direct sharing, even a limited form, is impossible. In order to share a block directly, both processes must have the same protection key value as the block. In addition, in order to reference the other private information that each process needs, all of the other blocks in each process's domain must have identical protection key values. Thus, each sharer of a shared block has complete access to all of the private information of all of the other sharers. This is clearly unacceptable.

If the primary memory subdivision mechanism is to be effective, users must be prevented from changing the subdivisions. They must also be prevented from directly reading or writing secondary memory. Thus, there must be a subdivision of the control processor's instruction set into two types: those that both the system and the users can execute and those that only the system can execute. These latter instructions are called the *privileged instructions*. In order to prevent users from executing privileged instructions the control processor must know whether it is executing a system or a user procedure.

The usual method for determining what kind of procedure the control processor is executing is to define two states in which it can execute; *user* and *supervisor* (called *slave* and *master* on some computers). The control processor can execute a privileged instruction only if it is in the supervisor state. Instructions or other mechanisms are then required for changing the state of the processor. The instruction for changing from supervisor to user state need not be a privileged instruction, since nothing will happen if it is executed when the processor is already in the user state. When a procedure is executing in the user state, it is normally prevented from changing the state of the processor. However, a user's procedure must be able to call the system when it needs some service from the system. When the system is called in the proper way, the state of the processor must be changed to supervisor state so that the system can perform properly. Normally, this change in state is effected automatically when a valid call to the system is executed by a user procedure. In addition, most interrupts result in a transfer of control to the system. In these cases the state of the processor must also be changed. With this kind of hardware, as long as the system always changes the processor state to user before transferring to a user program, protection can be made foolproof (provided the memory sub-division hardware is adequate).

In the IBM 360 all calls to the system must be done using the *supervisor call* instruction (SVC). When this instruction is executed, control is transferred to the system, and the processor's state is changed to supervisor. A different approach is taken in the Honeywell 645 and 6180 (the Multics computers). Since these computers have segmentation hardware, the segment table for a process contains a descriptor for each segment that the process can reference. The descriptor for each procedure segment contains a single bit, the *processor state bit*, which indicates whether the processor is to be in user or supervisor state when it is executing that procedure. Whenever control is transferred to any procedure segment, the state of the processor is automatically set according to the value of the processor state bit in the segment's descriptor.

As we will see later, there is additional mechanism that allows descriptors for both system procedure and system data segments to coexist in the same segment table along with the descriptors for the user's segments, yet user procedures cannot normally reference these system segments. The segment table is system data and the instruction for loading the descriptor base register is a privileged instruction. Thus, once started in user mode, there is no way for a user procedure to change the descriptors or change the location of the segment table. So there is no way that he can cause one of his procedures to execute in supervisor state. A significant advantage of this approach is that the system can be composed of many separate procedure segments, most of which can be set to execute in user state. Only those few system procedures that need to use the privileged instructions will be allowed to execute in supervisor state. This provides an extra measure of protection against accidental damage.

In order to use these hardware features to enforce protection the system must be able to identify each user. Only if the user is correctly identified can the system establish the proper domain of access and perform various functions on his behalf without violating the domains of the other users of the system. This implies two software requirements: the ability to validate the identity of a perspective user and controlled entry to the system procedures.

Validation of the identity of a perspective user is usually done by requiring the user to supply a secret password whenever he submits a job to the system or logs in from a terminal. This secret password is unique for each user and initially known only to him. If he does not tell his password to anyone, then no one else can successfully pretend that he is the user. A certain amount of physical security is required if the passwords are to be effective. If the user is signing on from a terminal, the system should turn off the printing mechanism during the time the user is typing his password; otherwise, it will be visible to anyone who happens to wander by. If the user is submitting a batch job on cards, the input card deck that defines his job should be submitted directly to computer center personnel at some kind of enclosed and guarded receiving station. The user password is included in the card deck. When the deck is submitted, the user should receive a receipt that he must present in order to receive the results of his job and his original input deck. For additional security the password card may be removed automatically by the system when the input deck is read (using a multistacking card reader) or manually removed by the computer center personnel.

Controlled entry to the system procedures is required if protection is to be guaranteed. If a user can transfer control to any arbitrary point in the system, then the system's behavior is unpredictable and thus not reliable. It might even be possible for a clever user to bypass the access control imposed by the system by transferring to the appropriate place in the system, a place that was never intended as a user entry point. In the IBM 360 the only instruction that the user can use to transfer to the system is the SVC instruction. Execution of this instruction always causes control to be transferred to a fixed location in the system. The R1 and R2 fields of the SVC instruction taken together contain a number that is an *interrupt code* corresponding to the system procedure that the user really wants to call. The state of the processor is switched to supervisor state when the transfer takes place. The system then examines the interrupt code of the SVC that caused the transfer. If it is a valid number corresponding to a procedure that users are allowed to call, the system passes the call on to that procedure. Controlled system entry when segmentation hardware is available is explored in the next section.

17.3 Segment Access Control

In order to discuss the control of access to segments in our model system we give a brief description of a slightly simplified version of the segmentation hardware of the Honeywell 6180 (the new Multics computer). Figure 17.5 shows the format and contents of a descriptor. All of the segments in a process are divided into two groups: system segments and user segments. The system bit in a segment's descriptor indicates the group to which the segment belongs. A system procedure segment may access any segment in the process, in either group. With one exception, a user procedure segment may not reference any segment unless it is a user segment, that is, a segment whose descriptor has its system bit equal to zero. The exception is a *gate segment*. Such a segment is a system segment with a nonzero G field. A valid system entry point is called a *gate*. Any system procedure may contain gates. If one does, then the G field in its descriptor indicates how many gates it has. The only restriction is that if a system segment has N gates, they must be the first N locations in the segment. Thus, if a user procedure tries to transfer to a system segment whose descriptor has a nonzero G field and the word number in the target address is less than the value of the G field, the transfer is permitted. If any of these conditions are not satisfied, then the transfer is not permitted.

The processor state bit controls only the use of the privileged instructions by dictating which state the control processor will be in when it is executing the corresponding procedure. Since the ability of a procedure segment to use priv-

S	G	PS	E	R	W	P	N	PT

S—system bit: 0 = user segment
        1 = system segment
G—number of gates
PS—processor state bit: 0 = segment executes in user state
        1 = segment executes in supervisor state
E—execute bit: 0 = segment cannot be executed
        1 = segment is procedure and can be executed
R—read bit: 0 = contents of segment cannot be read
        1 = contents of segment can be read
W—write bit: 0 = segment cannot be written into
        1 = segment can be written into
P—presence bit: 0 = page table does not exist
        1 = page table exists
N—length of segment
PT—address of page table

Figure 17.5 Format and contents of a descriptor.

ileged instructions is independent of its status as a system segment, system procedure segments will normally execute in user state, unless they need to use privileged instructions, thus decreasing the chances of accidental damage. In addition, the distinction between system and user segments allows all of the system procedures to be part of each process, thus greatly simplifying calls to the system and the referencing of user data by the system when it is carrying out a user request.

The E, R, and W fields control the type of access that is permitted to the corresponding segment. This access control is always effective, whatever the state of the control processor. Thus, not even a system procedure can normally write in a user's segment if the W field of that segment's descriptor is zero. Every memory reference by the control processor is checked against the access control bits in the target segment's descriptor. Any transfer-type instruction is blocked unless the execute bit is one, any fetch-type instruction is blocked unless the read bit is one, and any store-type instruction is blocked unless the write bit is one.

The hardware described above makes it possible to achieve our objectives of controlled sharing and protection in a flexible and efficient way. However, we need supporting software. In particular we need a way for users to specify which segments they wish to share, with whom they are to be shared, and the type of access that is to be permitted. This information needs to be retained by the system so that the user's wishes regarding access to each of his segments will be enforced until he explicitly changes the access. The segment manager must check this information when making a segment known to a process and construct the proper descriptor from it if access is permitted. In addition, we must identify the system segments and the number of gates in each of the system procedure segments.

The logical place for this information is in the directory entry for a segment. Recall that each segment has a directory entry somewhere in the hierarchy. Therefore, we extend the directory entries to contain the following information.

1. Name of the segment.
2. Length of the segment.
3. Type of segment: system or user.
4. Number of gates.
5. Execution state of the segment: supervisor or user.
6. Access control list.
7. Other information.

The *access control list* is a list of pairs of the form (user identification, access type). A user has no access at all to a segment unless his identification appears in one of the pairs in the access control list. If his identification does appear there, then he has access, but only of the type specified. This access may be any of the combinations: read, read-write, execute, read-execute, or read-write-execute (impure procedure). The owner of a segment is automatically entered on the access control list with read-write access when the segment is created. He may later explicitly change this to some other type of access for his own self-protection.

The entries of the segment manager in Chapter 15 that create and destroy segments must be modified so that a user cannot create a system segment or destroy any segment that does not belong to him. Only special individuals, such as the computation center's administrator, can be permitted to create or destroy system segments. His identification must be known to the system as having special privileges so that segment control will permit him to create or destroy system segments. Likewise, the entries that change a directory must be modified so that a user can only modify his own directories and then only make changes that do not violate security. For example, a user must not be able to change the type of any of his segments to system nor their execution state to supervisor.

A major addition to the directory manager is an entry that allows a user to add new entries to, or delete old entries from, the access control list of any of his segments. The segment manager also has to be modified to validate the access whenever it is called to make a segment known to a process. After the directory entry for the segment is found, the access control list in that entry must be searched for an access pair that contains the identification of the owner of the process in which the segment manager is currently executing. The identification of this user is found in the process control block for the process. A pointer to this process control block is found in the running list (see Chapter 14) at a location whose index is equal to the numbr of the processor that is executing the process. The number of a processor is always known to it. If a matching access control list entry is found the segment is made known to the process. If no matching entry is found, the segment manager returns to its caller, indicating that the requested segment could not be found. Notice that all the data needed for validating the request of this user and his process can be obtained from data bases whose contents can be easily and entirely controlled by the system.

17.4 Reliability

All of the protection features just described contribute to reliable operation of the system. There are two major additional functions that the system can perform to greatly improve the reliability of its operation: *backup* and *information repair*. Both of these contribute to recovery after a system failure. There is no possible way to guarantee that a system will never fail. So we prepare for this eventuality by providing means to help recover from a system failure. In actual practice the more reliable a system is, the more users assume it will not fail. Thus, the probability increases that they will be badly hurt when the system does fail. Therefore, the more reliable a system is, the greater the need for the system to take adequate steps to minimize each user's loss when the system fails. This is a complex problem, so we will only briefly indicate in what direction solutions lie.

Redundancy built into system data will help in both the early detection of failure and the repair of the damaged data. One form of redundancy is to include some form of parity or check code in each entry of a table or list. Periodic checking of this code permits early detection of failure. When properly selected check codes are used, it is often possible not only to tell what information is bad, but also to repair the damaged information, especially if the critical information is repeated elsewhere. Another example of where such redundancy can be introduced is the page map in a directory entry. Each word in the page map points to the location of a page in secondary memory. If the block in which the page is stored also contains a pointer to the page map entry for that page, we may be able to repair damage to either the page or the page map.

If damaged information cannot be repaired, it can at least be restored to its original value or to some recent previous value if the system automatically provides backup. To achieve automatic backup, the system must make a copy of each segment reasonably soon after it is modified. By reasonably soon we mean not so soon that the segment will be immediately modified again, but soon enough that the probability is extremely high that the copy is made before the system fails. A reasonable time to start making copies of a user's segments that have been modified is immediately after he has signed off after a work session (unless the work session was extremely long, in which case the copying might be started before he signs off). At this point the probability of additional modification of any of his segments in the near future is very low. Furthermore, any segments that were created for temporary use during the work session will have been destroyed, so we do not copy a lot of useless information.

To facilitate this automatic backup of segments, we must include additional information in each segment's directory entry. The information needed or useful is:

1. Date and time the segment was created.
2. Date and time the segment was last modified.
3. Date and time the segment was last copied for backup.
4. Location of last backup copy.

The backup copies of segments need to be put on some removable storage such as magnetic tape. In addition to a copy of the contents of a segment, a copy of all of the information in its directory entry (except the page map) must also be included on the backup tape. When the system fails, any information that cannot be repaired is restored by retrieving it from the backup tapes. This restoration procedure is very complex and time consuming. As time goes on, the backup tapes will accumulate many outdated copies of many of the segments. In order to keep the volume of information under control these outdated copies will have to be eliminated some way. One way to achieve this is to periodically dump the entire contents of the hierarchy. This complete dump contains a copy of all the currently existing segments and can serve as a base point for restoration of damaged information. Thus, all backup tapes created prior to the complete dump can be discarded.

EXERCISES

17.1 Modify the segment and page management of Chapter 15 so that it will work properly with shared segments. Shared procedure segments require their own private copy of the linkage section. How can this be accomplished?

17.2 Write two IL procedures that manage the access control list in a directory entry for a segment. The call

add(user_id, access, tree_name);

should add the pair (user_id,access) to the access control list for the segment whose tree name is the value of tree_name. The call

remove(user_id, tree_name);

should remove any entry for the user user_id from the access control list for the segment whose tree name is the value of tree_name.

17.3 Modify the segment management of Chapter 15 so that it enforces access control as described in this chapter.

Appendix A
Our 360 Machine Language

The machine language used for examples in this text is essentially a subset of the IBM 360. We call this subset *Our 360*. This appendix contains a brief description of Our 360 machine language. Appendix B contains a brief description of an assembly language based on Our 360.

1. The Memory

The basic addressible memory unit is 8 bits in size and is called a *byte*. A byte in memory is addressed by an integer i, where $0 \leqslant i < 2^{24}$. In this appendix we will refer to byte i of memory using the notation M_i. A group of 4 consecutively addressed bytes, the first of whose address is evenly divisible by 4, is called a *full word*. The 4 bytes in a full word are numbered 0, 1, 2, and 3 in the same order as the addresses of the bytes and are pictorially represented as

byte 0	byte 1	byte 2	byte 3

A full word is addressed using the address of byte 0. There are 32 bits in a full word. These bits are numbered 0, 1, ... , 31 in the same order as the addresses of the bytes and are pictorially represented as

```
┌──┬──┬──────────────────────┐
│  │  │                      │
└──┴──┴──────────────────────┘
 0  1                       31
```

In general, the leftmost bits (bytes), which have the lowest numbers, are called the *high-order bits (bytes)* and the rightmost bits (bytes), which have the highest numbers, are called the *low-order bits (bytes)*.

2. Data

There are three types of data. Each type is stored in a different format.

(i) Arithmetic data is stored in a full word as a 32-bit *signed integer*. Such an integer I is stored in the format:

```
┌──┬─────────────────────────┐
│ S│           N             │
└──┴─────────────────────────┘
 0  1                       31
```

where $0 \leqslant S \leqslant 1$ and $0 \leqslant N < 2^{31}$. If $I < 0$ then $S = 1$, otherwise $S = 0$. Bit 0 is called the *sign bit*. If $S = 0$ then $I = N$, otherwise $I = -(2^{31} - N)$. This is the *twos complement* representation for negative integers.

(ii) Logical data is stored in a full word. No special interpretation is placed on any of the bits in logical data. If interpreted as a 32-bit *unsigned integer* N, then $0 \leqslant N < 2^{32}$.

(iii) Character data is stored in a byte. A character is represented as an 8-bit unsigned integer:

N

$0 \leqslant N < 2^8$

0　　　　　7

The value of N for various characters is given in a table at the end of this appendix.

3. Registers

The control processor of Our 360 contains 16 general registers and one special purpose register called the *program status word* (PSW). The PSW is subdivided into several *fields*. Two of these fields are of interest in this appendix: the *instruction counter* (IC) and the *condition code* (CC).

(i) The general registers are each 4 bytes (32 bits) in size. The bytes and bits in a general register are numbered exactly like the bytes and bits in a full word. The general registers are used for operations on all types of data. They may also be used in making references to memory and in transfer of control. References to the general registers are written as $G_0, G_1, \ldots, G_{15}$.

(ii) The IC is 3 bytes (24 bits) in size. Its contents are interpreted as a 24-bit unsigned integer, which is the address of the next instruction to be executed.

(iii) The CC is 2 bits in size. Its contents are interpreted as a 2-bit unsigned integer, which indicates the result of the last comparison or computational instruction.

4. Instruction Types and Formats

There are four types of instructions, each having a different format. The following table gives the format and size of each type of instruction. Each instruction type is named by a single letter. The bits in each field are numbered from 0, starting with the leftmost bit in the field.

Instruction type	Size in bytes	Byte 0	Byte 1		Byte 2	Byte 3
R	2	OP	R1	R2		
X	4	OP	R	X	B	D
M	4	OP	R1	R2	B	D
S	4	OP	R	///////	B	D

0　　　　7　8　　11　12　　15　16　　19　20　　　　　31

The first part of interpretation of any instruction by the control processor is called the *instruction* fetch and consists of the following steps.

(i) The control processor reads (fetches) the 2 bytes $M_{IC} \ldots M_{IC+1}$. The first of these 2 bytes is the *operation code* (OP).

(ii) **if** OP is type R **then** IC : = IC+2;
        else begin Read 2 additional bytes $M_{IC+2} \ldots M_{IC+3}$;
        end;

5. Effective Addresses

For all instructions, except type R, one operand is specified by the *effective address* (EA). The control processor computes the value of EA from the contents of the D field of the instruction and the contents of certain general registers specified by the contents of the X and B fields of the instruction. The value of EA is a 24-bit unsigned integer. In the following definition of the value of EA, all arithmetic computation is assumed to be done modulo 24 bits, and all quantities involved are treated as nonnegative integers.

    if B=0 **then** base : = 0;
        else base : = G_B;
    if type of OP is X
        then if X=0 **then** index : = 0;
                        else index : = G_X;
        else index : = 0;
    EA : = D + base + index;

6. Description of Individual Instructions

This section contains a description of each individual instruction in Our 360 grouped according to the general function of the instruction. Each description has the format:

symbolic operation code	numeric operation code	instruction type	name of instruction

description of instruction's action

The numeric operation code is given in hexadecimal representation. Whenever an instruction references a full word, the value of EA must be evenly divisible by 4. If this is not true an interrupt occurs. The result of this interrupt is that the operating system gains control (see Appendix F).

In the following descriptions note that for many of the type X instructions there exists a companion type R instruction that performs the same operation. The only difference is the location of the second operand. The first operand of an instruction of either type is in a general register. The second operand of a type R instruction is also in a general register, while the second operand of a type X instruction is in primary memory.

A. DATA MOVEMENT INSTRUCTIONS

(i) *Load instructions*

IC   43 X insert character

The contents of byte 3 of G_R are replaced by the contents of M_{EA}.

L   58 X   load
G[R] := M[EA] ... M[EA + 3];
The contents of G_R are replaced by the contents of the full word
$M_{EA} ... M_{EA+3}$.

LR   18 R   load register
G[R1] := G[R2];
The contents of G_{R1} are replaced by the contents of G_{R2}.

LM   98 M   load multiple
G[R1] ... G[R2] := M[EA] ... M[EA + 4*K − 1];
The contents of the K general registers $G_{R1} ... G_{R2}$ are replaced by
the contents of the K full words $M_{EA} ... M_{EA+4*K-1}$. The value of K is
defined as
> **if** R1 ≤ R2
>> **then** K := R2 − R1 + 1;
>> **else** K := 16 − R1 + R2 + 1;

When R1 > R2 the ordering of the general registers is
$$G_{R1} ... G_{15}G_0 ... G_{R2}.$$

LA   41 X   load address
The contents of G_R are replaced by the value of EA (*not* the contents
of M_{EA}).

(ii) Store instructions

STC   42 X   store character
The contents of M_{EA} are replaced by the contents of byte 3 of G_R.

ST   50 X   store
M[EA] ... M[EA + 3] := G[R];

STM   90 M   store multiple
M[EA] ... M[EA + 4*K − 1] := G[R1] ... G[R2];
(See description of load multiple instruction in the preceding section
for additional comments.)

B. DATA MANIPULATION INSTRUCTIONS

(i) Arithmetic operations

A   5A X   add
G[R] := G[R] + M[EA] ... M[EA + 3];
The contents of the full word $M_{EA} ... M_{EA+3}$ are added to the contents
of G_R. After the addition the value of CC is set as follows:
> **if** G[R] = 0 **then** CC := 0;
> **if** G[R] < 0 **then** CC := 1;
> **if** G[R] > 0 **then** CC := 2;
> **if** overflow **then** CC := 3;

Overflow occurs when the result of the addition cannot be represented
by the 32 bits of G_R, that is, the result is either ≥ 2^{31} or < -2^{31}.
When overflow occurs, an interrupt may also occur (see Appendix F).

AR   1A R   add register
G[R1] := G[R1] + G [R2];
Set CC as described for the add instruction.

S   5B   X   subtract
         G[R] := G[R] − M[EA] ... M[EA + 3];
         Set CC as described for the add instruction.

SR   1B   R   subtract register
         G[R1] := G[R1] − G[R2];
         Set CC as described for the add instruction.

M   5C   X   multiply
         G[R] ... G[R + 1] := G[R + 1] *M[EA] ... M[EA + 3];
         The CC is unchanged. The result of the multiplication is a 64-bit
         signed integer. G_{R+1} contains the low-order 32 bits of the product, and
         G_R contains the high-order 32 bits. The value of R must be even.

MR   1C   R   multiply register
         G[R1] ... G[R1 + 1] := G[R1 + 1] *G[R2];
         (See description of the multiply instruction for additional comments.)

D   5D   X   divide
         G[R + 1] := G[R] ... G[R + 1] /M[EA] ... M[EA + 3];
         G[R] := remainder of the division;
         The CC is unchanged. The dividend is a 64-bit signed integer. The
         quotient and the remainder are 32-bit signed integers. G_{R+1} contains
         the quotient, and G_R contains the remainder. If the quotient cannot
         be represented by the 32 bits of G_{R+1}, no division takes place, and an
         interrupt may occur (see Appendix F). The value of R must be even.

DR   1D   R   divide register
         G[R1 + 1] := G[R1] ... G[R + 1] /G[R2];
         (See description of the divide instruction for additional comments.)

SLA   8B   S   shift left arithmetic
         i := remainder(EA/64);
         if i ⩾ 31 **then if** G[R] ⩾ 0 **then** G[R] := 0;
                                 else G[R] := -2^{31};
            else Left shift bits 1–31 of G[R] by i bits and fill bits vacated on
                  the right with 0;
         If a bit unlike the sign bit (bit 0) is shifted out of bit 1, an interrupt
         may occur (see Appendix F). If i < 31 the effect is equivalent to mul-
         tiplying G_R by 2^i.

SLDA   8F   S   shift left double arithmetic
         i := remainder(EA/64);
         Left shift bits 1–31 of G[R] and all 32 bits of G[R + 1] by i bits and
            fill bits vacated on the right with 0;
         Bits are shifted from bit 0 of G_{R+1} into bit 31 of G_R. If a bit unlike
         the sign bit (bit 0) of G_R is shifted out of bit 1 of G_R, an interrupt
         may occur (see Apepndix F). If i < 63 the effect is equivalent to mul-
         tiplying G_R ... G_{R+1} by 2^i. The value of R must be even.

SRA   8A   S   shift right arithmetic
         i := remainder(EA/64);
         if i ⩾ 31 **then if** G[R] ⩾ 0 **then** G[R] := 0;
                                 else G[R] := −1

else Right shift bits 1–31 of G[R] by i bits and fill bits vacated on the left with copies of the sign bit (bit 0);
If $i < 31$ the effect is equivalent to dividing G_R by 2^i.

SRDA   8E   S   shift right double arithmetic
i : = remainder(EA/64);
Right shift bits 1–31 of G[R] and all 32 bits of G[R+1] by i bits and fill bits vacated on the left with copies of the sign bit (bit 0) of G[R];
Bits are shifted from bit 31 of G_R into bit 0 of G_{R+1}. If $i < 63$ the effect is equivalent to dividing $G_R \ldots G_{R+1}$ by 2^i. The value of R must be even.

(ii) Logical operations

N   54   X   and
G[R] : = G[R] and M[EA] ... M[EA+3];
The contents of the full word $M_{EA} \ldots M_{EA+3}$ is bitwise logically **anded** to the contents of G_R. After the **anding** the value of CC is set as follows:

$$\text{if } G[R] = 0 \text{ then } CC := 0;$$
$$\text{else } CC := 1;$$

NR   14   R   and register
G[R1] : = G[R1] and G[R2];
Set CC as described for the and instruction.

O   56   X   or
G[R] : = G[R] or M[EA] ... M[EA+3];
Set CC as described for the and instruction.

OR   16   R   or register
G[R1] : = G[R1] or G[R2];
Set CC as described for the and instruction.

SLL   89   S   shift left logical
i : = remainder(EA/64);
if $i \geqslant 32$ **then** G[R] : = 0;
    else Left shift the contents of G[R] by i bits and fill bits vacated on the right with 0;

SLDL   8D   S   shift left double logical
i : = remainder(EA/64);
Left shift the contents of G[R] ... G[R+1] by i bits and fill bits vacated on the right with 0;
Bits are shifted from bit 0 of G_{R+1} into bit 31 of G_R. The value of R must be even.

SRL   88   S   shift right logical
i : = remainder(EA/64);
if $i \geqslant 32$ **then** G[R] : = 0;
    else Right shift the contents of G[R] by i bits and fill bits vacated on the left with 0;

SRDL   8C S shift right double logical
i : = remainder(EA/64);
Right shift the contents of G[R] ... G[R+1] by i bits and fill bits vacated on the left with 0;
Bits are shifted from bit 31 of G_R into bit 0 of G_{R+1}. The value of R must be even.

c. COMPARISON INSTRUCTIONS

(i) Arithmetic

C   59 X compare
if G[R] = M[EA] ... M[EA+3] then CC : = 0;
if G[R] < M[EA] ... M[EA+3] then CC : = 1;
  else CC : = 2;
The two operands are treated as 32-bit signed integers.

CR   19 R compare register
if G[R1] = G[R2] then CC : = 0;
if G[R1] < G[R2] then CC : = 1;
  else CC : = 2;
The two operands are treated as 32-bit signed integers.

(ii) Logical

CL   55 X compare logical
if G[R] = M[EA] ... M[EA+3] then CC : = 0;
if G[R] < M[EA] ... M[EA+3] then CC : = 1;
  else CC : = 2;
The two operands are treated as 32-bit signed integers.

CLR   15 R compare logical register
if G[R1] = G[R2] then CC : = 0;
if G[R1] < G[R2] then CC : = 1;
  else CC : = 2;
The two operands are treated as 32-bit signed integers.

d. CONTROL INSTRUCTIONS

BC   47 X branch on condition
i : = CC;
if bit i of R = 1 then IC : = EA;
Each bit in the R field of the instruction whose value equals 1 selects a condition. If any one of these conditions is true the contents of IC are replaced by the value of EA and a transfer of control occurs in the executing program.

BCR   07 R branch on condition register
i : = CC;
if bit i of R1 = 1 then IC : = bits 8 ... 31 of G[R2];

BAL   45 X branch and link
G[R] : = IC;
IC : = EA;

The instruction fetch occurs before the contents of G_R are replaced by the contents of IC, therefore, the value of IC will be the address of the byte immediately following the low-order byte of this instruction.

BALR   05   R   branch and link register
temp : = bits 8 . . . 31 of G[R2];
G[R1] : = IC;
if R2 $\neq$ 0 then IC : = temp;

SVC   0A   R   supervisor call
An interrupt occurs and control is transferred to the operating system (see Appendix F). The values of the R1 and R2 fields in the instruction are concatenated and made available to the system as an 8-bit *interrupt code*.

7. USASCII Character Codes

When manipulating character data, each character is represented by an 8-bit unsigned integer *code*. The following table gives the codes for the commonly used characters. The coding used is the USA Standard Code for Information Interchange (USASCII). Unfortunately, even though this coding is an official "standard," it is not used in many computers. In the table the codes are expressed in hexadecimal representation. Thus, each code is expressible by two hexadecimal digits. In the table the row in which the character appears determines the first hexadecimal digit and the column determines the second hexadecimal digit.

Second Digit

First Digit	0	1	2	3	4	5	6	7	8	9	A	B	C	D	E	F
0																
1																
2	SP	!	"	#	$	%	&	'	(	)	*	+	,	—	.	/
3	0	1	2	3	4	5	6	7	8	9	:	;	<	=	>	?
4	@	A	B	C	D	E	F	G	H	I	J	K	L	M	N	O
5	P	Q	R	S	T	U	V	W	X	Y	Z	[	\	]	^	—
6	`	a	b	c	d	e	f	g	h	i	j	k	l	m	n	o
7	p	q	r	s	t	u	v	w	x	y	z	{	|	}	~	
8																
9																
A																
B																
C																
D																
E																
F																

SP means (blank) space character.
 _ is the underline.
 — is the minus (hyphen).

Appendix B
Our 360 Assembly Language

This appendix is a brief description of the assembly language used in this text. It is basically a subset of the assembly language used in the IBM OS/360 and is compatible with the machine language for Our 360.

1. Structure of an Assembly Language Source Segment

An assembly language source segment consists of a sequence of *symbolic instructions* and *comments* that begins with a START instruction and terminates with an END instruction. Each comment or symbolic instruction is punched on a single card (or typed on a single line if using a typewriterlike device).

A comment has the form:

where *comment* is any string of characters. (When using cards, columns 73 to 80 are normally used for identification purposes.) Comments are ignored by the assembler except for printing them in the output listing.

A symbolic instruction has the form:

where *label* is either a single symbol or omitted, (SP) is one or more (blank) spaces, *operation* is a single symbol that must be a valid symbolic operation code, *operand* is composed of one or more symbolic expressions, and *comment* is any string of characters. There are two kinds of symbolic instructions: *symbolic machine instructions* and *pseudo instructions*. A symbolic machine instruction represents exactly one machine language instruction. A pseudo instruction may represent one or more data items, or may simply give information to the assembler without representing either instructions or data.

Corresponding to an assembly language source segment, the assembler generates a single object segment. The symbol appearing in the label field of the START instruction will be the name of the object segment. All of the machine language instructions and data that are specified by symbolic instructions in the source segment will be part of the object segment.

2. Symbolic Expressions

Symbolic expressions are used in the operand field of symbolic instructions to define values. In particular the values of the fields in the machine language instruction corresponding to a symbolic machine instruction are defined by symbolic expressions.

A. SYMBOLS AND THEIR DEFINITION

Symbols are one of the major components of symbolic expressions. A *symbol* is composed of from one to eight letters and/or digits, the first of which must be a letter. In a valid assembly language procedure each symbol that appears must be defined. When a symbol is defined, it has a value, V, where $0 \leqslant V < 2^{24}$. A symbol is defined if and only if it appears in the label field of exactly one symbolic machine instruction, EQU pseudo instruction, DC pseudo instruction, or DS pseudo instruction.

The value of a symbol appearing in the label field of a symbolic machine language instruction is the relative address of the first byte of the corresponding machine language instruction. The *relative address* of a byte in a segment is the location of that byte relative to the beginning of the segment. The value of a symbol appearing in the label field of a DC instruction is the relative address of the first byte of the corresponding data. The value of a symbol appearing in the label field of a DS instruction is the relative address of the first byte of the corresponding reserved space. The value of a symbol appearing in the label field of an EQU instruction is the value of the (symbolic) expression in the operand field of that instruction. This value will be a relative address if the value of the expression is a relative address. It may also be simply a value that has no special interpretation and is called an *absolute value*.

B. LITERALS AND CONSTANT DEFINITIONS

A constant definition has the form

$$t'constant'$$

where t is a letter that specifies the type of *constant*. There are four possible choices for t: "F", "C", "X", and "A".

(i) F

In this case *constant* is a string of decimal digits optionally preceded by a minus sign. The corresponding constant will occupy a full word of memory and will be stored as a twos complement binary integer. Its value, I, must be such that $-2^{31} \leqslant I < 2^{31}$.

(ii) C

In this case *constant* is a string of characters that does not include any unpaired quote characters ('). Each (consecutive) pair of quote characters ('')

in *constant* will be interpreted as a single quote character ('), which will be part of the corresponding constant. All other characters in *constant* will be included in the corresponding constant. If there are N characters in the corresponding constant, they will be stored in N consecutive bytes of memory and represented using the USASCII codes in Appendix A.

(iii) X

In this case *constant* is a string of hexadecimal digits. If there are N hexadecimal digits in *constant*, the corresponding constant will occupy ceiling(N/2) consecutive bytes of memory. Two hexadecimal digits will be stored in each byte, unless N is odd. If N is odd the right half of the last byte will be zero.

(iv) A

In this case *constant* is a symbol whose value is a relative address. When the assembled procedure executes, the corresponding constant will be the absolute address that corresponds to the relative address that is the value of the symbol.
A *literal* has the form

$$= constant\_definition$$

where *constant_definition* is a constant definition as defined in the preceding paragraphs. The assembler collects the constants corresponding to all of the constant definitions that appear in literals. These constants, with all duplicates eliminated, are appended to the end of the assembled procedure. A literal functions as a special symbol whose value is the relative address of the constant corresponding to the constant definition in the literal.

C. Form and Value of Symbolic Expressions

A *symbolic expression* is composed of symbols, decimal integers, and operators. A *decimal integer* is a string of one or more decimal digits. Its value, I, must be $0 \leqslant I < 2^{12}$. The permissible *operators* are "+" (addition), "−" (subtraction), "*" (multiplication), and "/" (division).

An *expression* is either a single literal or a sequence of one or more terms separated by the operators "+" and "−". A *term* is a sequence of one or more primaries separated by the operators "*" and "/". A *primary* is either a symbol (not a literal), a decimal integer, or the special symbol "*".

If an expression is a literal, the value of the expression is the value of the literal, that is, the relative address of the constant corresponding to the constant definition in the literal. Otherwise, the value of an expression is the sum and difference of the values of the terms in the expression, that is, the value of the expression

$$t_1 \pm t_2 \pm \ldots \pm t_n$$

is

$$\text{value}(t_1) \pm \text{value}(t_2) \pm \ldots \pm \text{value}(t_n)$$

The value of a term is the product and quotient of the values of the primaries in the term, that is, the value of

$$p_1 \, *_/ \, p_2 \, *_/ \ldots *_/ \, p_n$$

is

$$\text{value}(p_1) \, *_/ \, \text{value}(p_2) \, *_/ \ldots *_/ \, \text{value}(p_n)$$

If a primary is a symbol its value is just the value of the symbol. If it is a decimal integer, its value is the value represented by the decimal integer. If it is the special symbol "*", its value is the current value of the assembler's location counter, that is, the relative address of the instruction in which it appears, when it appears in a symbolic machine instruction, and the relative address of the next instruction, when it appears in an EQU pseudo instruction.

The value of an expression is always a nonnegative integer. This value may be a relative address, an absolute value, or invalid. The value of an expression is a relative address only if it is less than 2^{12} and the terms in the expression can be rearranged to have the form

$$address\_symbol \pm absolute\_expression$$

where *address_symbol* is a single symbol whose value is a relative address, and *absolute_expression* is an expression whose value is an absolute value. The only rearrangements of terms that are permitted are those that result from an arbitrary number of applications of the associative law for addition and subtraction, that is,

$$A * B - C + D$$

may be rearranged to any of the forms:

$$A * B + D - C$$
$$D - C + A * B$$
$$D + A * B - C$$

but not, for example, to any of the forms:

$$C - A * B + D$$
$$A * B - D + C$$
$$- C + A * B + D$$

The value of any literal is always a relative address. The value of the special symbol '*' is always a relative address.

The value of an expression is always absolute if it does not contain any symbols whose value is a relative address. The value of an expression that contains symbols whose values are relative addresses will be absolute if all of these symbols cancel each other. Cancellation occurs only if each such symbol is a term by itself and the expression can be rearranged (using only the associative law) to have the form

$$s_1 - s_2 + s_3 - s_4 + \ldots + s_{n-1} + s_n + non\_address\_expression$$

where each s_i is a symbol whose value is a relative address, *non_address_expression* contains no symbols whose value is a relative address, and all of the s_i occur in consecutive pairs having the form

$$s_{k-1} - s_k$$

where k is an even integer. For example, if the values of A, B, C, and D are relative addresses and the values of X and Y are absolute, the value of

$$A + X - B - C + Y + D$$

is absolute, since it can be rearranged to

$$A - B + D - C + X + Y$$

The value of a null expression is either a relative address or absolute, depending on the context. All other expressions whose value is neither a relative address nor an absolute value according to the above definitions have an invalid value. This includes expressions such as

$$A * B$$

where either one or both of the symbols A and B have values that are relative addresses.

3. Symbolic Machine Instructions

A symbolic machine instruction corresponds to a single machine language instruction. The symbol in the operation field of the symbolic machine instruction, called the *symbolic operation code*, defines the value of the numeric operation code in the corresponding machine instruction. The value of all the remaining fields in the corresponding machine instruction are defined by the expressions in the operand field of the symbolic machine instruction.

A. Definition of Machine Instruction Field Values

The format of a symbolic machine instruction's operand field depends on the type of the corresponding machine instruction. In Appendix A we distinguish four types: R, X, M, and S.

(i) R

The format of this type of machine instruction is

OP	R1	R2

0   7 8   11 12   15

The operand field of a corresponding symbolic machine instruction has the form

$$e_{R1} , e_{R2}$$

where both e_{R1} and e_{R2} are expressions whose value must be absolute and less than 2^4. The value of the R1 (R2) field in the corresponding machine instruction will be equal the value of the expression e_{R1} (e_{R2}).

(ii) X

The format of this type of machine instruction is

OP	R	X	B	D

0   7 8   11 12   15 16   19 20   31

The operand field of a corresponding symbolic machine instruction has one of the forms:

$$e_R , e_D (e_X , e_B)$$
$$e_R , e_D (e_X)$$
$$e_R , e_D$$

The expressions e_R, e_X, and e_B must have absolute values less than 2^4. The values of the R, X, and B fields in the corresponding machine instruction will be equal to the values of e_R, e_X, and e_B, respectively. The value of the expression e_D must be a relative address. In the third form, which omits e_X, the value of the X field in the corresponding machine instruction will be zero. In the second and third forms, which omit e_B, the value of the B field will be equal to the number of the default base register as defined by the USING pseudo instruction (see Section 4c following).

(iii) M

The format of this type of machine instruction is

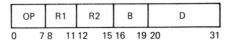

0   7 8   11 12   15 16   19 20   31

The operand field of a corresponding symbolic machine instruction has one of the forms:

$$e_{R1}, e_{R2}, e_D (e_B)$$
$$e_{R1}, e_{R2}, e_D$$

The expressions e_{R1}, e_{R2}, and e_B must have absolute values less than 2^4. The values of the R1, R2, and B fields in the corresponding machine instruction will be equal to these values. The value of the expression e_D must be a relative address, which will be the value of the D field in the corresponding machine instruction. In the second form, which omits e_B, the value of the B field will be equal to the number of the default base register as defined by the USING pseudo instruction.

(iv) S

The format of this type of machine instruction is

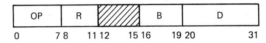

0   7 8   11 12   15 16   19 20   31

The operand field of a corresponding symbolic machine instruction has one of the forms:

$$e_R, e_D (e_B)$$
$$e_R, e_D$$

The expressions e_R and e_B must have absolute values less than 2^4, and the expression e_D must have an absolute value less than 2^{12}. The values of the R, B, and D fields in the corresponding machine instruction will be equal to these values. In the second form, which omits e_B, the value of the B field will be equal to the number of the default base register as defined by the USING pseudo instruction.

B. EXTENDED OPERATION CODES

An extended operation code is equivalent to a branch instruction (BC or BCR) with a particular mask value. The extended operation codes are defined in the following table.

Name	Extended operation code	Implied mask
Branch on high	BH	2
Branch on low	BL	4
Branch on not equal	BNE	7
Branch on equal	BE	8
Branch on not low	BNL	11
Branch on not high	BNH	13
Branch	B	15
Branch register	BR	15

The last entry corresponds to a BCR instruction, and all of the other entries correspond to BC instructions. The names are chosen to be meaningful, assuming that the extended operation follows immediately after a compare instruction (C). For example, the sequence of instructions

```
L     2,X
C     2,Y
BNL   RES
```

will branch to RES if $X \geqslant Y$, that is, if X is not lower than Y.

4. Pseudo Instructions

Pseudo instructions are used to define symbols, define constants, reserve space in the object segment, and provide the assembler with other information.

A. START AND END

These pseudo instructions must be the first and last symbolic instructions in an assembly language source segment. The START pseudo instruction has the format

$$seg\_name \quad \text{START}$$

The symbol in the label field, *seg_name*, will be the name of the object segment that results from assembling the assembly language source segment. The END pseudo instruction has the format

$$\text{END}$$

B. EQU

This pseudo instruction is used to define the value of a symbol. Its format is

$$symbol \quad \text{EQU} \quad expression$$

The value of the symbol in the label field is defined to be equal to the value of the expression in the operand field.

C. USING

This pseudo instruction is used to specify the number of the default base register. Its format is

$$\text{USING} \quad register\_nr$$

General register *register_nr* will be used as the default base register. This assignment is in effect for all succeeding symbolic machine instructions unless superseded by another USING pseudo instruction. A USING pseudo instruction does *not* load any value into the default base register. The assembly language source segment will have to include an additional instruction to load the base address of the procedure segment into the default base register, unless this address is already in the register when the procedure is entered.

D. DC AND DS

These pseudo instructions are used to define symbols, define constants, and reserve memory space. The format of a DS pseudo instruction is

$$symbol \quad \text{DS} \quad nt, \ldots ,nt$$

where each n is either an unsigned decimal integer or omitted, each t is one of the data type specifiers "F", "C", "X", or "A", and *symbol* is an optional label. The format of a DC pseudo instruction is

$$symbol \quad \text{DC} \quad nt'constant', \ldots ,nt'constant'$$

where n, t, and *symbol* are the same as in a DS pseudo instruction and each *constant* is a constant that is consistent with the immediately preceding *t*.

In both pseudo instructions each *nt* implies a certain number of bytes of memory, *s*, as defined by the following formulas. The integer *n* (if *n* is omitted then by convention $n = 1$) is the *replication count* and specifies *n* replications of the basic space unit, *k*, defined in the following table.

t	Basic space unit for *nt'constant'*	Basic space unit for just *nt*
F	$k = 4$	$k = 4$
C	$k = $ number of characters in *constant*	$k = 1$
X	$k = $ number of hexadecimal digits in *constant*	$k = 1$
A	$k = 4$	$k = 4$

The total number of bytes corresponding to a single occurrence of *nt* or *nt'constant'* is:

t	Total space
F	$s = n*k$
C	$s = n*k$
X	$s = \text{ceiling}((n*k)/2)$
A	$s = n*k$

A DS pseudo instruction causes space to be reserved in the object segment as specified by each occurrence of an *nt*, in the sequence in which they occur. The contents of this space after loading the object segment is undefined, that is, the values stored there are unpredictable. A DC pseudo instruction causes space to be reserved in the object segment as specified by each occurrence of an *nt'constant'*, in the sequence in which they occur. In addition, for each *nt'constant'* the corresponding space will be initialized with *n* replications of the constant corresponding to *consant*.

The space in a segment corresponding to the operand subfields of a single DS or DC pseudo instruction may not be contiguous due to the alignment requirement implied by the various *t*'s in the subfields. The space corresponding to an *nt* or *nt'constant'* must begin at an address evenly divisible by 4 if *t* is "F" or "A". If *t* is "C" or "X" the corresponding space may begin at any address. The space corresponding to a subfield will be contiguous to the space corresponding to the preceding subfield, if possible. If not, the minimum number of bytes required to achieve proper alignment will be skipped over. The contents of these skipped bytes after loading the object segment are undefined. Bytes may also be skipped preceding the space corresponding to the first subfield in order to achieve the required alignment. This will depend on the memory requirements of the preceding symbolic instructions.

If a symbol appears in the label field of a DS or DC pseudo instruction, that symbol is defined to have a value equal to the relative address of the first byte of space corresponding to the first operand subfield. This address is determined *after* any skipping of bytes required for the alignment implied by the first operand subfield.

Appendix C
Definition of the Instran Language

The Instran Language is an approximate subset of the PL/1 language as defined in the IBM publication, "IBM Systems/360, PL/1 Reference Manual," Form Number C28-8201. The definition of the semantics of those PL/1 features included in the Instran Language is basically the same as the definition of the corresponding features in PL/1. The following paragraphs specify the Instran Language by formally defining its syntax. Those restrictions that are not expressible in the syntax specification are described in the notes following the productions.

In the productions, which specify the syntax, the following notational conventions are used. Nonterminals (names of syntactic classes) are written in lowercase letters. The brackets "{ }" are used to enclose a sequence of one or more items, all of which must occur exactly once. The brackets "[ ]" are used to enlcose a sequence of one or more optional items, that is, they all occur exactly once or they are all omitted. The three large dots "•••" are used to indicate that the preceding item or bracketed sequence of items may be repeated an indefinite number of times in succession. For example,

$$item \bullet \bullet \bullet \quad \text{and} \quad \{item\} \bullet \bullet \bullet$$

Both indicate one or more occurrences of *item*, while

$$[item\text{-}1\ item\text{-}2] \bullet \bullet \bullet$$

indicates zero or more occurrences of the sequence

$$item\text{-}1\ item\text{-}2$$

The productions have the form

$$nonterminal ::= alternate\text{-}1 \mid alternate\text{-}2 \mid \ldots \mid alternate\text{-}n$$

where *n* is greater than zero. An alternate may be any sequence of terminals, nonterminals, and bracketed sets of alternates.

The Productions

The sentence symbol for the Instran Language is *procedure*. The productions are organized into groups. The productions in each group are related to each other.

procedure :: = entry-name : PROCEDURE [(dummy-argument-list)];
    head body END;
dummy-argument-list :: = identifier [,identifier] • • •
head :: = [declaration] • • •
body :: = { [label :] statement } • • •
entry-name :: = identifier
label :: = identifier

declaration :: = DECLARE { simple-variable | structure } ;
simple-variable :: = identifier [dimension] type [storage]
dimension :: = (integer)
type :: = FIXED | character | bit
character :: = CHARACTER (integer)
bit :: = BIT (integer)
storage :: = EXTERNAL | INTERNAL

structure :: = 1 identifier [dimension] [storage] { ,element} • • •
element :: = integer { scalar | substructure }
scalar :: = identifier type
substructure :: = identifier { ,element } • • •

statement :: = group | assignment | go-to | if | return | call | null
group :: = DO ; body END ;
assignment :: = variable = expression ;
go-to :: = GOTO label;
if :: = IF b-expression THEN true-statement [ELSE false-statement]
return :: = RETURN;
call :: = CALL identifier [(argument-list)] ;
null :: = ;
true-statement :: = statement
false-statement :: = statement

argument-list :: = expression [, expression] • • •

expression :: = a-expression | s-expression | b-expression
a-expression :: = a-expression { + | − } a-term | a-term
a-term :: = a-term { * | / } a-primary | a-primary
a-primary :: = a-variable | a-constant | (a-expression) | − a-primary
a-variable :: = variable
a-constant :: = integer

s-expression :: = s-expression cat s-primary | s-primary
s-primary :: = s-variable | s-constant | (s-expression)
s-variable :: = variable
s-constant :: = bit-constant | character-constant
cat :: = ||

b-expression :: = b-expression or b-term | b-term
b-term :: = b-term & b-primary | b-primary
b-primary :: = b-variable | b-constant | (b-expression) | ¬ b-primary | relation
b-variable :: = variable
b-constant :: = bit-constant
relation :: = a-expression { = | ¬ = | < | < = | > | > = } a-expression
or :: = |

variable :: = identifier [.identifier] • • • [(a-expression)]
bit-constant :: = ' { 0 | 1 } • • • 'B
character-constant :: = ' [data-character] • • • '
integer :: = digit • • •
identifier :: = letter [letter | digit] • • •

digit :: = 0 | 1 | 2 | 3 | 4 | 5 | 6 | 7 | 8 | 9
letter :: = A | B | C | D | E | F | G | H | I | J | K | L | M | N | O | P | Q | R | S |
          T | U | V | W | X | Y | Z | _
data-character :: = any-character-except-quote-mark | ' '

constant :: = integer | bit-constant | character-constant
operator :: = + | − | * | / | = | ¬ = | > | > = | < | < = | & | or | cat
punctuation :: = (|) | , | : | ; | ' | .
key-word :: = DECLARE | FIXED | CHARACTER | BIT | EXTERNAL |
             INTERNAL | GOTO | IF | THEN | ELSE | RETURN |
             PROCEDURE | END | DO | CALL
comment :: = /* [any-character-string-not-containing-*/] */
string :: = CHARACTER | BIT
type-word :: = FIXED | CHARACTER | BIT
commas :: = , | ;
any :: = any-terminal-or-nonterminal

Restrictions and Notes

1. Integer constants are 10 or fewer digits and their value must be non-negative and less than 2^{31}.
2. Character or bit constants are 4095 or fewer characters.
3. Identifiers are 12 or fewer characters.
4. The underscore "_", which is called the break character, is considered to be a letter.
5. The upper bound of an array must be less than 4096 (the lower bound is always zero).
6. The integers preceding the elements in a structure declaration define the levels of the structure. All elements at the same level must have the same level number. Each inner-level element must have a level number exactly one greater than its immediately enclosing level.
7. The space character is not permitted interior to an identifier, integer, bit constant, or key word.
8. One or more spaces are permitted between any consecutive pair of components in any of the productions, except as restricted by 7 above.

9. One or more spaces are required between any consecutive pair of the following items: identifier, key word, and constant.

10. A comment may occur anywhere a space may occur and serves the same delimiting function that a space does.

11. Key words are reserved and may not be used for any other purpose.

12. The pair of characters '' is interpreted as a single ' when appearing as a data character in a character constant.

13. The last group of productions defines nonterminals that do not appear in any other productions. They are defined so that the names of these syntactic classes may be used in the text.

14. An a-variable is a variable that was declared with data type FIXED.

15. An s-variable is a variable that was declared with data type CHARACTER or BIT.

16. A b-variable is a variable that was declared with data type BIT.

Appendix D
Instran Tables

The major tables used by Instran are described in this appendix. For each table the form and contents of an entry in the table are described and an Instran declaration is given for the table. The permissible values are often given in comments embedded in the declaration. The letters enclosed in parentheses following the table name are the code letters used in tokens which reference entries in the table.

1. Identifier Table (I,K)

Contains one entry for each unique identifier that appears in the source procedure. In addition, the first 14 entries contain the key words of the Instran Language.

DECLARE IDENT (100) CHARACTER (12);

2. Constant Table (C)

Contains one entry for each distinct constant appearing in the source procedure.

```
DECLARE 1 CONSTANT (30),
            2 VALUE FIXED, /* value or pointer to value */
            2 TYPE CHARACTER (1), /* = 'F' if integer
                                        'C' if character string
                                        'B' if bit string
                                        'A' if address constant */
            2 LENGTH FIXED, /* length of string */
            2 ADDRESS FIXED; /* offset from base of constant block */
```

If the entry is for an integer, VALUE is the actual value of the integer. If the entry is for a string, VALUE is a pointer to the actual value. For an address constant this value is the index of the symbol's identifier in the identifier table. Character string values are stored in C_STRINGS and bit strings are stored in B_STRINGS. The pointer is the index in C_STRINGS or B_STRINGS of the first character or bit of the value. The first entry of the constant table (index = 0) is preset to contain the integer zero, while the second entry is preset to contain the integer one.

3. Character String Constants

 Storage area for character string constants.
 DECLARE C_STRINGS (200);
The individual character string constants are stored contiguously in C_STRINGS as substrings.

4. Bit String Constants

 Storage area for bit string constants.
 DECLARE B_STRINGS (200);
The individual bit string constants are stored contiguously in B_STRINGS as substrings.

5. Punctuation-Operator Table (P)

 Contains an entry for each operator and punctuation in the Instran Language.
 DECLARE 1 PUNCT (20),
     2 NAME CHARACTER (2), /* character or pair of characters
       used for the terminal, e.g., < = */
     2 OPERATION FIXED; /* index in operation table of
       corresponding macro operation */

6. Symbol Table (S)

 Contains one entry for each simple variable, array, structure, label, and structure element that appears in the source procedure.
 DECLARE 1 SYMBOL (100),
     2 NAME FIXED, /* index in identifier table of name */
     2 TYPE CHARACTER (1), /* = 'F' if fixed
         'C' if character
         'B' if bit
         'L' if label
         'S' if structure
         'U' if undefined */
     2 LENGTH FIXED, /* length of string */
     2 STORAGE CHARACTER (1), /* = 'I' if internal
          'T' if text
          'E' if external
          'D' if dummy argument
          'S' if structure element
          'U' if undefined */
     2 ARRAY BIT (1), /* = 0 if scalar
         1 if array */
     2 UPPER FIXED, /* upper bound for subscript */
     2 PARENT FIXED, /* index in symbol table for parent of this
       structure element */
     2 BROTHER FIXED, /* index in symbol table of next

structure element with same level number as this structure element */
2 SON FIXED, /* index in symbol table of first element of this structure or substructure */
2 DEFINED CHARACTER (1), /* = 'D' if defined
'M' if multiple definitions
'U' if undefined */
2 ADDRESS FIXED; /* offset from base of appropriate block or position in argument list if a dummy argument */

The offset ADDRESS for labels is relative to the text block. For internal variables the offset is relative to the internal storage block. For external variables the offset is the offset of the corresponding address constant and is relative to the constant block.

7. Stack

Each entry in the stack is a token.
DECLARE 1 STACK (100),
2 TYPE CHARACTER (1), /* token type,
= 'K' if key word
'I' if identifier
'C' if constant
'P' if punctuation-operator table
'S' if symbol table
'N' if node list */
2 INDEX FIXED; /* index of entry in table indicated by TYPE field */

8. Phrase Table

Each entry is a token of the phrase most recently matched by a reduction rule.
DECLARE 1 PHRASE (10),
2 TYPE CHARACTER (1),
2 INDEX FIXED; /* see #7 (stack) for description */

9. Operation Table (R)

Contains one entry for each macro operation that can appear in the node list.
DECLARE OPERATION (25) FIXED; /* index in macro definition table of first instruction in the macro definition corresponding to this operation */

10. Label Table (L)

Contains one entry for each internal label generated by Instran.
DECLARE LABEL (50) FIXED; /* offset from base of text block of labeled instruction */

11. Node Table (N)

Each entry contains three tokens that form a node definition (macro instruction). The first token of each macro instruction has its TYPE field equal to 'R' (the operation table). The tokens for the operands of the macro instruction follow consecutively in order. These operand tokens have TYPE fields that are not equal to 'R'.

```
DECLARE 1 NODES (500),
          2 OPERATION,
            3 TYPE CHARACTER (1), /* token type,
                            = 'R' for operation table */
            3 INDEX FIXED, /* index in operation table */
          2 LEFT,
            3 TYPE CHARACTER (1), /* token type,
                            = 'S' if symbol table
                              'C' if constant
                              'L' if internal label
                              'N' if node table */
            3 INDEX FIXED, /* index in table indicated by TYPE */
          2 RIGHT,
            3 TYPE CHARACTER (1), /* same as for LEFT */
            3 INDEX FIXED; /* same as for LEFT */
```

12. Instruction Table (I)

Contains one entry for each machine instruction in the object language.

```
DECLARE 1 INSTRUCTION (40),
          2 NUMERIC FIXED, /* numeric machine instruction code */
          2 TYPE CHARACTER (1); /* operand type */
```

13. Temporary Table (T)

Contains one entry for each temporary needed by the object procedure.

```
DECLARE TEMPORARY (20) FIXED; /* offset from base of temporary
          storage block */
```

14. Assembly Code

Each entry is an assembly code instruction in the object procedure.

```
DECLARE 1 CODE (400),
          2 INSTRUCTION FIXED, /* index in instruction table of
                machine instruction code */
          2 R FIXED, /* contents of R field */
          2 X FIXED, /* contents of X field */
          2 B FIXED, /* contents of B field */
          2 TYPE CHARACTER (1), /* operand token type,
                          = 'S' if symbol table
                            'C' if constant
                            'L' if internal label table
                            'T' if temporary table */
```

2 INDEX FIXED; /* index of operand entry in table indicated by TYPE field */

15. Macro Definitions

Each entry is one line of a macro definition. The definitions vary in length and are terminated by a special entry. The beginning of each definition is indicated in the macro operation table.

DECLARE 1 MACRO (400),

2 INSTRUCTION FIXED, /* index in instruction table of machine instruction code */

2 OPERAND FIXED; /* operand of instruction,

= 0 if temporary

1 if first operand of macro instruction

2 if second operand of macro instruction */

An entry with both fields (INSTRUCTION and OPERAND) equal to zero terminates a macro definition.

Appendix E
Instran Reduction Rules and Actions

The Instran reduction rules are written in the form

label: phrase → replacement ; actions @next_label

where all of the parts are optional except for "→" and ";". In the phrase part of a rule, lowercase words are names of nonterminals defined in the grammar for the Instran Language found in Appendix C. The replacement part of a reduction rule is either the single symbol "#", which means delete the phrase with no replacement, or a sequence of symbols of the form "P_i ... P_j", where P_i represents the ith token in the matched phrase. The sequence of tokens corresponding to the P_i's will replace the phrase. If the replacement part of the reduction rule is empty, the phrase is not replaced. The actions are all named, and each action is described later in this appendix. Lines beginning with "/*" are comments.

The Reduction Rules

/* The P group parses a PROCEDURE statement.

P1: → ; lexical lexical lexical
P2: identifier : PROCEDURE → P_0 ; begin_procedure lexical @P3
P21: → ; error
P3: (→ # ; lexical @P6
P4: ; → # ; lexical @D1
P5: → ; error

/* Parse dummy argument list.

P6: identifier → # ; define_dummy lexical @P7
P61: → ; error
P7: , → # ; lexical @P6
P8:) → # ; lexical @P4
P9: → ; error

/* The D group separates a simple variable declaration from a structure declaration.

D1: DECLARE → # ; lexical @D3
D2: → ; lexical @S1

D3: identifier → ; define lexical @VD1
D4: 1 → # ; lexical @SD1
D5: → ; error

/* The VD group parses a simple variable declaration.

VD1: (→ ; lexical lexical @VD5
VD2: string → # ; set_type lexical lexical lexical @VD7

/* Parse data type.

VD3: type-word → # ; set_type lexical lexical @VD9
VD4: → ; error

/* Parse dimension attribute.

VD5: (integer) → # ; set_dimension lexical @VD2
VD6: → ; error

/* Parse string attribute.

VD7: (integer) → # ; set_length lexical lexical @VD9
VD8: → ; error

/* Parse storage class.

VD9: identifier storage ; → # ; set_storage lexical @D1
VD10: identifier ; any → P_0 ; @D1
VD11: → ; error

/* The SD group parses a structure declaration.

SD1: identifier → ; define_structure lexical lexical @SD3
SD2: → ; error
SD3: (any → ; lexical lexical @SD6
SD4: identifier storage , → # ; set_storage lexical lexical @SD8
SD41: identifier , any → P_0 ; @SD8
SD5: → ; error

/* Parse structure dimension.

SD6: (integer) → # ; set_dimension lexical lexical @SD4
SD7: → ; error

/* Parse structure element.

SD8: integer identifier → P_0 ; set_element lexical @SD10
SD9: → ; error

/* Parse substructure element.

SD10: commas → ; set_substructure @SD16

/* Parse basic element.

SD11: string → # ; set_type lexical lexical lexical @SD14
SD12: type-word → # ; set_type lexical @SD16
SD13: → ; error
SD14: (integer) → # ; set_length @SD12
SD15: → ; error

/* Check end of structure.

SD16: any , → # ; lexical lexical @SD8
SD17: any ; → # ; lexical @D1
SD18: → ; error

/* The S group parses a statement.
/* Labeled statement.

S1: identifier : → # ; label lexical lexical

/* Assignment statement.

S2: identifier (→ ; compile_expression @E1
S3: identifier . → ; compile_expression @E1
S4: identifier = → ; compile_expression @E1

/* Other statements.

S5: GOTO identifier → P_0 ; compile_branch lexical @E1
S6: IF any → P_0 ; lexical compile_expression compile_if @S14
S7: RETURN ; → # ; compile_return lexical lexical @E3
S8: CALL identifier → P_0 ; lexical compile_expression @E1
S9: ; any → P_0 ; lexical @E3
S10: DO ; → P_1 ; lexical lexical @S1

/* End of group.

S11: DO END ; → # ; lexical lexical @E3

/* End of procedure.

S12: PROCEDURE END ; → ; end_procedure
S13: → ; error

/* Then clause of IF statement.

S14: THEN → ; lexical lexical @S1
S15: → ; error

/* The E group parses the end of a statement including the end of then
    and else clauses.

E1: any ; → # ; lexical lexical @E3
E2: → ; error

/* End of a then clause and the beginning of an else clause.

E3: THEN ELSE any → P_1 P_0 ; compile_then lexical @S1
E4: any THEN any any → P_1 P_0 ; end_if @E3

/* End of an else clause.

E5: any ELSE any any → P_1 P_0 ; compile_else @E3
E6: → ; @S1

The Actions

The remainder of this appendix contains, in alphabetical order, the definition of each action referenced in the reduction rules. Each action is an argumentless procedure. All communication is carried out through named external variables. The principal ones are the stack S and the array P. At the time an action is executed, the replacement specified in the matched reduction rule has been carried out, that is, the matched phrase has been copied into the array P and the contents of the stack modified as specified in the replacement part of the rule. In the action definitions, S_i refers to the ith token from the top of the stack after the stack has been modified (S_0 refers to the token at the top of the stack). P_i refers to the ith token in the matched phrase, which has been copied into the array P.

When an entry is to be made in the node table, we will simply write it in the form

$$\{t_1 \ldots t_n\}$$

This shorthand will mean, "Add the sequence of tokens $t_1, \ldots, t_n$ to the node table as its next entry." The t_i will be either actual tokens, such as R_{EQU} (a token pointing to the operation table entry for EQU) and L_i (a token pointing to the ith entry in the label table), or names of variables whose value is a token such as S_j or P_i.

1. begin_procedure

   Save the identifier pointed to by the token in P_2 as the name of the procedure's entry name.
   dummy_count := 0;
   $\{R_{\text{ENTER}}\}$

2. compile_branch

   Change the token in P_0 to point to the symbol table entry for the label whose name is the identifier pointed to by the token currently in P_0.
   $\{R_{\text{BRANCH}} \, P_0\}$

3. compile_expression

   Make the node table entries for the expression beginning with S_1 and ending with the first semicolon or keyword that follows in the source input. When finished, S_1 will contain a single token that is the result of the expression, and S_0 will contain a token for the key word or terminal that terminated the expression.

4. compile_if

   Generate a new internal label whose token is L_i.
   $\{R_{\text{FALSEB}} \, S_1 \, L_i\}$
   $S_1 := L_i;$

5. compile_return

   $\{R_{\text{RETURN}}\}$

6. compile_then

   Generate a new internal label whose token is L_j. S_2 contains a token of the the form L_i.
   $\{R_{\text{BRANCH}} \, L_j\}$
   $\{R_{\text{EQU}} \, S_2\}$
   $S_2 := L_j;$

7. define

   Make an entry in the symbol table for the variable whose name is the identifier pointed to by the token in S_0. If an entry already exists for an entity with the same name that is not a structure element (STORAGE $\neq$ 'S') and not a dummy argument (STORAGE $\neq$ 'D'), mark this existing entry as multiply defined and return. If an identically named entry already exists for a dummy argument then in that entry

DEFINED : = 'D';

Otherwise make a new entry.

NAME : = INDEX from token in S_0;
TYPE : = 'U';
STORAGE : = 'U';
DEFINED : = 'D';
All other fields in entry : = 0;

The token in S_0 is then modified to point to the corresponding entry in the symbol table; that is, the token type is changed to 'S', and the INDEX field of the token is set to the index of the corresponding entry in the symbol table.

8. define_dummy

Make an entry in the symbol table for the dummy variable whose name is the identifier pointed to by the token in P_0. If an entry for an entity with the same name already exists, mark this existing entry as multiply defined and return. Otherwise make a new entry.

NAME : = INDEX from token in P_0;
TYPE : = 'U';
STORAGE : = 'D';
DEFINED : = 'U';
ADDRESS : = dummy_count;
All other fields in entry : = 0;
dummy_count : = dummy_count + 4;

9. define_structure

Make an entry in the symbol table for a structure whose name is the identifier pointed to by the token in S_0. If an entry already exists for an entity with the same name that is not an entry for a structure element and not an entry for a dummy argument, then mark the existing entry as multiply defined and return. If an identically named entry already exists for a dummy argument, then in that entry

DEFINED : = 'D';
TYPE : = 'S';

Otherwise make a new entry.

NAME : = INDEX from token in S_0;
TYPE : = 'S';
STORAGE : = 'U';
DEFINED : = 'D';
All other fields in entry : = 0;

The token in S_0 is then modified to point to the corresponding entry in the symbol table. Initialize the pointers used for linking structure elements.

old_level : = 2;
parent_element : = INDEX from token in S_0;
last_element : = INDEX from token in S_0;

10. **end_if**

> P_3 contains a token of the form L_j.
> $\{R_{EQU}\ P_3\}$

11. **end_procedure**

> The end of pass one of Instran.
> $\{R_{END}\}$
> **call pass_two;**

12. **error**

> A generic action that is executed when an error has been detected. An appropriate error comment is made and translation proceeds if possible. Normally, each separate occurrence of this action would be a different action procedure.

13. **label**

> Make an entry in the symbol table for the label whose name is the identifier pointed to by the token in P_1. If no entry already exists for an entity with the same name that is not a structure element, then make a new entry.
> NAME := INDEX from token in P_1;
> TYPE := 'L';
> STORAGE := 'T';
> DEFINED := 'D';
> All other fields in entry := 0;
> If an identically named entry already exists that has TYPE = 'L' and DEFINED = 'U', then, in that entry,
> DEFINED := 'D';
> If the symbol table contains any other identically named entry that is not a structure element, then, in that entry,
> DEFINED := 'M';
> Modify the token in P_1 so that it points to the symbol table entry for the label.
> $\{R_{EQU}\ P_1\}$

14. **lexical**

> This action procedure is the lexical analyzer, which assembles a token for the next basic element in the source input and puts it on the top of the stack. The basic elements in the Instran Language are identifiers, constants, key words, and operator-punctuation symbols.

15. **set_dimension**

> The UPPER field of the symbol table entry pointed to by the token in S_0 is set equal to the integer pointed to by the token in P_1. The ARRAY field of the same entry is set equal to 1.

16. **set_element**

> Make an entry in the symbol table for a structure element whose name is the identifier pointed to by the token in S_0. A new entry is made even if there exist other entries for entities having the same name.

NAME : = INDEX from token in S_0;
TYPE : = $'U'$;
STORAGE : = $'S'$;
DEFINED : = $'D'$;
All other fields in entry : = 0;
The token in S_0 is modified to point to the new entry in the symbol table. Update the structure linking pointers.

this_element : = INDEX from token from S_0;
new_level : = integer pointed to by token in P_1;

This structure element is added to the structure's tree representation, which is embedded in the symbol table, by linking its symbol table entry to the entries for the other elements of the structure that are already in the symbol table.

if new_level = old_level
    then if last_element = parent_element
        then SYMBOL.SON[parent_element] : = this_element;
        else SYMBOL.BROTHER[last_element] : = this_element;
    else begin
        if new_level > old_level then Error;
        if new_level ≤ 1 then Error;
        if last_element = parent_element then Error;
        repeat
            old_level : = old_level − 1;
            last_element : = parent_element;
            parent_element : = SYMBOL.PARENT[last_element]
        until old_level = new_level;
        SYMBOL.BROTHER[last_element] : = this_element
    end;
SYMBOL.PARENT[this_element] : = parent_element;
last_element : = this_element;

17. set_length

The LENGTH field of the symbol table entry pointed to by the token in S_0 is set equal to the integer pointed to by the token in P_1.

18. set_storage

The STORAGE field of the symbol table entry pointed to by the token in P_2 is set equal to the storage code corresponding to the key word pointed to by the token in P_1.

19. set_substructure

SYMBOL.TYPE[this_element] : = $'S'$;
old_level : = old_level + 1;
parent_element : = this_element;

20. set_type

The TYPE field of the symbol table entry pointed to by the token in S_0 is set equal to the type code corresponding to the key word pointed to by the token in P_0.

Appendix F
Our 360 I/O Processor and Interrupts

The I/O processor controls all of the input and output. It and the control processor normally operate concurrently. Communication between them is limited and is mostly through exchange of information in the primary memory. The following diagram illustrates the relationship between the control processor (**CP**), the primary memory (**PM**), the I/O processor (**IOP**), the control units (**CU**), and the devices (**D**).

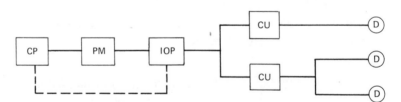

The solid lines indicate paths for both data and control signals. The broken line indicates a path that is limited to control signals. Using this path, the control processor can tell the I/O processor to start some I/O and can test the I/O processor's current status. Using this same path, the I/O processor can *interrupt* the control processor, thus indicating that it has terminated some I/O. All other information is passed between the I/O processor and the control processor through the primary memory.

The I/O processor controls the various devices through the control units. More than one control unit can be attached to the I/O processor, and more than one device can be attached to a control unit. All input read from any device into the primary memory and all output written from the primary memory to any device always passes through both the I/O processor and the device's control unit.

The action of the I/O processor and the interrupt mechanism described here is not a subset of the IBM 360. While retaining the flavor and spirit, it is simpler and it corrects some awkward problems which exist in the IBM 360.

1. The Basic Interrupt Mechanism

The I/O processor uses the computer's I/O interrupt feature to inform the control processor that it has terminated some input or output. The interrupt mechanism that it uses is a slightly augmented version of the basic interrupt mechanism. The augmentations will be described in a later section. In this section we describe the basic interrupt mechanism.

The current state of the control processor is retained in the program status word (PSW). As noted in Appendix A, one of the fields in the PSW is the instruction counter (IC), which contains the address of the next instruction. To change the instruction execution sequence (i.e., to branch), the contents of the IC are replaced by a new address. When an interrupt occurs, the entire current contents of the PSW are exchanged with a new PSW. Thus, the state of the control processor at the time of the interrupt is saved, and a branch takes place. This action automatically occurs whenever an interrupt condition occurs, unless that type of interrupt is disabled.

There are five types of interrupts: *program, SVC, I/O, timer,* and *trouble.* A program interrupt is caused by conditions such as attempting to execute an illegal instruction, attempting to reference a full word with an effective address that is not evenly divisible by four, an overflow, or attempting to do an improper division. An SVC interrupt occurs when the SVC instruction is executed; an I/O interrupt is caused by a signal from the I/O processor; a timer interrupt occurs when the value in the timer is decremented to zero; and a trouble interrupt occurs when the I/O queue is full.

Corresponding to each interrupt type there is a fixed area in primary memory called the *interrupt exchange area* into which the current contents of the PSW are stored (called the *old PSW*) and from which the *new PSW* is loaded. These fixed memory areas are:

Absolute memory address (in hex)	Contents
0–7	Old PSW for program interrupt
8–F	New PSW for program interrupt
10–17	Old PSW for SVC interrupt
18–1F	New PSW for SVC interrupt
20–27	Old PSW for I/O interrupt
28–2F	New PSW for I/O interrupt
30–33	I/O queue length
34–37	I/O queue index
38–3F	Old PSW for timer interrupt
40–47	New PSW for timer interrupt
48–4B	Timer
4C–4F	Unused
50–57	Old PSW for trouble interrupt
58–5F	New PSW for trouble interrupt

Since there is a separate old-new PSW pair for each type of interrupt, a different branch address may be provided for each type.

In addition to the CC and the IC, the PSW contains additional information relating to interrupts. The format of the PSW is:

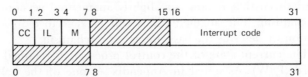

Here, IL is the length of the last instruction executed before the interrupt occurred, while IC is the address of the instruction that would have been executed next if the interrupt had not occurred. The *interrupt code* further identifies the exact cause of the interrupt. M is a mask that controls the disabling of interrupts. The bits in M have the following correspondence.

Bit	Interrupt type
4	Timer
5	I/O
6	Program: overflow
7	Program: improper division

When the corresponding bit equals one, interrupts of the indicated type are disabled. SVC and trouble interrupts cannot be disabled, and only program interrupts caused by an overflow or an improper division can be disabled.

When an interrupt condition occurs, the control processor completes the execution of the current instruction, if it is able to do so (for example, it cannot complete the execution of an illegal instruction). Then the current contents of the PSW are stored as the old PSW in the interrupt exchange area corresponding to the type of interrupt and the associated new PSW is loaded. Before the old PSW is stored, the interrupt code and the IL are set to the appropriate values. This exchange of PSW's causes a branch to that part of the operating system responsible for the corresponding type of interrupt. The new PSW may contain ones in any of bits 4–7, thus disabling some or all of the interrupts.

2. I/O Tasks

The operation of the I/O processor is based on the concept of an *I/O task*. An I/O task consists of one or more device control operations, input or output of one or more records of data, or a mixture of both. The composition and duration of an I/O task is determined by the program that initiates it. All that is required is that the sequence of operations in an I/O task can proceed to termination without intervention by the initiating program, unless some trouble arises. Once the I/O processor starts an I/O task, the control processor has no further control over any of the operations in that I/O task.

The I/O processor is capable of concurrently performing more than one I/O task. Each I/O task is defined by an *I/O program*, which is stored in primary memory. An I/O program consists of a sequence of I/O processor *commands*.

For each I/O task there is a *device control block* (DCB) that associates the I/O program that defines the task with the device that the I/O task is using. In addition, at certain times, information concerning the status of a task is stored in its device control block.

A device control block has the format:

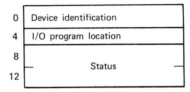

The *device identification* specifies both a *channel* and a device that is attached to that channel. The I/O processor has several channels connecting it to the control units. Each channel is an independent path for the transfer of data and control information. More than one control unit can be attached to a single channel, and this attachment is fixed. Normally, only one I/O task may be using a given channel. Thus, the number of concurrently active I/O tasks is limited by the number of channels.

3. I/O Processor Commands

The I/O program that defines an I/O task consists of a sequence of *commands*. There are three commands: READ, WRITE, and CONTROL. All commands have the format:

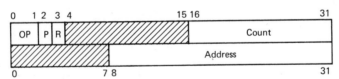

where OP is the operation code (READ, WRITE, or CONTROL). The P field (the *program end flag*) must be zero in all commands except for the last command in an I/O program.

The READ and WRITE commands transfer data between a device and primary memory. The *count* field specifies the number of bytes to be transferred, and the *address* field specifies the base address of a primary memory area into which the data is read or from which the data is written. An I/O program always reads or writes an integral number of records. Each READ or WRITE command specifies a full record or less than a record. When the R field (the *record end flag*) of a command is zero, that command specifies part of the same record that the next command in sequence specifies. When the record end flag is one, the next command in sequence, if any, specifies a different record.

A sequence of WRITE commands that specifies a single record will always write a record whose length is exactly equal to the sum of their count fields. A sequence of READ commands that specifies a single record will never read more than one record, even if the sum of their count fields, N, is larger than the length of the record. If N is equal to or greater than the length of the record, all of the record is read. In the latter case the record is read into the first N memory locations specified by the count and address fields. The contents of the remaining memory locations are unchanged. If N is less than the length of the record, only the first N bytes in the record are read, and the remaining bytes in the record are skipped.

The CONTROL command passes control information to a control unit. The I/O processor combines both the count and address fields into a single number and sends it to the control unit, which controls the device specified in the device control block. The meaning of this information is specific to each device.

4. Control Processor I/O Instructions

There are only two instructions that the control processor can use to reference the I/O processor: STARTIO and TESTIO. The effective address defined by each of these instructions is the location of the device control block for some I/O task.

The STARTIO instruction requests the I/O processor to begin execution of the specified I/O task. The control processor does this by sending a start signal and the address of the device control block for the I/O task to the I/O processor. After the control processor has finished executing this instruction, the CC will indicate if execution of the I/O task was successfully started. The following codes are used to indicate the result.

Value of CC	Result
0	Execution of I/O task successfully started
1	Channel or device addressed is being used by another task
2	Unusual condition; status stored
3	I/O processor inoperative

Only when the CC is zero will the I/O processor continue to execute the I/O task. This execution is independent of the control processor and continues until the task has been finished, unless some trouble is encountered. In all other cases the I/O processor makes no further attempt to execute the I/O task until another STARTIO instruction is executed by the control processor.

When an unusual condition is encountered by the I/O processor in trying to start an I/O task, status information is stored in the device control block for the task. This information will identify the specific problem. When the I/O processor is inoperative, all that is known is that the control processor failed to receive any kind of response from the I/O processor within a reasonable time after a start signal was sent to it.

The TESTIO instruction requests the I/O processor to store information concerning the status of the indicated I/O task into its device control block. The control processor does this by sending a test signal and the address of the device control block for the I/O task to the I/O processor. Execution of a TESTIO instruction does *not* cause the I/O processor to stop execution of any I/O task. After the control processor executes this instruction the CC will indicate if the status has been stored and if the channel or device are busy. The following codes are used.

Value of CC	Result
0	Status successfully stored
1	Channel busy; status also stored
2	Device busy and channel free; status also stored
3	I/O processor inoperative; status not stored

Even though the status information indicates if the channel and device are busy, this information is duplicated in the CC setting. This permits rapid testing for these common conditions.

The status information is stored in the form:

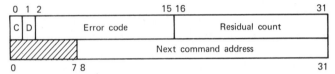

The flags C and D, if equal to one, indicate that the channel or device, respectively, is busy. The *error code* identifies any errors that have occurred since the status for the specified I/O task was last stored. The *next command address* is the location of the next command to be executed in the I/O program that defines the specified I/O task. If the I/O program was completed before the status was stored, the next command address will be eight larger than the address of the last command that was executed. If a READ or WRITE command was in progress when the status was stored, or the last command executed was a READ command that read less than the number of bytes specified in the count field of the command, the *residual count* will equal the (positive) difference between the number of bytes specified in the command and the number actually read or written.

5. I/O Interrupts

An I/O interrupt functions like the basic interrupt mechanism with some additional functions. Whenever an I/O task being executed by the I/O processor terminates normally or encounters some difficulty that the I/O processor cannot resolve, the I/O processor sends an interrupt to the control processor. When the control processor receives an interrupt signal from the I/O processor and I/O interrupts are enabled, an interrupt occurs exactly as described in Section 2. If I/O interrupts are disabled, no interrupt occurs. However, a counter is incremented in the control processor. This counter records the number of interrupt signals which have been received, but for which no interrupt has yet occurred.

In addition to the action described above, before sending an interrupt signal to the control processor, the I/O processor makes an entry on the I/O queue that identifies the I/O task responsible for the interrupt and stores the status of that task in its device control block. An I/O queue entry has the format:

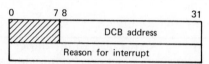

where *DCB address* is the address of the device control block for the responsible I/O task. Both the status and the queue entry are stored, even if I/O interrupts are disabled. That is, as far as the I/O processor is concerned, an I/O interrupt always occurs. Only the control processor knows if an interrupt actually occurred.

Because the I/O processor may concurrently execute several I/O tasks, several additional interrupts may be signaled by the I/O processor while I/O interrupts are disabled during the processing of a prior interrupt. The I/O queue is provided so that interrupt information for several I/O tasks can be held until the corresponding interrupts can be processed. Each entry in the queue corresponds to one I/O task. When there are queue entries for interrupt signals that have not yet been acted on, an I/O interrupt will occur as soon as I/O interrupts are enabled. This will happen each time I/O interrupts are enabled until all queue entries have been processed.

Whenever an I/O queue entry is made, it is stored in primary memory at the location whose address is the current value of the *I/O queue index* in the I/O interrupt exchange area (see Section 1). After the queue entry has been made, the queue index is automatically incremented by eight in anticipation of the next interrupt. The incremented value of the queue index is compared with the current value of the *I/O queue length,* which is the address of the last word reserved for the I/O queue. If it is greater, a trouble interrupt occurs, since the I/O queue is full and the system will be in trouble if further I/O interrupt conditions occur. It is up to the system to reset the I/O queue index whenever it has processed, or at least removed, all of the I/O queue entries.

Index